Insects Causing Damage to Plants

NIPA® GENX ELECTRONIC RESOURCES & SOLUTIONS P. LTD.
New Delhi-110 034

About the Authors

Mr. H, Lewin Devasahayam after completion of his post graduate degree course from the Banaras Hindu University, Varanasi joined the Tamil Nadu Agricultural University, Coimbatore in 1960 and served the University for 30 years in various capacities and retired as Associate Professor of Plant Pathology in 1990. He has published more than 50 research articles in various leading Indian and Foreign journals. He is the author of the books entitled 'Practical manual of Entomology', 'Elements of Entomology', 'Illustrated Plant Pathology' and co-author of the books 'Crop diseases' and 'Diseases of spice crops', and all the above books have been published by NIPA, New Delhi. He is also the co-author of the book entitled 'Plant Ecology' published by M/S. Agri-Biovet Press, New Delhi.

Dr. L. Darwin Christdhas Henry obtained his B.Sc.Degree in Botany from the Madras University Chennai. He continued his under graduate and post graduate degree course in Agriculture at Allahabad Agricultural Institute. He joined the Faculty of Agriculture, Annamalai University He has completed his doctorate degree, specializing in the field of edible mushrooms. He has published a number of research articles in National and International Journals. He has authored the books entitled 'Crop Diseases – Identification, Treatment and Management' and 'Diseases of spice crops', both published by NIPA, New Delhi and also the book entitled 'Plant Ecology' published by M/S. Agri-Biovet Press, New Delhi. He is also the co-author of the book entitled 'Illustrated Plant Pathology' published by NIPA, New Delhi. Apart from his wide exposure to different agricultural crops, he has specialized in the cultivation of different species of edible mushrooms.

An Illustrated Handbook
Insects Causing Damage to Plants

H. Lewin Devasahayam
Associate Professor (Retd.)
Tamil Nadu Agricultural University
Coimbatore, Tamil Nadu

L. Darwin Christdhas Henry
Associate Professor (Retd.)
Faculty of Agriculture
Annamalai University
Annamalainagar, Tamil Nadu

NIPA® GENX ELECTRONIC RESOURCES & SOLUTIONS P. LTD.
New Delhi-110 034

NIPA® GENX ELECTRONIC RESOURCES & SOLUTIONS P. LTD.

101,103, Vikas Surya Plaza, CU Block
L.S.C.Market, Pitam Pura, New Delhi-110 034
Ph : +91 11 27341616, 27341717, 27341718
E-mail:newindiapublishingagency@gmail.com
www: www.nipabooks.com

For customer assistance, please contact
Phone:+ 91-11-27 34 17 17
Fax: + 91-11-27 34 16 16
E-Mail: feedbacks@nipabooks.com

Print ISBN: 978-81-19103-14-0

ebook ISBN: 978-81-19103-19-5

Composed and Designed by NIPA.

Preface

Insects are ubiquitous, in other words they are omnipresent and are found everywhere all over the world. They are present in the highest mountain peaks, ice bergs, deep oceans, thick forests, scorching deserts, darkest caves and even in oil wells, where no other living organisms can survive. Insects possess special innate adaptations to withstand any kind of environment and so they can survive under any extremes of adverse environmental conditions and are capable of perpetuation unabated.

According to available evidences, the first insects evolved in marine or coastal habitats and took to land at around the same time as plants appeared some 440 million years ago. But the first humans emerged in Africa only around 2.0 million years ago. As such man has intruded into the domain of insects. To satisfy his basic requirements of food, clothing and shelter, man started raising crops. In the meanwhile, insects have deterred from their normal feeding habits and started damaging the crops raised by man and the produce obtained from the crops. From that time onwards, a conflict has arisen between man and the insects. Man by virtue of his ingenuity and resourcefulness has adapted various means such as, physical, mechanical, chemical, biological and other means to destroy insects, which have turned into crop pests or which he considers to be harmful in any way. However, in spite of all his efforts, man is not able to control the harmful insects completely and the struggle is going on and on.

In general insects are free living and a vast majority of them are harmless. Only some of them have turned into pests, while a few are beneficial to man in various ways. Among the beneficial a few are productive insects and directly provide products of commercial value. Of these the most important ones are the honeybees that supply honey and bee wax, the silk worms that give silk, the lac insect that provide lac and the cochineal insects that supply certain dye substances. Desert locusts, grasshoppers, termites and some other insects are being used as food by man in different countries. Certain substances of medicinal value are obtained from some insects. Bee venom is obtained from honeybees. Parasitoids and predators are helpful insects as they destroy certain harmful insect pests. Insects are the major agents that aid in the pollination of higher plants. Some insects such as, *Drosophila* flies are extensively used in the field of genetical studies.

However, some insects, which are termed as insect pests, damage cultivated and wild plants. A few blood sucking insects attack humans and livestock. Further, some insects are known to cause or transmit several fungal, bacterial and viral diseases in plants, humans and livestock.

Acquiring knowledge about insects gives an idea about their survival potential. Ability to adapt themselves to any kind of adverse environments, small size, presence of exoskeleton, diversity food habit, high fecundity and controlled reproduction, different developmental stages under hormonal control, different modes of regeneration, greater mobility, migratory habit, defense mechanism, sensory appendages, diverse respiratory systems and zenith of evolution are all acquired traits of insects, which have made it possible for them to survive and reproduce successfully.

Studies on the general aspects about insects, their morphology, anatomy, development and metamorphosis, classification etc. help to understand more about insects, especially the harmful and harmless insects, their feeding habits, how they ingest and digest food, how the blood is circulated to each and every part of the body, how they respire, how they develop and multiply, as well as to identify the various insects already identified and also to identify hitherto unidentified insects and place them in their respective taxonomic position.

All information obtained from such studies are ultimately employed to evolve suitable and appropriate measures to control harmful insects and also to explore the possibilities of utilizing the activities of beneficial insects for the benefit of mankind.

The text, which has been compiled base on the Syllabus of Agriculture Degree course will not only satisfy all the needs of undergraduate students of Agriculture, but also will be of assistance and help to Post Graduate Students of Entomology and Zoology, and also to the elite public who are interested to know more about insets. Further, it will serve as a valuable reference book for those who are employed in the Department of Agriculture.

The authors are thankful to friends and colleagues, who have inspired and encouraged them in their endeavor to complete the book. The authors wish to express their sincere thanks to M/S. New India Publishing Agency, 101-Vikas Surya Plaza, Pitam Pura, New Delhi for the excellent manner in which the book has been brought out.

H. Lewin Devasahayam
L. Darwin Christdhas Henry

Contents

Preface v
List of Figures xi
Glossary of Scientific Terms xvii
1. Insects 1
General Aspects About Insects 1
Characters of Organisms Belonging to Phylum - Arthropoda 2
Position of Insects in the Animal Kingdom 3
Dominance of Insects Over Other Living Organisms 4
Orientation and Tropism 7
Adaptation of Insects for Survival, Offence and Defense 9
Beneficial Insects 17
Major Productive Insects 17
Minor Productive Insects 21
Helpful Insects 22
Insect Ecology 23
Environmental Factors Affecting Insects 24
i) Abiotic factors 24
ii) Biotic Factors 27
Habitats of Insects 32
Types of reproduction in insects 33
Insect Pest 34
Biotic or Natural Balance 35
2. Morphology of Insects 37
Integument or Exoskeleton 38
Stages in the Development of Cuticle 41
Cuticular processes, Appendages or Outgrowths 42
Body Regions of Insects 43
The Head 43
Important Structures of the Head 45
Eyes 45

Antennae 46
Mouthparts 49
Thoracic Appendages 60
Legs 60
Muscles and Locomotion 61
Modifications of the legs of insects 62
Prolegs 64
Wings 64
Movement of Wings of Insects 65
Development of Wings 66
Modification of wings 66
The Wing Coupling Mechanism 67
Wing Venation 69
Abdomen 71
External Genitalia of Female Grasshopper 72
External Genitalia of Male Grasshopper 73

3. Anatomy of Insects 75
The Alimentary System or Digestive System 76
Process of Digestion 81
The Excretory System 83
The Circulatory System 84
The Nervous System 88
The Endocrine System 91
The Exocrine System 94
Types of Pheromones 94
The Respiratory System 97
The Reproductive System 102
The Female Reproductive Organs 103
Oogenesis 105
Fertilization 106
The Male Reproductive Organs 106
Types of Reproduction 108
Sense Organs in Insects 110
Sound Producing Organs 119
Light Producing Organs 120

4. Development and Metamorphosis of Insects......................................121

Embryonic Development 121
Post Embryonic Development and Metamorphosis................................ 122
Different Types of Metamorphosis 123
Types of Larvae 126
The pre-pupa or pharate pupa 128
The Pupa 129
The Adult or Imago 130

5. Classification...133

Classification of Living Organisms 133
Classification of Phylum - Arthropoda 134
Class - Insecta 136
Sub class - Apterygota or Ametabola... 136
Order - Protura (Protrurans)... 136
Order - Thysanura (Silver fish and Bristle tails).................................... 138
Order - Collembola (Spring tails) 138
Sub class - Pterygota or Metabola 138
Division - Exopterygota or Heterometabola; 138
Sub division - Hemimetabola 138
Order - Odonata (Dragon flies and Damsel flies) 138
Order - Ephimeroptera (Mayflies, Day flies etc.) 140
Order - Plecoptera (Stone flies) 142
Sub class - Perygota or Metabola 142
Division - Exopterygota or Heterometabola; 142
Sub Division - Paurometabola 142
Order - Dictyoptera (Cockroaches and Mantids).................................... 143
Order - Orthoptera (Grasshoppers, Locusts, crickets etc.) 145
Order - Plasmida (Stick insects, Leaf insects etc.) 149
Order - Dermaptera (Earwigs) 150
Order - Embioptera (Web spinners, Embids etc.) 151
Order - Isoptera (Termites, White ants etc.) 153
Order - Psocoptera (Psocids, Book lice, bark lice etc.) 157
Order - Mallophaga (Bird lice, Elephant lice etc.) 158
Order - Siphunculata or Anoplura (Sucking lice) 158
Order - Hemiptera (Bugs, Water scorpions, Hoppers, Aphids etc.) 168

Order - Thysanoptera (Thrips) 176
Sub class - Pterygota or Metabola 178
Division - Endopterygota or Holometabola 178
Order - N europtera (Alder flies, lace wings, ant lions) 178
Order - Mecoptera (Scorpion flies) 181
Order - Lepidoptera (Butterflies, skippers and moths) 182
Order - Trichoptera (Caddis flies) 196
Order - Diptera (True flies and moosquitoes) 197
Order - Siphonaptera or Aphaniptera (Fleas) 205
Order - Hymenoptera (Ants, bees, wasps) 206
Order - Coleoptera (Beetles, weevils) 215
Order - Strepsiptera (Stylopids) 231
Appendices........ 233
Colour Plates........ 237
References........ 241
Index........ 243

List of Figures

Fig. 1a. Percentage of insect species in Phylum - Arthropoda 3
Fig. 1b. Number of species in each of the Classes of Phylum - Arthropoda 4
Fig. 1c. Percentage of insects in relation to total living organisms 4
Fig. 2. Body regions of a typical insect (e-g) Grasshopper (Lateral view) 37
Fig. 3. Body regions of a typical insect (e-g) Cockroach 39
Fig. 4. Structure of the integument of an insect 40
Fig. 5. Moulting in an insect 42
Fig. 6. Morphology of the head of an insect (e-g) Grasshopper 43
Fig. 7. Types of antennae in insects 47
Fig. 8. Biting and chewing type of mouthparts 51
Fig. 9. Piercing and sucking type of mouthparts (Bug type) 53
Fig. 10. Piercing and sucking type (Mosquito type) 54
Fig. 11. Siphoning type of mouthparts 55
Fig. 12. Chewing and lapping type of mouthparts (e-g) Honeybee 57
Fig. 13. Sponging type of mouthparts (e-g) Housefly 58
Fig. 14. Types of legs in insects 61
Fig. 15. Mechanism of the basic wing strokes in flight of an insect 65
Fig. 16. Wing coupling structures 68
Fig. 17. Wing of a Hymenopteron showing the margins and angles 70
Fig. 18. A hypothetical insect wing showing the arrangement of veins 70
Fig. 19. External genitalia of female grasshopper 73
Fig. 20. External genitalia of male grasshopper 74
Fig. 21. Arrangement of inner organs in insects (e-g) Grasshopper 75
Fig. 22. T. S. of the crop 77
Fig. 23. Longitudinal section of the gizzard 78
Fig. 24. T.S. through the stomach 79
Fig. 25. T.S. of the rectum 80
Fig. 26. The salivary system 80
Fig. 27. The alimentary system (e-g) Grasshopper 82
Fig. 28. The circulatory system (e-g) Grasshopper 87
Fig. 29. The nervous system (e-g) Grasshopper 90
Fig. 30. Spiracle, tracheae and tracheoles 98
Fig. 31. Vertical, longitudinal section of the head and thorax 99
Fig. 32. Vertical, longitudinal section of abdomen 100
Fig. 33. The respiratory system (e-g) Cockroach 101
Fig. 34. Internal reproductive organs of a female (e-g) Grasshopper 103
Fig. 35. Longitudinal section of an ovariole 104
Fig. 36. Internal genitalia of a male insect (e-g) Grasshopper 106
Fig. 37. Longitudinal section of testes follicle (e-g) grasshopper 107
Fig. 38. A campaniform sense organ 112

Fig. 39. A scolophore of the integument type 112
Fig. 40. Longitudinal section of the base of the antenna showing Johnston's organ 113
Fig. 41. Auditory and sound drums of male cicada 114
Fig. 42. Tympanum in fore leg of locust 114
Fig. 43. Tympanum of a locustviewed from within 114
Fig. 44. Tympanum on the first abdominal segment of grasshopper 114
Fig. 45. Chemosensory peg organ in the antenna of grasshopper 115
Fig. 46A. Compound eye adapted for apposition image 119
Fig. 46B. Compound eye adapted for superposition image 119
Fig. 47. Gradual metamorphosis (e-g) Grasshopper 124
Fig. 48. Incomplete metamorphosis (e-g) Dragon fly 124
Fig. 49. Complete metamorphosis (e-g) Butterfly *(Danais chrysipus)* 125
Fig. 50. Types of larvae of insects 126
Fig. 51. Silver fish *(Lepisma saccharina)* 137
Fig. 52. Spring tail *(Isotoma minos)* 138
Fig. 53. Dragon fly 140
Fig. 54. Damsel fly 140
Fig. 55. May fly *(Ephemera vulgata)* 141
Fig. 56. Stone fly (*Perla* species) 142
Fig. 57. Cockroach *(Periplaneta americana)* 144
Fig. 58. Praying mantis *(Mantis religiosa)* 145
Fig. 59. Long-horned grasshopper *(Mecopoda elongata)* 147
Fig. 60. Cricket *(Gryllus domesticus)* 147
Fig. 61. Mole cricket *(Gryllotalpa africana)* 148
Fig. 62. Grasshopper *(Hieroglyphus banian)* 149
Fig. 63. Leaf insect (Phyllium scythe) 150
Fig. 64. Stick insect *(Carausius* species*)* 150
Fig. 65. Earwig *(Euborellia stali)* 151
Fig. 66. Web spinner *(Oligotoma saundersi)* 152
Fig. 67. Ground inhabiting termite *(Odontotermes obesus)* 156
Fig. 68. Book louse *(Liposcellis transvaalensis)* 157
Fig. 69. Bird louse *(Osborniella guiraensis)* 158
Fig. 70. Human head louse *(Pediculus humanus capites)* 159
Fig. 71. Buffalo louse *(Haematopinus tuberculatus)* 159
Fig. 72. Pond skater *(Gerris spinolae)* 162
Fig. 73. Water strider *(Gerris remigis)* 162
Fig. 74. Cholam earhead bug *(Calocoris angustatus)* 163
Fig. 75. Bed bug *(Cimex hemipterus)* 163
Fig. 76. Red cotton bug *(Dysdercus cingulatus)* 164
Fig. 77. Rice earhead bug *(Leptocorisa varicornis)* 164
Fig. 78. Predatory bug *(Canthecona furcellata)* 165
Fig. 79. Predatory bug *(Andrallus spinidens)* 165
Fig. 80. Banana lacewing bug *(Stephanitis typicus)* 166
Fig. 81. Giant water bug *(Sphaeroderma molestum)* 166
Fig. 82. Giant water bug *(Belostoma indicum)* 166
Fig. 83. Water scorpion *(Ranatra elongata)* 167
Fig. 84. Water scorpion *(Laccotrephes maculatus)* 167
Fig. 85. Water boatman *(Micronecta striata)* 167
Fig. 86. Cicada of the plains *(Platypleura machinoni)* 169
Fig. 87. Spittle bug *(Clovia puncta)* 169

Fig. 89. Mango hopper *(Amritodes atkinsoni)* 170
Fig. 90. Rice green leaf hopper *(Nephotettix virescens)* 170
Fig. 91. Rice brown plant hopper *(Nilaparvata lugens)* 171
Fig. 92. Sugarcane hopper *(Pyrilla perpusilla)* 171
Fig. 93. Citrus white fly *(Aleurocanthus spiniferus)* 172
Fig. 94. Tobacco aphid *(Myzus persicae)* 173
Fig. 95. Apple wooly aphid *(Eriosoma lanigera)* 173
Fig. 96. Rice mealy bug *(Ripersia oryzae)* 174
Fig. 97. Mango mealy bug *(Drosicha mangiferae)* *174*
Fig. 98. Cottony cushion scale *(Icerya purchasi)* 174
Fig. 99. Indian lac insect *(Laccifer lacca)* 175
Fig. 100. Green scale of coffee *(Coccus viridis)* 176
Fig. 101. Hard scale of rose (*Aonidiella auranii)* 176
Fig. 102. Rice thrips (*Stenchaetothrips biformis*) 177
Fig. 103. Chilli thrips (*Scirtothrips dorsalis*) 177
Fig. 104. *Mantid fly* (Mantispa rugicollis) 179
Fig. 105. Green lacewing fly *(Chrysopa cymbela)* 179
Fig. 106. Thread lacewing fly *(Nemoptera sinuata)* 180
Fig. 107. Ant lion *(Palpares infirmus)* - Larva 180
Fig. 108. Scorpion fly *(Leptopanorpa furcata)* 181
Fig. 109. The lemon butterfly *(Papilio demoleus)* 185
Fig. 110. *Danaise* butterfly *(Danaise chrysippus)* 185
Fig. 111. The cabbage butterfly *(Pieris brassicae)* 186
Fig. 112. Anar butterfly *(Viracola isocrates)* 186
Fig. 113. The blue butterfly *(Lampides boeticus)* 186
Fig. 114. Groundnut leaf miner *(Aproaerema modicella)* 187
Fig. 115. Rice grain moth *(Sitotroga cerealella)* 187
Fig. 116. Black-headed caterpillar *(Opisina arenosella*) 188
Fig. 117. Citrus leaf miner *(Phylocnistis citrella)* 188
Fig. 118. Diamond-back moth *(Plutella maculipennis)* 188
Fig. 119. Coffee red stem borer *(Zeuzera coffeae)* 188
Fig. 120. Stem and bark borer of cashew *(Indarbela quadrinotata)* 189
Fig. 121. Coconut slug caterpillar *(Contheyla rotunda)* 189
Fig. 122. Rice meal moth *(Corcyra cepalonica)* 190
Fig. 123. Sugarcane early shoot borer *(Chilo infuscatellus)* 190
Fig. 124. Brinjal shoot borer *(Euzophera perticella)* 190
Fig. 125. Brinjal shoot and fruit borer *(Leucinodes orbonalis)* 191
Fig. 126. Red gram spotted pod borer *(Maruca testulalis)* 191
Fig. 127. Sugarcane top borer *(Scirpophaga nivella)* 191
Fig. 128. Bottle gourd plume moth *(Sphenarches caffer)* 192
Fig. 129. The mulberry silkworm *(Bombyx mori)* 192
Fig. 130. Tea looper *(Biston supressaria)* 193
Fig. 131. Horned caterpillar *(Acherontia styx)* 194
Fig. 132. Hawk moth *(Herse convolvuli)* 194
Fig. 133. Sunnhemp hairy caterpillar *(Utitheisa pulchella)* 194
Fig. 134. Groundnut red hairy caterpillar *(Amsacta albistriga)* 194
Fig. 135. American cotton boll worm *(Helicoverpa / Heliothis armigera)* 195
Fig. 136. The gram cutworm *(Agrotis ipsilon)* 195
Fig. 137. The snake gourd semilooper *(Plusia signata)* 195
Fig. 138. The tobacco caterpillar (*Spodoptera litura*) 195

Fig. 139. Castor semilooper *(Achaea janata)* ... 195
Fig. 140. The fruit sucking moth *(Othreis materna)* ... 195
Fig. 141. Rice yellow hairy caterpillar *(Psalis pennatula)* ... 196
Fig. 142. Rose hairy caterpillar *(Euproctis fraterna)* ... 196
Fig. 143. Malaria mosquit *(Anopheles culicifacies)* ... 198
Fig. 144. Filariasis mosquito *(Culex pipens fatigans)* ... 198
Fig. 145. Dengue mosquito *(Aedis aegypti)* ... 198
Fig. 146. Gingelly gallfly *(Asphondylia sesami)* ... 199
Fig. 147. *Rice gall fly (Pachydiplosis oryzae)* ... 199
Fig. 148. Sand fly (*Phlebotomus minutus)* ... 200
Fig. 149. The common gad fly *(Tabanus triatus)* ... 201
Fig. 150. The cucurbit fruit fly *(Bactrocera* (*Dacus*) *cucurbitae)* ... 202
Fig. 151. The red gram pod fly *(Melanagromyza obtusa)* ... 203
Fig. 152. The common eye fly of man *(Siphunculata funicola)* ... 203
Fig. 153. Parasite of sugarcane top boreror *(Sturmiopsis inferens)* ... 203
Fig. 154. The common house fly *(Musca domestica)* ... 204
Fig. 155. The stable fly *(Stomoxys calcitrans)* ... 204
Fig. 156. The dog fly *(Hippobosca capensis)* ... 204
Fig. 157. The horse fly *(Hippobosca maculata)* ... 204
Fig. 158. The rat flea *(Xenopsilla cheopis)* ... 205
Fig. 159. The mustard saw fly *(Athalia lugens proxima)* ... 207
Fig. 160. Ichneumonid parasitoid *(Isotima javensis)* ... 208
Fig. 161. Bracon parasitoid *(Bracon brevicornis)* ... 209
Fig. 162. Chalcid parasitoid *(Brachymeria nephantidis)* ... 209
Fig. 163. Pupal parasitoid *(Trichospilus pupivora)* ... 210
Fig. 164. Egg parasitoid *(Trichogramma chilonis)* ... 210
Fig. 165. Bethylid larval parasite *(Perisierola nephantidis)* ... 211
Fig. 166. Red tree ant *(Oecophylla smaragdina)* ... 212
Fig. 167 - Black ant *(Camponotus compressus)* ... 212
Fig. 168. Social wasp *(Vespa orientalis)* ... 213
Fig. 169. Digger wasp *(Ampulex compressa)* ... 213
Fig. 170. Leaf cutter bee *(Megachile lanata)* ... 214
Fig. 171. The Indian honey bee *(Apis indica)* ... 214
Fig. 172. Tiger beetle *(Cicindela sexpunctata)* ... 216
Fig. 173. Predatory beetle *(Scarites indicus)* ... 217
Fig. 174. Water scavenger beetle *(Hydrophilus olivaceus)* ... 218
Fig. 175. Sugarcane white grub *(Holotrichia consanguinea)* ... 219
Fig. 176. Coconut rhinoceros beetle *(Oryctes rhinoceros)* ... 220
Fig. 177. Shining leaf chafer *(Anamola biharensis)* ... 221
Fig. 178. Rose flower cutting beetle *(Oxycetonia versicolor)* ... 221
Fig. 179. Dung roller beetle *(Heliocopris bucephalus)* ... 221
Fig. 180. Jewel beetle *(Buprestes parsina)* ... 222
Fig. 181. Click beetle *(Agrypnus fuscipes)* ... 222
Fig. 182. 'Khapra' beetle *(Trogoderma granarium)* ... 223
Fig. 183. Carpet beetle *(Anthrenus voras)* ... 223
Fig. 184. Wood boring beetle *(Stegobium paniceum)* ... 224
Fig. 185. Lesser grain beetle *(Rhizopertha dominica)* ... 225
Fig. 186. Saw-toothed grain beetle *(Oryzaephilus surinamensis)* ... 225
Fig. 187. Lady bird beetle *(Epilachna vigintioctopunctata)* ... 226
Fig. 188. Lady bird beetle *(Brumus suturalis)* ... 226

Fig. 189. Lady bird beetle *(Rodolia cardinalis)* 226
Fig. 190. Red flour beetle *(Tribolium castaneum)* 227
Fig. 191. Blister beetle *(Lytta tenuicollis)* 227
Fig. 192. Mango stem borer *(Batocera rufomaculata)* 228
Fig. 193. Grapevine stem girdler *(Sthenias grisator)* 228
Fig. 194. The pulse beetle *(Bruchus chinensis)* 229
Fig. 195. Red pumpkin beetle *(Raphidopalpa foveicollis)* 230
Fig. 196. Coconut red palm weevil *(Rhynchophorus ferrugineus)* 231
Fig. 197. Ash weevil *(Myllocerus pusthulata)* 231
Fig. 198. Mango leaf twisting weevil *(Apoderus tranquebaricus)* 231
Fig. 199. Stylopid *(Eoxenos laboulbeni)* 231

Glossary

Abdominal legs. Fleshy legs found in the abdomen of caterpillars bearing characteristic arrangement of crochets.

Acephalus larva. Larva with a reduced or no head capsule and appendages. The larva is also called a maggot (e-g) Larva of housefly and many other Dipterous insects.

Aedaegus. The male intromittent organ, refers to the penis plus accessory structures, used to introduce sperms into the female genitalia.

Aestivation. Dormancy during a hot or dry season.

Alate. Winged form in the life cycle of certain insects (e-g) Aphids.

Amphigonus. Having separate males and females within a species.

Antennae. Mobile, joined pair of appendages that articulate in the head in front or between the eyes of insects and serve as feelers.

Apodous larva. Degenerate type of larva, common in weevils, wasps bees and flies, in which the body appendages are completely suppressed and legs are totally absent and body poorly sclerotized (e-g) Larva of Dipteron flies such as house fly, blow fly.

Appendages. External limbs or similar structures arising from segments of the head, thorax or abdomen.

Apterous. Wingless form in the life cycle of certai insects (e-g) Aphids.

Atrophy. Decrease in size of a tissue organ or part after full development has been attained; a condition in which the affected cells undergo degenerative and autolytic changes, become smaller and have a lessened functional capacity.

Binomial nomenclature. The system of classification of organisms as advanced advanced by Carolus Linnaeus consisting of the name of a genus and the name of a species coupled together.

Bot. The larva of certain flies that are parasitic in the body of mammals.

Brachypterous. Winged adult insects; Having wings; In the case of insects such as, aphids, the comparatively short winged forms; The condition of having short wings that that do not cover the entire abdomen (e-g) Brown plant hopper.

Campodeiform larva. Larva more or less fusiform with depressed body, well sclerotized prognathous head, long thoracic legs and a pair of cerci. They are the most primitive larval forms and have no compound eyes or ocelli (e-g) Neuroptera, some Coleoptera, Strepsiptera and Trichoptera.

Carabeiform larva. Larva shaped like the larva of a carabid beetle, that is elongate, flattened and with well developed legs and filaments lacking at the end of abdomen.

Cardia. Valvular apparatus connecting the oesophagus and intestine.

Caterpillar. The polypod or eruciform larva of Lepidoptera or saw fly is generally called as caterpillar; Larval stage of butterflies and moths, known as eruciform larva, with distinct head capsule, chewing mouthparts, thoracic legs, abdominal prolegs and cylindrical in shape.

Cercus (pl. Cerci). One pair of appendages at the posterior end of the abdomen, usually long, filamentous and extending beyond the wings and abdomen.

Chitin. A complex nitrogenous carbohydrate forming the main skeletal body of the arthropods.

Chrysalis (pl. chrysalids). A naked or open pupa; Not enclosed or covered (e-g) Pupa of a butterfly

Cocoon. A silken case woven by a full-grown larva or a fibrous case made by a larva or an earthen case in the soil made by a larva in which it pupates.

Copulation. Sexual act between the male and female resulting in fertilization.

Corbicula. The pollen basket in bees located on the hind tibia, which is hardened on both the sides with fringes of long curved hairs for collecting pollen.

Cornicle. A pair of dorsal tubular structures extending from the dorsal area of the abdomen in aphids.

Crawlers. The early stage nymphs of mealy bugs, which have functional legs and can crawl about short distances.

Crochets. Small hooked spines located at the tips of the prolegs of butterfly and moth larvae.

Dengue. A virus disease of man marked by fever and severe pain in head, eyes, muscles and joints that is transmitted by certain mosquitoes.

Detrivorous. Organisms feeding on organic debris of plant or animal origin.

Deutonymph. The third instar nymphal stage in the development of mites and ticks.

Digonic hermaphroditism. Both sperms and ova are simultaneously produced by different parts of the same gonad.

Dioecious. Having the male and female sex organs in different individuals; Any one individual being either male or female.

Diurnal. Active during the day time; Habitually flying by day only.

Dormancy. A state of quiescence or inactivity.

Ecdysis / Moulting. The process of shedding the skin (Exoskeleton) periodically by insects and other arthropods.

Ecdysone. The moulting hormone of metamorphosis.

Ectoparasite. A parasite that lives externally and feeds from the host by sending in specialized organs or mouthparts to ingest nourishment from the host, but does not kill the host outright; A parasite feeding on the host from the exterior.

Egg case. The case or covering formed or secreted by an insect to contain or hold together the egg mass as a whole (e-g) Praying mantis.

Egg pod. Eggs laid in a compact cluster arranged in a vertical, spiral manner inside a capsule; A capsule enclosing the egg mass (e-g) Eggs of grasshoppers and locusts.

Egg raft. A compact mass of floating eggs consisting of 200-300 eggs, which are laid in water (e-g) Cu*lex mo*squito.

Elytra (Sing. Elytron). The hardened, modified, veinless, leathery or chitinous forewings of Coleopteran beetles and weevils that serve as protective covering to the membranous hind wings and commonly meeting in a straight line down the middle dorsum in repose.

Embryo. The young one before emerging from the egg.

Encephalous larva. Larva with a distinct, well developed, sclerotized head capsule, mouthparts and antennae (e-g) Larva of mosquito.

Endemic pest. A pest permanently established in a moderate or severe form in a defined area or commonly a country or part of a country.

Endoparasite. A parasite, which enters a host, lives within the host and feeds from the host.

Entognathous. Having the mandibles and maxillae retracted into spouches in the head **as in** the case of Protura, Collembola and Diptera

Entomophagous. Organisms, which feed on insects or their parts; Insectivorous.

Enzyme. A protein produced by living cells that can catalyze a specific biochemical reaction such as, digestion of food etc.

Ephimeral. Short-lived.

Epidemiology. The study of disease development and spread throughout a host population.

Epithelium. A layer of cells lining the internal or external surface of an organ or tissue.

Eruciform larva (Polypod larva). A larva with a more or less cylindrical body, a well developed head and with both thoracic legs and abdominal prolegs; A caterpillar (e-g) Larvae of butterflies and moths).

Exarate pupa (Free pupa, pupa libra or decticous pupa). Pupa remains uncovered, without puparium; All the appendages such as, the wings and legs are free from the rest of the body and the pupa resembles somewhat the adult condition (e-g) Beetles.

Exuviae. The cast off skin of nymphs or larvae at the time of metamorphosis.

Fecundity. The number of eggs produced, regardless of heir fertility.

Femur. The single-piece part of the leg in-between the trochanter and tibia; In jumping legs as in the case of grasshoppers, the femur is much enlarged and very strong and helps the insect to jump.

Filariasis. A human disease caused by the filarial nematode - Wu*chereria bancrofti tr*ansmitted by the mosquito - Cu*lex fatigans;* Also called as Elephantiasis.

Forma specialis (f.sp.). A group of races and biotypes of a pathogen species differing in some physiologic properties that can infect only plants within a certain host genus or species; A special form.

Frass. The wet or dry saw dust-like excrement of borers usually evident at their exit holes in fruits and other plant parts.

Frassy cocoon. A type of cocoon constructed by the larva with frass and saliva (e-g) Op*isina arenosella.*

Free pupa. A pupa in which the appendages and limbs are not fused to the outer covering.

Fringed wings. Wing margins fringed with long setae or hairs (e-g) Wings of thrips.

Genitalia. The sexual organs and associated structures; The external sexual organs.

Genus. A taxonomic category containing a single species or a monophyletic group of species, which is separated from other taxa of the same rank by a decided gap.

Gravid female. Adult female that is capable of laying eggs; Adult female insect containing fertilized eggs.

Gregarious. Forming aggregations; Living in groups.

Grubs. The larvae of Coleoptera and Hymenoptera are generally called as grubs.

Guiding ring. A cuticularized structure that surrounds and guides the stylet.

Haematophagous. Feeding on blood.

Haematophagus. Blood suckers; Feeding or subsisting on blood.

Hairy cocoon. A type of cocoon constructed with body hairs (e-g) Wooly bear.

Halteres / Balancers. Knob - shaped, modified hind wings of Diptera, which act as balancing organs

Hemelytra. Wings in which the basal half is thick and leathery, while the distal half is membranous. (e-g) Fore wings of Heteropteran bugs.

Hemicephalous larva. The head capsule and appendages are considerably reduced and the head is retracted into the thorax and the head capsule is in-between that of Encepalous and Acephalous larva (e-g) Brachycera, Tipulidae.

Hermaphroditism. Individuals functioning both as males and females.

Heterogenesis. A form of reproduction in some insects, which has sexual and asexual or parthenogenetic forms or some cyclic form of alteration of generations.

Hibernation. State of inactivity; Passing the winter in sleep or inactivity.

Histological changes. Changes in the cellular structure of tissues.

Honeydew. A sugary exudate secreted by sucking insects such as, aphids, white flies, scale insects and plant hoppers; Excess sugary plant sap secreted from the hind end of the gut by sap sucking insects.

Hormone. An internally secreted chemical messenger substance produced by endocrine tissues (glands), which influences or activates other organs or physiological processes within the same organism; A chemical secreted in one part of the body that regulates the function of some other part of the body.

Hyperparasitoid. An insect parasite of another primary parasitoid, also known as secondary parasitoid.

Hyperplasia. A plant overgrowth due to increased cell division; An abnormal increase in the number of cells of a tissue.

Hypertrophy. A plant overgrowth due to abnormal cell enlargement; An abnormal increase in the size of dells of a tissue.

Hypognathous. Having the head vertical and the mouth directed downward (e-g) Orthoptera

Hypoplasia. A defective or incomplete development of an organ, tissue or organ system.

Imago. The adult, Sexually matured form of an insect.

Immature stage. The sage of an insect prior to the imago (Adult form)

Ingestion. Taking in food; Eating.

Instars. Stages between moults.

Integument. The outer covering or skeleton of insects and other invertebrates; In insects and other arthropods, the integument is primarily composed of chitin.

Isthmus. Relatively narrow portion of the esophagus between the metacarpus and basal bulb.

Juvenile hormone. A small carbon chain hormone secreted by the corpora allata into the haemolymph that maintains the immature form of an insect during early moults.

Juvenile stage. Any or every post-embryonic stage in the development of an organism that precedes the sexually mature adult stage such as, egg, larval and pupal stages of insects.

Lacerate. Tearing of the tissues and causing injury.

Larva aculeata. Larva with dense, fur-like hairs.

Larva cornuta. Larva with fleshy horns or processes.

Larva furcifera. Larva with furcate processes.

Larval parasitoid. Parasitoid that deposits its eggs in or on the host larvae. The progeny of the parasitoid emerges from the host larvae, parasitizes and destroys the host larvae

Larviparous. Giving birth to larvae directly instead of laying eggs as in some Diptera

Longevity. Life span of an individual.

Longicorn. Having the antennae as long or longer than the body as in the case of Cerambycid beetle; Long-horned.

Looper. Caterpillar of the Family- Geometridae, Order - Lepidoptera with only one pair of abdominal prolegs in addition to the terminal claspers and which moves by looping its body; also called as a measuring worm (e-g) Daincha borer.

Maceration. Chewing without ingestion of the chewed material.

Macropterous. Winged; In insects such as aphids , the comparatively long-winged forms, which are mostly migratory in nature.

Maggots. The larvae of a fly (Diptera) are generally called as maggots; Vermiform larva devoid of true legs and without a well developed, distinct head capsule, the head-end of which is pointed and the posterior end somewhat flattened and appears as if cut off at the tip.

Malformation. Part of a plant or parts of a plant that is deformed, badly formed or badly shaped.

Metabolism. The entire process or series of changes or transformation of ingested food into tissue and cell substances and of these latter into waste products.

Metamorphosis. The series of changes in the bodily form, which most insects undergo as they develop into adulthood.

Mimicry. The ability of an animal to imitate or mimic another species of animal or plant.

Mosaic disease. A disease characterized by variegated pattern of green and yellow on the foliage of plants caused by certain viruses.

Moulting. Casting off the outer skin to facilitate further growth; Shedding or casting off of the cuticle.

Mycoplasma. Obligate plant parasitic pleomorphic mollicute (Prokaryote without cell wall) found in the phloem tissue; Microorganisms intermediate in size between viruses and bacteria, possessing many virus-like properties and ultramicroscopic.

Naiads. The aquatic ymph of a dragon fly, may fly or other hemimetabolous insects.

Naked pupa. A pupa not enclosed in a cocoon or other covering.

Necrosis. Death of a definite area of plant tissue.

Negatively phototrophic / Photonegative. Natural instinct to shun away from artificial light source.

Nocturnal. Active during the night time.

Nomenclature. The term nomenclature is derived from two Latin words viz., 'Nomen'= name and 'Calare'= to call, which literally means to call by name.

Non-persistent virus. A stylet-borne virus; A virus that does not undergo part of the development within the body of its vector.

Nuclear polyhedrosis virus (NPV). A virus that causes viral diseases of insects, mainly the larvae of certain Lepidoptera and Hymenoptera, characterized by the formation of polyhedral inclusion bodies in the nuclei of the infected cells.

Nuptial flight / Mating flight. A flight, which may be very long or short that precedes copulation between an adult male and female.

Nymph. An immature stage (instar), which is similar to the adult except for the absence of functional wings and sexual organs.

Obligate parasite. A parasite that can only lead a parasitic mode of life on a host and which cannot exist in any other way.

Obtect pupa. The pupa in which the appendages and body are completely united or fused by a hardening of the outer skin, termed the theca (e-g) Moth pupa.

Oesophagus. A tubular part of the fore gut between the mouth and the crop.

Oligopod larva. Larva having a well marked head, thorax and abdomen. Thorax bearing three pair of thoracic legs but no abdominal legs (e-g) Larva of beetles.

Omnivorous. Feeding on many different kinds of both plant and animal food.

Ootheca. A hardened egg case or covering formed by the secretion of accessory genital glands or oviducts as in cockroaches.

Opisthognathous. Having the front of the head deflected backward, so that the mouthparts project to the rear as in certain Hemiptera.

Osmaterium. Fleshy, tubular, eversible processes producing a penetrating odor, capable of being projected through a slit in the prothoracic segment of certain papilionid caterpillars or from openings elsewhere in the bodies of certain other forms.

Oviparous. Reproduction by eggs laid by a female.

Ovoviviparous. In the case of some species of insets, the eggs are retained in the genital tracts of the female parent until they are ready to hatch and so hatching occurs just before or as the eggs are being laid.

Parasite. An organism deriving all or a portion of its nutrition from another organism (host) while permanently or temporarily attached to or remain within the host.

Parasitoid. Entomophagous insects, which live on or in the body of another insect (host) or in or on any stage of the host from which they get protection and nourishment at least during any stage of their life cycle.

Parthenogenesis. Development of eggs without being fertilized; Reproduction from eggs without fertilization by sperm cells.

Pathogen. A disease causing organism such as, fungus, bacteria, virus, protozoa and such other organisms.

Pharate pupa (Prepupa). The non-feeding, quiescent stage between the larval and pupal period; Usually uring this period the cocoon or puparium is formed; In male coccids and thrips, the first instar in which external wing pads appear is sometimes termed prepupa.

Pheromone. Chemical substance that attracts members of the same species or one sex of that species, especially insects and nematodes.

Phloem. Food conducting and food storing tissue in the vascular system of roots, stems and leaves.

Phyllophagous. Feeding upon leaf tissues.

Phytophagous. Herbivorous; Plant eating.

Polyphagous pests. Pests infesting more than one unrelated species of plants.

Polypod larva (Eruciform larva). Larva having well marked head, a thorax of three segments, each bearing a pair of clawed legs and an abdomen of ten segments bearing five pairs of fleshy, hooked legs called prolegs (e-g) Caterpillar of butterfly.

Positive phototrophic / Photopositive. Natural instinct that makes organisms to move towards artificial light source.

Predator. An organism that lives by preying upon other living organisms.

Prepupa. A quiescent stage between the larval period and the pupal period; Also the last nymphal instar of thrips.

Proctodaeal feeding. Contents of the rectal pouch supplied as food by the workers in a termite colony to soliciting young nymphs, queen and soldiers.

Prolegs. The fleshy abdominal legs found in caterpillars bearing a characteristic arrangement of crochets; Also called as abdominal legs, pseudo legs, false legs or prop legs.

Pronotum. Dorsal surface of the prothorax.

Propagative virus. The circulative viruses that multiply in the bodies of their vectors (e-g) Rice tungro virus; Circulative virus.

Proteolytic enzyme. Protein splitting enzyme; Enzymes capable of splitting proteins into other compounds during metabolism.

Prothoracic shield. A thick, hard, protective layer of of skin covering the prothorax of larvae like a shield.

Protonymph. The first nymphal stage of a Tetranychid mite. It is also called the larval stage.

Pseudocoel. A false coelom or body cavity not lined with an epithelium of mesodermal origin containing a liquid in which various internal organs are suspended.

Pterygotes. Insects, which have wings; Winged insects.

Pulverulent. Powdery, dusty, covered with very minute powder-like scales.

Pupa. A resting stage in insect metamorphosis between larva and young adult, during which movement and feeding ceases and a great change in form occurs.

Pupa adecticous / Pupa complex. The appendages remain in separate sheaths and are only partly free as in the case of mosquito pupa (Tumbler) or adhere to the body and the abdomen is capable of movement as in the case of most Lepidopterous and Dipterous pupa or enclosed in last larval skin or puparium as in the case of some Dipterous pupa.

Pupa coarctata. The last larval skin is not cast off but formed into a puparium The insect inside cannot be seen. The adult insect emerges from the puparium by making a circular opening on the pupal case by means of special structures as in the case of pupa of house fly stable fly etc.

Pupa nuda. A naked pupa, which lies free of any attachment.

Pupa obtecta. The legs and wings are bound down to the body by moulting fluid and can be seen externally and the pupa is active . Adult insect emerges from the pupal case through a 'T'- shaped slit on its dorsal surface as in the case of mosquito, sand fly and horse fly pupa or the appendages are tightly oppressed to the body and are covered by an external skin and the abdomen is capable of movement. Such pupa known as a chrysalis has a posterior process or cremaster, which may be furnished with apical hooks or anchorage as in the case of some Lepidopterous pupa.

Puparium / Pupal case. After the last moult of the larva, the larval skin is not cast off, but becomes hardened and chitinous to form a case or covering.

Quiescent. A state of relative biological inactivity induced by physical factors such as, low temperature or other adverse environmental conditions.

Race. Sub group or biotype within a species or variety, which are morphologically identical but differs from other races by virulence, symptom expression or host range.

Rachis. Elongated main axis of an inflorescence.

Radicle. Part of the plant embryo that develops into the primary root.

Relative humidity (RH). The ratio of the amount of water present in a sample of air to that which could be contained in the same volume, if it were saturated.

Reproductive potential. The maximum possible rate of increase in population (Multiplication) of an organism; This is also called 'Reproductive capacity' or 'Biotic potential'.

Reticulate. Like a net; Net-like.

Rhizome. Mostly horizontal, jointed, fleshy, elongated, usually underground stem.

Rhizosphere. The soil immediately surrounding the roots of living plants; The area in the soil immediately surrounding plant roots and influenced by them.

Rosette. A short, circular cluster of leaves, that gives the plant a bunchy appearance; Short, bunchy habit of plant growth.

Scaly wings. Wings covered with small, colored scales, which are unicellular, flattened outgrowths of the body wall and are responsible for color as in wings of butterflies and moths.

Scarabeiform larva. A grub - like, stout, sub-cylindrical, 'C' - shaped, sluggish larva with a well developed head, short and diminutive thoracic legs , but no abdominal legs and has soft, fleshy body (e-g) Larvae of some families of Coleoptera

Scientific name. A Latinized name of a species internationally recognized; The scientific name of a species consists of the generic and a species name; Scientific names are always written in 'Italics'.

Scutellum. A triangular - shaped sclerite located behind the pronotum in Homoptera, Hemiptera and Coleoptra;

SDwarms. A large numbr of insects moving together simultaneously.

Sedentary. Stationary without any bodily movement; Remaining in a fixed location.

Semilooper. A caterpillar in which only one or two pairs of the abdominal legs are wanting and when in movement only small loops are formed (e-g) Castor semilooper.

Seminal vesicle. A portion of the male reproductive tract functioning in the temporary storage of sperms.

Serpentine mines. Winding larval feeding tunnels in leaves or stems of plants made by some species of Lepidoptera and Diptera.

Sessile. Permanently attached.

Seta (pl. Setae). A slender hair - like outgrowth of the integument.

Sexual dimorphism. Species in which the two sexes differ morphologically.

Slug caterpillar. Thick, short, stout and fleshy caterpillar with a short and retractile head and the body covered with poisonous *spines called* **'scoli'; Also** called as platyform larva (e-g) Parasa lepida.

Social insects. Insects that ive in groups or colonies generally with differentiated castes or or forms with specific functions in the colony (e-g) Termites, honey bees.

Species. Groups of interbreeding natural populations that are reproductively isolated from other such groups.

Spermatheca. An enlarged portion of the female gonad between the oviduct and uterus functioning in the storage of sperms.

Spermatocyte. A cell giving rise to sperm cells or spermatozoa.

Stadium (pl. Stadia). The time interval between moults in a developing insect. Stomodaeal feeding. A mixture of salivary secretion and regurgitated intestinal contents supplied as food by the workers to young nymphs, queen and soldiers in the colony.

Stomodeum. The anterior portion of the digestive system formed by the invagination of the body wall.

Stridulate. To make a noise by rubbing two surfaces together.

Stubble. Crop residue left over in the field after the harvest of the crop.

Stunting. The reduction in height of a vertical axis, resulting from a progressive reduction in the length of successive internodes or a decrease in their numbers.

Stylet. Stiff, relatively long, slender, hollow, tube-like feeding organ of plant parasitic nematodes or sap-sucking insects such as, aphids, hoppers etc.

Stylet borne virus. Virus that is carried only in the stylet and is not persistent in the vector; The infectivity of the vector is rapidly los unless there is a renewed access to a fresh virus source (e-g) Banana bunchy top virus; Groundnut rosette virus etc.

Sub-species. An aggregate of phenotypically similar populations of a inhabiting a geographic sub-division of the range of a species and differing taxonomically from other populations of the species.

Suckers. Young plants arising from the mother corm as side shoots as in the case of banana plants.

Sucking insects. Insects that suck blood, plant sap, nectar, or other liquid food.

Susceptibility. The state of being readily affected or acted upon by an injurious agent.

Suture. Impressed line where two body sections meet or where two elytra meet when closed as in the case of Coleoptera.

Syngonic hermaphroditism. Hermaphroditic reproduction in which both sperms and eggs are produced by the same gonad.

Synonym. A different name given to a species or genus previously named and described.

Systematics. Derived froms Greek word 'systema' meaning systems of classification. It may be defined as scientific study of kinds and diversity of organisms and of any or all relationships among them; The science of diversity of organisms.

Taxis. A directed response involving the movement of a living organism towards or away from a stimulus.

Taxonomy. It is derived from two Greek words viz., "Taxis' = Arrangement; 'Nomos' = law; It may be defined as 'The theoretical study of classification, including its basis principles, procedures and rules'; The arrangement of species and groups into system, which exhibits their relationship to each other and their place in a natural classification; The science of classification and naming of species; Arrangement of species and groups into a system, which shall exhibit their relationship to each other and their places in a natural classification.

Termitarium. The earthen structure or nest inhabited by a termite colony.

Terrestrial. Strictly living on the ground; Usually living on dry land.

Testis. Portion of the male reproductive system.

Thermotropism. The response or reaction of an organism to temperature changes.

Thigmotropism / Thibmotaxis. Response or reaction to touch or contact.

Tibi**al spur. A l**arge spine on the tibia usually located at the distal end of the tibia.

Tissue. A group of cells of similar structure, which performs a special function.

Tolerance. A basis for resistance in which a host shows an ability to grow and reproduce or to repair injury, despite supporting a pest population equal to that, which may cause damage to a more susceptible host

Toxicity. The capacity of a compound to produce injury; Ability to poison or to interfere adversely with vital processes of the organism by physico-chemical means.

Tracheal gills. Larval stages of some species of insects, which are quatic to some extent, breathe under water through gills, which are outgrowths of the tracheae; Respiratory organs of certain aquatic insects consisting of the outgrowths of the body wall containing numerous closed tracheoles into which dissolved oxygen can diffuse from the surrounding water.

Trail marking pheromone. An intraspecific chemical messenger produced by foraging ants and termites to indicate source of requisites to other colony members.

Transient phase. The stage of development, which lasts for only a short time.

Translocation. Transfer of nutrients or virus through the plant; Movement of chemical substances within a plant.

Transmission. The transfer or spread of a virus or other pathogen from one plant to another.

Transovarial transmission. The ability of a virus vector to transmit the virus to its progeny by means of eggs.

Trench fever. A non-fatal disease of human beings characterized by sudden onset of fever, headache, dizziness and pain in muscles and bones caused by *Ricketsia quintana* transmitted by the body louse through its faeces.

Tritonymph. The third stage nymph of a Tetranychid mite.

Trypanosomiasis. The disease caused by the presence in the body of a protozoan parasite of the Genus - *Trypanosoma,* marked by fever, anemia and redness of the skin, transmitted by the Tse tse flies; Also known as Africal sleeping sickness.

Tubercles. Small, wart - like protuberances on the surface of the body.

Tumblers. The aquatic pupal stage of mosquitoes.

Typical host. A host in which pathogenic microorganism or parasite is commonly found; Natural host.

Vagina. The cuticle lined canal leading from the uterus to the vulva.

Variety. A loosely defined taxonomic category below the level of sub-species usually possessing some visibly different character.

Vascular bundle. Strand of conductive structures, usually composed of xylem and phloem ; In leaves the small vascular bundles are called veins.

Vector. A living organism such as, insect mite, bird, higher animal, nematode, parasitic plant and human able to carry and transmit a pathogen, especially an insect, nematode etc. able to transmit a virus.

Vessel. A xylem element or series of such elements whose function is to conduct water and mineral nutrients.

Vestigial. Having the nature of a degenerate or atrophied organ, more fully functional in an earlier stage of development.

Virulence. State of being pathogenic.

Viruliferous. A vector containing a virus and capable of transmitting it.

Viviparity. A type of reproduction in some species of insects in which the embryonic development is completed within the body of the female parent and hence instead of laying eggs, young ones oe nymphs are laid.

Weevils. Insects belonging to Order - Coleoptera with prominent snout in the adult insects.

White ears. White-colored, dried up, chaffy ear heads invariably without any normal grains, produced as a result of stem borer attack, which cuts off the inner portion of the culm. The white ears come off quite easily when pulled; Also called as white head.

Wrigglers. The aquatic larval stage of mosquitoes.

Xylem. A plant tissue consisting of tracheids, vessels 'parenchyma cells and fibres; Wood.

Yellow fever. An acute infectious disease caused by a virus transmitted by Aedes aegiptii and marked by fever, jaundice and albumin and globulin in the blood.

Yellows. A type of plant disease characterized by pronounced chlorosis; A plant disease characterized yellowing and stunting of the host plant.

Zoophagous. Feeding on material of animal origin.

Zygote. A fertilized egg.

1

Insects

General Aspects About Insects

'Insects' are almost ubiquitous. They are found all over the world, in the icy regions of the poles, deserts, forests, mountain peaks, oceans and even in oil wells. They can adopt themselves to any climatic conditions and can live in any habitat where food is available. Insects are mostly small, air-breathing organisms of the animal world. Most of the insects live freely in nature. Many species are wild and harmless to human beings. Many are beneficial because of their role in the ecosystem. Some are a part of the food chains of wild creatures, while others through predation and parasitism hold in check many other species of insects. Insects also play a vital role in pollination. Relatively only a few species are harmful to man and are his chief competitor on earth. The harmful insects besides destroying crops cultivated by man, damage stored grains and other stored products. They do harm to animals, birds etc. raised by man by parasitism, as well as by spreading various diseases. Likewise they do harm to human beings also. They are also responsible for destruction of many household articles. Harmful insects are generally called as **'insect pests'**. The name **'Pest'** is derived from the Latin word **'Pestis'**, which means causing destruction.

According to evidences, insects came into existence in this world about 250 - 500 million years ago and have been a part of our earth for ever. Dragon flies have been on the earth for more than 300 million years and grasshoppers existed even before the Dinosaur age, On the other hand humans appeared in this world only about a million years ago. As such, only man had intruded into the domain of insects. Man started cultivating crops to meet his requirements of food, clothing and shelter. Simultaneously insects had gradually changed their original food habits and started feeding and destroying the crops raised by man. From that time onwards a rapport had started between the insects and man. Man consider insects as his prime enemy and by virtue of his ingenuity man had adopted various ways and means to destroy the harmful insects. Despite all his efforts, man had not been successful in eliminating the harmful insects and the struggle is going on even now.

Characters of Organisms Belonging to Phylum - Arthropoda

The name - **Arthropoda** has been derived from two Latin words viz., **'Arthro'** = jointed and **'Poda'** = foot. Thus the name - Arthropoda means organisms with jointed foot.

1. The body of organisms belonging to this Phylum is bilaterally symmetrical. The various parts of the body such as, mouth parts, antennae, eyes, legs, wings, reproducrory organs etc. are all symmetrically arranged on the left and right sides of the median.
2. The body is made up of several segments arranged serially one behind the other. In this respect, the Arthropods resemble earthworms, millipedes, centipedes etc. The segments are fused to form three distinct parts viz., the **head, thorax** and **abdomen**. Some segments of the body or all the segments of the body may have appendages such as, one pair of legs in each segment and wings in some segments. These simple, unspecialized appendages in each segment in the primitive state, during the process of evolution have become specialized to suit various specific functions such as, eating, walking, jumping, flying and a great variety of other activities.
3. Jointed foot of Arthropods are appendages, which are composed of a number of segments, each capable of special movements according to the nature of their living habits.
4. The body of Arthropods is covered by a non-living **exoskeleton** made up of chitin and protein. The exoskeleton gives not only protection, but also a mechanical advantage in that the way the muscles are attached to the appendages and the body wall resulting in great strength in proportion to their size.
5. The exoskeleton is secreted by the epidermis. This skin is periodically moulted as long as the organism continues to grow in size.
6. The muscular system is highly developed with numerous separate muscles comprising of finely striated and quickly contracting fibres. The way the muscles are attached to the appendages and the body wall results in great strength in proportion to their size. Even the small ants are capable of carrying heavy loads in their mandibles.
7. The digestive tract is complete. The mouth parts are provided with lateral jaws. The mouth parts may be adopted for biting and chewing or for piercing and sucking liquid food.
8. Cilia and flagella are absent.
9. The body is made up of **haemocoel**. The **Caelom** encloses only the excretory organs and the reproductory organs. Except this cavity, the rest

of the body is filled with blood and all other organs are floating in the blood. However, the coelom is greatly reduced.

10. Excretion is by means of **malpighian tubules**, which empty into the alimentary system.
11. The nervous system consists of the brain, dorsal ganglia over the mouth, a ring or collar round the gullet and a ventral chain of ganglia joining the paired the nerve cord.
12. The sexes are generally separate and the males and females are easily differentiated. Fertilization is internal and most of the Arthropods undergo a more or less extensive metamorphosis, which is accompanied by frequent moultings of the inelastic, chitinous exoskeleton.

Position of Insects in the Animal Kingdom

According to estimates 2.16 million animal species have been recorded throughout the world in the Animal - Kingdom as on the year 2022 (Animal -Wikipedia, NET). This is divided into 10 or 12 Phyla. Among the Phyla an important one is Phylum - Chordata. Under this Phylum, all the higher vertebrates such as, animals, birds, lizards, snakes, crocodiles, turtles, sharks, bony fishes, frogs etc are included. All other numerous species of invertebrates of the lower orders are included in the other Phyla. Of these, largest number of organisms including insects are classified under the Phylum - Arthropoda. In the Phylum - Arthropoda there are about 1.05 million species of insects under Class - Insecta, which constitutes about 86 per cent of species in the Phylum-Arthropoda **(Fig. 1a),** while under Class - Arachnida and Class - Crustacea there are about 60,000 and 25,000 species respectively and in Class - Chilopoda and Diplopoda put together, there are only about 10,000 species **(Fig. 1b)**. The main characteristics of organisms belonging to important classes of Phylum - Arthropoda are furnished in Appendix-1.

Further compared to all living organisms, both plants and animals put together (species-wise), insects alone account for just over 50 per cent **(Fig. 1c)**.

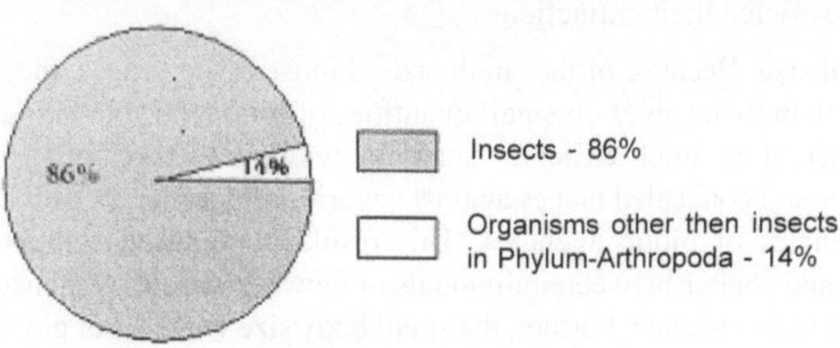

Fig. 1a. Percentage of insect species in Phylum - Arthropoda

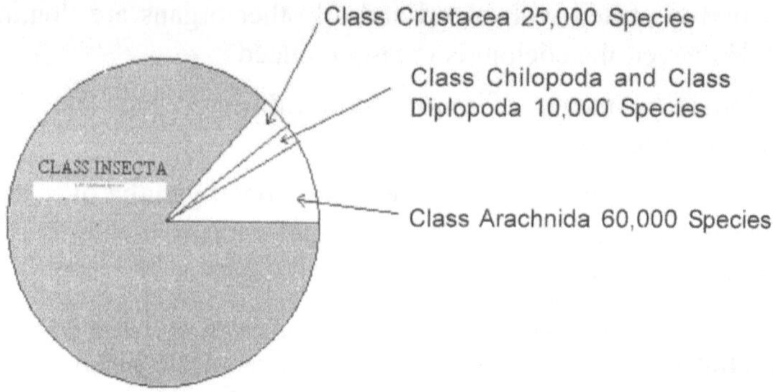

Fig. 1b. Number of species in each of the Classes of Phylum - Arthropoda

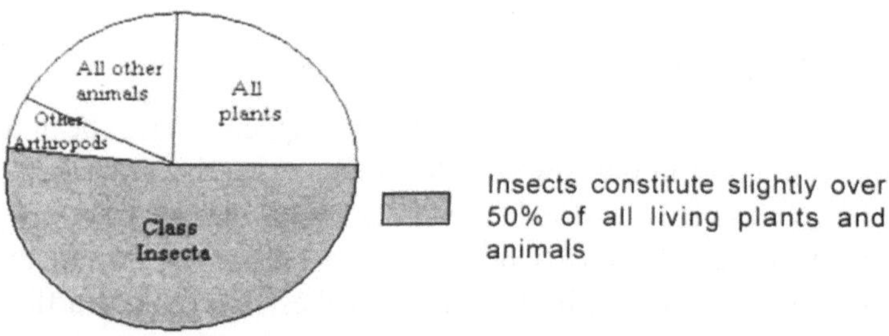

Fig. 1c. Percentage of insects in relation to total living organisms

Dominance of Insects Over Other Living Organisms

Insects possess several characters, which are not found in other living organisms, as such they adapt themselves to any kind of adverse environmental conditions, survive, exist, multiply and thus outnumber all other living organisms.

1. **Ability to adapt themselves to any kind of adverse environments.** Since their evolution insects have acclimatized themselves to live and carry on life under any type of adverse environmental conditions. This extraordinary capacity to adapt themselves to any kind of stress over millions of years has resulted in a state of perfection of their lives, which has restricted their extinction.

2. **Small size.** Because of the small size of most of the insects, they are able to sustain themselves on small quantities of food available to individuals. Further, it is much easier to small individuals to take refuge in small niches and concealed places against vagaries of weather, as well as against exigencies of biotic agencies. This results in reduced competition for food and shelter between individuals or between species and ensures their continued existence. Further, the small body size with less of physiological functions maximizes utilization of energy.

3. **Presence of exoskeleton.** The exoskeleton of insects is well developed and sufficiently hard to withstand extremes of unfavorable conditions such as, severe cold or drought. The exoskeleton forms a protective outer covering of the insect body, which helps to protect them from their natural enemies and other pathogenic organisms. Further, the exoskeleton is akin to the true skeleton of vertebrates on to which the body muscles are attached and enable the insects to perform their activities more efficiently.
4. **Diversity in food habits.** Among the insects there are **'monophagous'**, **'oligophagous'** and **'polyphagous'** ones. This specificity is not only to different plant species, but also to the different parts of the same plant species such as, foliage, stems, roots, fruits etc. For this reason the mouth parts of insects are also well modified for different modes of feeding. The polyphagous insects can feed and live on a number of collateral and alternate host plants in the absence of their primary host. Due to such diversity in food habits, the competition for a particular food becomes less. Further, the food habits of phytophagous, carnivorous and haematophagous insects vary considerably and as such the competition for specific type of food is very much reduced. Under such circumstances, a single insect species or a number of species can coexist simultaneously in a habitat.
5. **High fecundity and controlled reproduction.** The large number of eggs laid by gravid female insects, as well as the rapid rate of development together makes insects very prolific. The Lepidopterans, Coleopterans, Hymenopterans etc. have high fecundity and are capable of laying hundreds of eggs at a time. The cottony cushion scale insect - *Icerya purchasi* lays 500 - 1400 eggs at a time. The eggs of most of the insects are well protected and can withstand unfavorable environmental conditions and also escape from natural enemies. The aphids usually lay about a hundred young ones. Such high level of fecundity helps the insects to tide over natural calamities such as, floods, earthquakes, drought etc. and provides the insects more chances of survival. Such advantages possessed by the insects through the ages of evolution have helped them from the possibility of extinction. The social species of wasps, bees and termites have developed a high degree of control on reproduction. In a colony, all females except one or a few are rendered as sterile females. This controlled reproduction not only reduces wasteful reproduction, but also provides for division of labor.
6. **Different developmental stages under hormonal control.** The two different stages viz., the larval (feeding stage) and the adult (reproductive stage) in several insect species are quite different not only in their morphological characteristics but also in their requirements of both food and habitat and as such their internal competition for food and shelter gets

completely eliminated. In the absence of such competition, the chances of their survival are enhanced.

7. **Different modes of regeneration.** In many insect species, both sexual and asexual modes of regeneration are possible so that proliferation of their progenies is not hindered at any given time Even in the event of any natural calamities, their progenies are available for perpetuation. Aphids can reproduce sexually and give birth to offsprings. However, in the absence of males or when conditions are not favorable they are capable of reproducing 'parthenogenetically'. Further, when there is dearth of food in a particular habitat or when there is over population some insect species such as, aphids and hoppers produce macropterous forms, which can fly to other places for survival.
8. **Greater mobility.** Among the invertebrates, insects are capable of flying. This ability to fly enables them to move quickly from one place to another place for feeding and breeding. Because of their ability to fly, insects can easily escape from their natural enemies. Insects are also provided with different kinds of legs for the purpose of running, climbing, digging etc., which also help them to escape from their enemies. The hexapod mode of locomotion gives the insects high degree of stability while running.
9. **Migratory habit.** Like birds, many insects have the tendency to migrate. When living conditions become unsuitable or when there is lack of food, they migrate enmass to distant places, where living conditions are favorable and where there is enough supply of food.
10. **Defense mechanism.** In the course of evolution, insects have developed many defense mechanisms such as, protective coloration, mimicry, defensive chemicals etc. to safeguard themselves from their enemies and survive.
11. **Sensory appendages.** During the process of evolution, insects have perfected their appendages, which are vital for their existence. The antennae serve as the most important organs of feeling. The legs are not only useful for locomotion, but also useful for different functions such as, digging, clinging, jumping, swimming, for catching prey etc. in different species. The leg segments also serve as sensory and auditory organs. These sense organs forewarn insects of impending danger from their enemies and help them to escape from their enemies.
12. **Diverse respiratory system.** Insects have overcome problems of terrestrial respiration and can utilize atmospheric oxygen more efficiently. Insects are provided with lateral air openings called spiracles at the rate of two per body segment through which atmospheric air enters into a net work of air

tubes or tracheae. The tracheae ramify further into a net work of smaller tubes or tracheoles that carry and supply fresh air to every part of the insect body. This system, which is independent of the circulatory system found in higher animals, is more efficient and serves to adapt insects to different environmental conditions and enhance their survival potential.

13. **Zenith of evolution.** Insects survived through 259 million years or more on this planet and in the process of evolution have overcome the problems of terrestrial respiration, water conservation and locomotion in a terrestrial environment. The hardened exoskeleton, adaptive legs and wings, the ability to utilize atmospheric oxygen, their ability to sustain and adjust with nature, live and reproduce under any adverse conditions have made insects the most successful of all creatio.

Orientation and Tropism

The behavior of insects consists of responses to various stimuli. A response is the result of impulses that are set up by the stimuli, which travel over the nervous system to the effectors. The way an insect responds or reacts is determined by the pattern of pathways over which the impulses travel. The responses in many cases are automatic and are inherited over the generations. The internal stimuli and tropic reactions are mostly caused by environmental influences. The response to stimuli results in movement towards or away from the stimuli. Such orientation with respect to stimuli are called **'taxes'**. Several types of tropisms are noticed in insects.

i) **Phototropism - 'Response to light'.** Many insects are either attracted or repelled by artificial light. This reaction is termed **'phototaxis'**. Insects like the grubs of *Trogoderma granarium*, which show negative reaction to artificial light are said to be **'negative phototrophic'** or **'photo negative'**, while the moths of *Amsecta moorei*, which are attracted to artificial light are called **'photo positive'** or **'positive phototrophic'**. The photo negative insects are not affected by exposure to red light, whereas photo positive insects prefer to gather at violet light. This reaction may often be different in the various growth stages of the insect. The maggots of housefly are photo negative, while the adult flies are attracted by light. Bed bugs are photo negative at all growth stages.

ii) **Chemotropism - 'Response to chemicals'.** Positive response to certain chemicals often serves the insects in locating food, sites for oviposition, in finding a mate etc. A chemical producing a negative response is called a **'repellent'**. Certain flies and moths are **'chemo positive'** to some chemicals, oils, flesh etc. Many female insects oviposit only on their hosts after ascertaining the smell of the host plant. Chemotropism has many practical applications in pest control such as, use of poison baits, attractants and repellents, pheromone traps etc. Sorghum shoot fly adults *(Atherigona soccata)* can be attracted and killed by using fish meal traps. Fruit flies *(Dacus dorsalis* and *D. zonatus)* are attracted by methyl eugenol. Molasses attracts *Spodoptera* caterpillars and is used in poison baits to kill the larvae. Pheromone traps are used to attract the male moths of tobacco caterpillar or cut worms *(Spodoptera litura), cotton boll worm (Helicoverpa/Heliothis armigera)* etc.

iii) **Geotropism - 'Response to gravitational force'.** Larvae of many insects exhibit both geo positive and geo negative response at different growth stages. During the feeding stages, they crawl up, feed on the aerial parts of plants and exhibit **'geo negative'** response. However, they exhibit **'geo positive'** response during pupation. Fruit flies *(Dacus* sp.*)*, cutworms *(Spodoptera)* etc. exhibit positive geotropism during pupation.

iv) **Anemotropism - 'Response to wind'.** Many insects tend to orient themselves against air currents and wind. The butterflies found in large numbers during May and November tend to orient against the wind. Dragon flies show an **'anemo positive'** response. Unlike the air currents near the ground, the upper air currents are strong and rather constant in direction. Many insect species are known to spread to new localities or new countries through air currents. Young nymphs of San Jose scale, some other scale insects, aphids etc. are known to be carried away by air currents to distant localities or even to other countries.

v) **Rheotropism - 'Response to water current'.** Many aquatic insects swim upstream , thus exhibit **'positive rheotropism'**, whereas some other insects are carried by water currents to distant places.

vi) **Thigmotropism - 'Response to touch'.** Some insects move away or retract when touched, while some insects such as, cockroaches move into some type of hiding places. Some insects such as, larvae of *Heliothis armigera* curl and fall down, thus exhibiting **'negative thigmotropism'**.

vii) **Thermotropism - 'Response to temperature'.** Temperature controls the development, fecundity, feeding, movements and dispersal of many insects. Various insects have different ranges of temperature at which they thrive. Every insect has a preferred temperature range. Some insects thrive at high temperatures, while some others require low temperatures for their survival. Mosquitoes are more abundant in places having a temperature range of 21.0° - 27.0° C, but rare in places where the temperature reaches 44.5° C and above. Rice weevil *(Sitophilus oryzae)* cannot tolerate high temperatures beyond 32.0° C. When the temperature goes high they migrate to the upper layers, which are generally cooler. Honey bees respond positively and leave the hive at high temperatures and bright light, but respond negatively by remaining in the hive at low temperatures even at the same light intensity.

viii) **Heliotropism - 'Response to sunlight'.** The growth and moulting of insects are accelerated or retarded according to their requirement of sunlight. Generally an insect that lives normally in a well lit environment is adversely affected if subjected to subdued light or darkness and vice versa. Silk worms *(Bombyx mori)* develop faster in light than in darkness. The grubs of Khapra beetle *(Trogoderma granarium)* develop more rapidly in sunlight. The moths of the spotted boll worms of cotton - *Earias* species and the groundnut red hairy caterpillar moths *(Amsacta albistriga)* lay most of their eggs during periods of darkness or twilight, thus exhibiting negative heliotropism. Many moths are **'helionegative'** and are active only during the night time.

ix) **Hydrotropism - 'Response to moisture'.** Some insects tolerate high humidity, while some others are harmed by high humidity. Saturated condition of moisture is directly injurious for the development and viability of cotton spotted boll worm *(Earias fabia)*. Rainfall also affects the life of insects in different ways. High rainfall affects cutworms *(Spodopyera liura)* by increasing the moisture content of the soil in which the pest lives. If there is more than 12.5 cm. rainfall during May - July, the consequent increase in soil moisture is unfavorable to the pest and it is forced to come out of the soil and fall a ready prey to their parasites and predators. Rainfall plays an important role in inducing emergence of several species of insects from the soil, especially black ants and termites. The winged stages of these insects, though ready for emergence some time during the summer, would not actually come out until after a good shower of rain. Another instance of similar behavior is afforded by the hairy caterpillars *Amsacta albistriga* and *Amsacta moorei*, which also emerge from their pupa in the soil only after a few good showers.

Adaptation of Insects for Survival, Offence and Defense

Protection from natural enemies and adverse climatic conditions is necessary to safeguard insects and to this effect insects possess a great variety of protective structures and mechanisms. Above all, insects require food for their development and multiplication without which they cannot survive.

i) **Food.** Each insect species has certain nutritional requirements for the completion of its life cycle. If the required food for any particular species gets reduced in a particular locality, that species will gradually disappear from the area even though other environmental conditions may be favorable. Under normal conditions there may be sufficient food for the insect to live on and continue its race. In the event of sudden increase in the population of a pest, the density of its population becomes too high to be supported by the food available in the area. Under such conditions some individuals of the species either migrate to other suitable areas or starve if the species has limited power of dispersal such as many coccids. According to the nutritional requirements the insects are characterized as **'omnivorous'**, **'carnivorous'** and **'herbivorous'**. Those insects, which feed on many cultivated and wild plants are termed **'polyphagous'** (e-g) Locusts, grasshoppers etc. If the feeding activities of a particular species of insects are confined to only to plants of one botanical family it is known as **'oligophagous'** (e-g) Cabbage butterfly - *Pieris brassicae*. When a single species of plants constitute the food of a particular species of insects, it is termed as **'monophagous'** (e-g) Egg plant fruit borer (*Leucinodes orbonalis*).

Quantity of food available in an area determines the population of an insect species. Every insect multiplies up to the limit of its food supply, if other environmental factors are favorable. Further, different vegetation zones harbor different insect communities.

ii) **Inherent rate of increase.** Every organism has an inherent urge or capacity to survive, reproduce and multiply in numbers. This capacity is referred to as **'biotic potential'**. The biotic potential of a species depends upon its fecundity, sex ratio, number of generations per unit time and its ability to withstand adverse environmental factors and maintain its numbers.

Some insects such as, white ant queen and house flies lay large number of eggs. Some insects reproduce very rapidly. The mustard aphid - *Lipaphis erysime* has over 40 generations in a year. House flies, *Drosophila* flies, aphids etc. reproduce very fast. However, survival potential is the net result of biotic potential and environmental resistance. In this respect the polyphagous insects are better fitted than the oligophagous and monophagous insects, because they have more nutritional potential. Some termites can live on cellulose alone with the help of microbes, which live in their bodies. Insects such as, moths, gryllids, grasshoppers, cockroaches, cutworms etc. work at night and thus avoid many of their potential enemies.

iii) **Protective structures.** Most of the insects possess well sclerotized exoskeleton, which affords protection to the insects. The cuticle is often prolonged into bristles, spines, hairs or scales, which protect the insects from mechanical injury, from excessive heat or cold, from evaporation of water from the body or from natural enemies. Several caterpillars have offensive weapons in the form of urticating hairs. In certain caterpillars the hairs with venoms irritate or poison their enemies that touch the hairs. Some beetles have evertible glands, which they turn inside out and project their secretion outside their bodies to repel their enemies. The body fluids of blister beetles and some other insects are corrosive or poisonous. Many insects, especially the bugs have repulsive odors. The Reduvid, Corixid and Notonectid bugs have piercing mouthparts. Some such as, the larvae of citrus butterfly - *Papilio demoleus* give off bad odors or bitter secretions from evertible glands on their head, the sudden protrusion of which may have frightening effect on certain enemies.

iv) **Protective constructions.** Silk is much used by many insects for building shelters. Gingelly shoot webber *(Antigastra catalaunalis)*, castor shoot and capsule borer *(Dichocrosis punctiferalis)* groundnut leaf folder *(Stomopteryx subsecivella)*, rice leaf folder *(Gnaphalocrosis medinalis)* etc. construct nests by webbing together leaves and remain securely inside the nests. To protect themselves or their young ones insects build many types of constructions. The bark borers construct long, winding galleries with silken thread, frass and bark tissues. In some cases, the larval stage is protected by some type of cases into which they can retract in case of danger as in rice case worm *(Nymphula depunctalis)*, coconut bag worm *(Monatha alpipes)* etc. Many moths construct cocoons to protect the pupal stage. Some insects construct earthen cocoons for pupation and many insects pupate deep inside the stems or fruits. Mealy bugs and many aphids have waxy coating over their bodies. The scale insects live inside a hard, scaly covering. Ants construct earthen mounds and termites build soil pyramids for their safety. Banana lace wing bug *(Stephanitis typica)* lays eggs on the tips of elevated, stiff threads. Cockroaches lay their eggs inside thick egg pods. Locusts and grasshoppers lay their eggs in egg pods deep under the soil. Dragon fly larvae cloth themselves with a covering of debris to disguise their appearance. The larvae of lace wing bugs and tortoise beetles of sweet potato *(Aspidiomorpha miliaris)* pile their own excrement and shed skins on their backs for concealment. Some insects cover themselves with skins of their prey. In many flies the skin of the last larval instar , instead of being shed is retained on the pupa as a very hard and thickened case, which is known as the **'puparium'**.

v) **Protective size, form and color.** Size may have protective value for some insects against certain enemies. Extremely small insects may be overlooked by larger enemies and large insects may appear too formidable for certain of their smaller enemies. Some of the larger insects found are the tropical silk worm moths such as, *Antheris, Attacus* etc., which have wing span of 25.0 - 30.0 cm. Other giants among insects are some of the stag, dung and longicorn beetles, grasshoppers and stick insects. Some of the wood-boring larvae also reach 12.5 - 15.0 cm. in length. The smallest of insects are found among some beetles, flies, leaf hoppers, ants, parasitic wasps etc.

Some insects such as, the puss moth caterpillar - *Cerura liturata,* the sphinx caterpillar - *Acherontia styx* etc. assume threatening postures to frighten their enemies. Some harmless insects have color patterns of stinging wasps, bees, butterflies and bugs, which are distasteful or unpleasant to their enemies. Such color patterns termed **'warning coloration'** mislead their enemies, which avoid them.

Another important device for protection is known as **'mimicry'** by which one species imitates the appearance of another better protected species and thereby sharing its immunity from destruction; the former is known as the **'mimic'** and the latter as **'model'**. Examples of mimicry are found chiefly among some butterflies, moths, bugs, wasps and flies. Many butterflies and some other insects, which are edible to birds, frogs, toads etc. mimic as other butterflies or insects in shape, color or both, which are poisonous or bitter in taste and unpalatable. Some flies mimic as wasps, which have stings.

Some larvae sit erect, sway about and look like snakes. The sphinx moth *(Acherontia styx)*, while feeding looks like a humming bird and the hornet moth *(Aegeria rapiformis)* resembles a stinged hornet. Mimicry is categorized as **'Batesian mimicry'**, in which a vulnerable species imitates another that is not palatable or better protected by its warning coloration; as **'Mullerian mimicry'**, in which two or more well defined unpalatable species have developed the same pattern of warning coloration and as **'parasitic mimicry'** in which the host mimics as a parasite.

Some beetles such as, the blue death-feigning beetle *(Asbolus verrucosus)* when faced by enemies drop down and feign death. A few insects have protective or cryptic coloration or assume the color pattern of the surroundings. Thus the leaf and stick insects have protective resemblance and almost merge with their surroundings and are difficult to detect unless they move. The leaf butterfly - *Kallima* species, which resembles the dead leaf on which it rests is difficult to detect. Grasshoppers in general are inconspicuous in their surroundings and thus are able to feed uninterrupted without fear from their enemies.

vi) **Protective habitations.** Insects, which burrow into the soil or tunnel into trunks of trees or inhabit fermenting inorganic substances or swim in water or live on the skins or inside the bodies of animals gain a greater degree of security from variations in temperature, excessive evaporation, storms, parasites, predators etc.

vii) **Protective behavior.** Some insects take shelter or hide at the approach of certain enemies. Others depend on their legs and wings for escape. The stinging insects belonging to Hymenoptera such as, honey bees and wasps take the offensive when danger threatens them. Some insects, which have no specific weapons for offence, threaten to fight and may succeed if they resemble well protected insects. Many insects are nocturnal and confine their activities only during the nights.

viii) **Diapause.** This is a state of developmental arrest, wherein the dormant stage occurs due to the temporary stoppage of growth, which results from reduced metabolism and enhanced resistance to adverse environmental conditions. Diapause, because of very low temperatures is termed as **'hibernation'** and due to high temperatures is termed as **'aestivation'**. It is an important adaptation to overcome the unfavorable seasonal climatic conditions for continuous multiplication and to preserve the species. The diapause may occur in any growth stage, from egg to adult. Besides adverse environmental conditions, lack of food may also induce diapause. The red hairy caterpillar of groundnut undergoes aestivation as pupa in summer and emerges as adult moth only after the receipt of first showers. The mango hoppers - *Idiocerus* species hibernate in the adult stage on the stems and among leaves and become active in spring.

ix) **Intraspecific relationship.** Association of the two sexes of the same species, parental care, community associations among social insects, assemblage etc. are some other important adaptations of insects.

'Parental care' is exhibited in varying degrees among various groups of insects. All insects take care that the eggs are laid where conditions are favorable for hatching and where there is ample food for the emerging young ones. Some insects take further precautions and lay eggs in protected places or cover them with foreign materials or prepare brood nests to protect them from natural enemies. The solitary mud wasps *(Eumenes conicus)* actually provide food beforehand in their nests for the use of the young larvae. Dung beetles *(Heliocopris bucephalus)* roll dung of animals and carry this to their underground nests to serve as food for their future larvae. Some earwigs such as, *Labidura riparia* brood over their eggs in their underground nests. Insects like the burying beetles

(Necrophorus species*)* extend their parental care to the feeding of young ones with predigested meat. *Belostoma* giant water bugs carry the eggs on their back. Some flies stick their eggs to the bodies of biting flies, with the result they are carried to mammals they parasitize. Normally the female is involved in this relationship, but in some insects the males also cooperate in this function. Some aquatic cockroaches carry their eggs under their fore wings. Some flies stick their eggs to the abdomen of mosquitoes and *Stomoxys* stable flies with the result, they are carried to mammals, which they parasitize.

x) **Social insects.** Many insects occur in groups and these groups differ in the factors responsible for bringing the individuals together. Some groups are simply the result of a positive reaction by many individuals to the same stimulus (e-g) insects attracted to light or to a dead animal. Some may result from the simultaneous emergence of many adults as in may flies. The larvae of many Lepidoptera, which hatch from a given egg mass may remain together during most of their larval stage as a result of their reaction to a common food supply. Ants, termites, honey bees and some wasps live in much more integrated colonies called societies and these societies have some special features. Among social insects, the intraspecific relations are for the good of the colony. Besides large number of individuals living together with their mother in a common abode, polymorphism or caste differentiation of its members is evident. Besides polymorphism, division of labor is involved among the members for the welfare of the community as a whole. For example, in white ants , the queen, which is the sexually mature female entirely devotes her time to egg laying, often at the rate of one egg per second and is generally described as egg laying machine. The workers are very efficient laborers and attend to all sorts of duties including the care of eggs, young ones and the royal pair, building and cleaning the termitarium, growing of fungus gardens and procuring food for all the individuals of the colony. The soldiers protect the colony. Similar organization exists among social ants. Among social bees, there is further specialization and the sterile female workers according to their age perform all the duties of the colony except reproduction.

The workers in some ant colonies are of different sizes and the activities of these different sized workers may differ. The larger ones act as soldiers; the intermediate sized workers do most of the foraging and regurgitate food to other workers and the smaller ones are concerned chiefly with tending the brood, nest construction and other chores inside the nest.

Social insects maintain a means of communication between themselves. The chief communication in ants is by means of **'pheromones'.** In the case of honey bees, a worker upon returning to the hive after finding nectar source communicates information about this source to the other workers by means of **'characteristic dances'**.

Social insects have a home base, the nest and the individuals that forage away from the nest return back without any difficulty. Social insects form new colonies either by swarming as in the case of honey bees, where the queen along with a group of workers leave and establish a new colony and a new queen is developed in the group that did not leave or by the parental forms leaving and starting a new colony as in some ants.

Insect societies are often associated with various other organisms such as, other insects, mites and sometimes fungi. Some of these may be predators or parasites, while some are scavengers and the others guests, which are not only tolerated, but actually fed by the host insects. Many social insects cultivate fungi inside their nests, which they use as food. Many ants tend aphids and feed on the honey dew secreted by them.

Another important aspect of the bond governing social behavior is the mutual exchange of nourishment between adult and young ones. Thus, wasp larvae exude saliva in exchange for food supplied by the workers. This type of behavior is met with in termites also, but not in bees, perhaps because of the availability of pollen and nectar from external sources.

There are some differences between the societies of termites and those of the Hymenoptera. Termites have a child labor society in that young insects from about the second instar onwards must take care of themselves and remain as workers for the rest of their lives, except those that become soldiers or reproductive forms. The Hymenoptera society is built on adult labor and the helpless young ones are cared for by the adult workers.

The majority of insects are solitary and live for themselves in-between them and the social insects are those, which assemble or are gregarious during a part or whole of their lives. Desert locusts have gregarious and solitary phases. The Embiidae live in silken tunnels in colonies of over 20 individuals in each. Living in such groups is probably for protection. Some coccinellid beetles and fruit flies congregate for hibernation under dry creepers or other host plants.

Intraspecific relationship has yet another aspect. Under normal conditions of stabilization there is enough food and space for a particular species in a particular locality where it multiplies. When the density of population becomes high, the overcrowding proves harmful to the species as competition for food and space results, which may affect the individuals inhabiting the habitat. Too much overcrowding and exhaustion of food often results in **'cannibalism'.** This is particularly so among the predators such as, preying mantis. Even in polyphagous insects such as, red cotton bug and locust hoppers cannibalism is noticed under such conditions. Insect pests of stored products such as, *Tribolium* red flour beetles often start feeding on their own eggs, when the population density becomes high. Cut worms *(Spodoptera litura)* also exhibit this tendency. Competition for space is important in some groups of insects as in the case of wood borers. In bark borer beetles, usually competition occurs mostly between the larvae and sometimes between the larvae and the adults.

xi) **Interspecific relationship.** A large number of insects live at the expense of others and cause injury of varying degrees to the latter. The insects, which depend for their lodgings or sustenance or both on other animals are known as **'parasites'.** Numerous insects pass their early developmental stages inside the bodies of other insects as **'internal parasites'**, **'endoparasites'** or **'parasitoids'** and most of these are found in the groups of wasps and flies. The adult female parasitoid usually lays its eggs on or inside the body of the insect host. The egg hatches into a larva, which feeds on the internal contents of the host body, pupates and emerges as an adult parasitoid. In the mean while, the host, though it survives the attack for a time, gradually dies. Frequently the parasitic habit of some insects becomes economically important, when the host happens to be a pest of economic importance. Several wasps and flies act as natural enemies in nature and control many caterpillar and beetle pests of crops such as, rice, millets, cotton, sugarcane, coconut etc.

In contrast to this there are some insects, which act as hunters, capture and feed on other insect victims. These are known as **'predators'.** The only marked difference between parasitism and predatory habit is that in the former the death of the host is a slow and gradual torture, while in the latter, the death of the prey follows almost immediately after the capture by the predator. Well known examples of predatory insects are beetles of different kinds including the tiger, ground and lady bird beetles, the dragon flies, robber flies, antlions, the mantis, some stink bugs, the giant water bug, hornets and a few aquatic insects.

Some parasitoids are sometimes parasitized by some other insect species and these insects are called **'hyperparasites'.**

In addition to these important inter relations among insects as parasitoids and predators, there are instances of insects living as associates, mess mates, guests, slaves or friends with one another. These types of inter relationship between organisms of different species, which live in a more or less intimate association is termed **'symbiosis'**. Such associations may be between two insect species or between an insect and an animal, microorganism or plant. One species may feed on the food collected or stored by another species, thus indirectly harming the latter. This type of symbiosis is common among white ants, wasps, bees and ants and is known as **'social symbiosis'**. When both the symbionts are benefitted by the association, the phenomenon is termed **'mutualism'**. Termites and the flagellate protozoans inside their guts show mutualism. Termites feed on wood, which is practically cellulose, which cannot be digested without the aid of flagellates present in their stomach. The waste products after the cellulose has been digested serve as food for the white ants. Similarly some species of ants white ants, wood boring weevils and wasps cultivate certain types of fungi, which can grow only under the environment and care provided by the host. The fungi in return provide nutritious food to these insects. Ants, some aphids and coccids also exhibit this type of relationship. The ants provide shelter to these insects, which in turn supply honey dew to the ants. Some scarabeid beetles and staphylinids are accommodated in the nests of termites. Honey bees visit flowers for obtaining nectar and pollen for their own use and in doing so, they incidentally bring about cross pollination.

Beneficial Insects

Though many of the insect species are injurious to man, his crops and his livestock, not all insects are harmful. Some insect species are beneficial to man in many ways. Some are productive insects and yield materials directly useful o man. Of these the most important ones are the honeybees, the silk worms and the lac insects. Some other insects are useful to man indirectly in many ways.

Major Productive Insects

1. **Honey bees.** Honey bees are one of the most important productive insects in that they provide honey and bee wax. The practice of rearing honeybees under artificial conditions in bee hives for obtaining honey collected by the bees on a commercial scale is called **'Apiculture'** and the location where the hives are maintained is called an **'Apiary'**.

Of the different species of honey bees, the Indian honey bee *Apis indica*, which live normally in cavities of tree trunks, hollow of rocks, cavities in walls, wooden boxes, packing cases etc. is amenable for domestication and is reared in modern bee hives under closed conditions. The average yield of honey ranges from 2.0 - 2.5 Kg. in the plains and 3.5 - 4.5 Kg, in the hills per hive per year.

The European honey bees *Apis mellifera* is reared in many European countries, United States of America and a few other countries. It gathers a much higher quantity of honey, which is 9 – 10 times more than the Indian honey bee. It has been introduced to India and is successfully reared mostly in Punjab, Jammu and Kashmir on a large scale.

The bees get activated in the morning by about 7.00 or 8.00 AM, depending upon the sunshine and temperature, and the worker bees go out for foraging and collect pollen, nectar. propolis and water and carry them to the hive and make repeated trips till sunset. The bees that go out first and find out new sources of these materials are called **'searches bees'** or **'scout bees'**. They return to the hive and communicate the message to young foraging bees by means of definite patterns of dancing. They perform a rather circular dance to indicate the direction of food source. The bees usually forage within a distance of about 100 meters radius from the hive, but may go up to a distance of 1.5 KM. They are capable of flying at speeds up to 30 KM/hour and forage actively at temperatures ranging from 25° - 27°C.

Pollen is collected from flowers. The workers load both the pollen baskets fully and return to the hive. A bee can carry up to 35 per cent of its body weight of pollen in one trip and may make about 6,000 trips in a day and collect 500 - 1,000 gm. of pollen. Nectar is collected from the flowers and stored in the crop, where it is mixed with saliva. The enzyme invertase present in the saliva converts the sucrose in the nectar into dextrose and laevulose. The bee returns to the hive and regurgitates the contents of the stomach into the comb cells, which are covered by flat, air-tight caps when the cells are full. A worker makes about 19,000 trips a day for collecting nectar. The bees also collect propolis, a resinous material exuded by buds, leaves and other parts of some trees and shrubs and carry it to the hives in the pollen baskets, which is used to carry out repairs of cracks and crevices in the combs. Water required by the colony is also collected from wells, streams, channels and from plants.

Tamarind trees, neem trees, eucalyptus trees, prosophis trees, glyricidia trees etc. provide large quantities of nectar. Sorghum, maize, lesser millets, rose, palmyra, castor, pomegranate, sugar beet, several vegetable crops, flower plants, many weeds etc. are good sources of pollen. Banana, citrus plants, apple, pear, cotton, guava, mango, coconut, gingelly, sunflower, mustard, cucurbits, lady's finger, onion, many flower plants etc. are good sources of both nectar and pollen.

Honey and its uses. 'Honey' is an aromatic, viscid, sweet product of conversion of nectar of flowers by the honey bees. It contains about 20 per cent moisture, 60 - 75 per cent invert sugars viz., dextrose and levulose, about 5.0 per cent sucrose, Minerals such as, iron, copper, manganese, magnesium sodium, potassium, calcium, silica and phosphates, acids such as, acetic butyric, citric formic, lactic, maleic and succinic acids, some amino acids, enzymes such as invertase, diastase and phosphatase and vitamins A, B and C. A pure sample of honey weighs 1.4 Kg. per litre.

Honey is a rich, readily assimilable, energy giving food and with milk forms a perfect and complete food. It is so highly nutritious that 1.0 Kg. of honey is equivalent to 5.0 Kg. of milk, 1.5 Kg. of meat, 40 oranges or 50 eggs in food value. Honey is used as a laxative and a blood purifier and as a curative for ulcers in the tongue, sore throat. Use of honey is advocated for treating stomach and intestinal ulcers. It is used as a carrier in many Ayurvedic and Unani medicines. Honey is used as an ingredient in many skin and beauty lotions and also used in the manufacture of chewing gums.

'Bee wax' is an important bye product of bee keeping industry. It is used in the manufacture of several items of cosmetics such as, beauty creams, lipsticks, beauty lotions, ointments and polishes for shoes, floors and furniture. It is also used in the manufacture of lubricants, paints, varnishes, inks, electrical insulating equipments, candles etc.

2. **Silk worms.** Another important and valuable product obtained from insects is silk, which is commercially produced on a large scale from silk worms. In India, Karnataka, Bengal, Assam, Punjab and Kashmir are the main states where silk worm is grown on a large scale. In Tamil Nadu, silk worms are reared on a commercial scale in parts of Coimbatore and Salem districts.

Commercial silk is obtained from four species of silk worms viz., The mulberry silk worm *(Bombyx mori)*, The Indian Tassar silk worm *(Antheraea paphia)*, The Indian Muga silk worm *(Antheraea assamia)* and The Indian Eri silk worm *(Philosamia ricini)*. The art of rearing silk worms for the production of silk on a commercial scale is known as **'sericulture'**. A single caterpillar is capable of producing 650 - 1,300 metres of a single silk thread. Silk thread is elastic, resistant and a non-conductor of heat and electricity, has good tensile strength and nearly as strong as steel.

The mulberry silk worm is completely domesticated and is never found in the wild state. It feeds primarily on mulberry leaves and is the most important of all silk worms and produces very fine quality silk. Silk thread is reeled from processed silk worm cocoons. A processed dry cocoon weighs about 1.5 - 1.9 gram. The chrysalis forms 60 per cent and the silk matter 40 per cent of the weight of the cocoon. About 5,500 cocoons are required to yield 1.0 Kg. of raw silk and to produce 1.0 Kg. of silk, the silk worms have to consume about 1.0 ton of mulberry leaves.

The Eri silk worm is reared on a large scale in the states of Assam and West Bengal and in Bihar, Orissa, Uttar Pradesh and Tamil Nadu on a small scale. This is also a domesticated in sect. It is a polyphagous insect. Its primary food is castor, but it feeds on the leaves of tapioca, papaya and a few other plant species. The silken thread is not as glossy as the mulberry silk and is not a continuous one and hence not reelable . About 1,600 cocoons are required to get 1.0 Kg. of silk.

The Tassar silk worm is not a domesticated one and is usually found in the forest areas and feeds on Dalbergia, Shorea, Terminalia, Zizyphus, Ficus etc. The silken thread is continuous and hence it is reelable. The silk is white and bright but coarse.

The Muga silk worm is amenable for domestication, but are mostly found in the forest areas and feeds on a variety of forest tress. The cocoons are mostly collected from the forest areas. The silken thread is a continuous one and hence it is reelable.

Uses of silk. (i) Silk is primarily used in the textile industry for the manufacture of glistening and long lasting silk garments (ii) Silk is used by the defense services for he manufacture of parachutes because of their hard and tensile strength (iii) Silk thread is used as guts in surgical suters (iv) The chrysalis obtained after extraction of silk are dried and used as manure because of their rich protein content (v) The fatty matter obtained from the cocoons are used in the manufacture of soap.

3. **Lac insects.** Another important product of commercial value obtained from insects is 'lac'. As in the case of all scale insects, which secrete a hard encrustation over their body, the lac insects *(Laccifer lacca / Kerria lacca)* secrete a brown, hard, resinous encrustation, that is the true lac from which commercial shellac is manufactured. Ion nature, the insect is found growing on a number of trees in India, Myanmar, Burma and Thailand. India produces about 42,000 tons of lac per year, which is nearly 65 per cent of the world's output. Bihar accounts for 40 per cent of India's production. Madhya Pradesh, West Bengal, Orissa, Assam and Uttar Pradesh are the other major lac producing states, while Tamil Nadu, Telugu Desam, Karnataka, Punjab and Maharashtra states also produce lac to a lesser extent. Lac insects grow on a number of host trees, however Palas *(Butea monosperma),* Kusum *(Schleichera oleosa)* and Ber *(Zezyphus mauritiana)* are the best suited hosts in India. Red gram, Babul and Sal are also other hosts.

The lac encrustation is scrapped from the tree branches and this scrapped material is known as **'raw lac'** or **'scrapped lac'** or **'stick lac'**. The raw lac is processed and from this commercial **'shellac'** is obtained.

Uses of shellac. Shellac is mainly used in the manufacture of French polish, floor polish, industrial varnish, book varnish, bangles, printing inks, electrical insulators, nail polish, cosmetics, hair dye, sealing wax, laminated paper boards, plastic moulded articles etc. Shellac is used in photographic and engraving industries. It is also used as etch primer, for painting sides of ships, for stiffening felt hats, for filling hollows of ornaments, for backing of mirrors, coating of pills. It is also used in many Ayurvedic medicines, both for external application as well as for internal administration.

4. **Dye yielding insects.** The bright, beautiful **'carmine red cochineal dye'** is obtained from the cochineal insects, *Dactylopius coccus* and *D. tomentosus*, which live on *Opuntia* species such as, *Opuntia coccinellifera* in Mexico and other countries and *Opuntia dillenii* in India. The dye is actually the dried bodies of the above insects. When the insects are fully developed, they are brushed off the host plant, killed by hot water or sun drying and the dried material is marketed.

Cochineal contains 10 per cent **'carminic acid'** and is used for coloring beverages, cakes and sweets, for dyeing materials such as, wool, silk and leather, as a cosmetic and also in medicine for the treatment of whooping cough.

Minor Productive Insects

a) **Insects producing galls.** Infestation by some insects produces galls in the host plant parts. Most galls are produced by plant mites, gall midges and gall wasps. Such galls may yield certain materials of medicinal value or other useful products such as, dyes, inks etc. The Aleppo galls on oak trees provide certain materials, which have astringent and tonic properties. The dye obtained from Aleppo galls is used for dyeing wool and hair. The dyes and inks obtained from insect galls are quite permanent in nature. Most of the insect galls contain high quantities (up to 30 - 70 per cent) of tannic acid.

b) **Insects as food.** Different kinds of insects serve as food for human beings. Locusts and grasshoppers, crickets, cicadas, large ants, eggs of some large water bugs, grubs of some cockchafer beetles, caterpillars of some Saturnid moths, larvae and pupae of bees and some wasps etc. are consumed by man in different countries. The honeydew secreted by aphids and scale insects is used as a sweetener in some Middle East countries. Termites are used as a highly nutritious food in some parts of our country. Chrysalids of silk worm serve as a protein-rich food. Many insects serve as food for birds, fishes and animals.

c) **Insects producing medicines.** Some insects produce specific medicines. **'Allantoin'** is a substance obtained from the secretion of fly maggots, which has properties of healing old, deep wounds. **'Cantharidin'**, a substance obtained from the blister beetles *(Lytta vesicatoria)* is administered internally for treating urinary diseases and it acts as a vesicant and counter irritant when applied externally. **'Apis'** is another specific medicine extracted from honeybees and and is used against certain urinary diseases, Diphtheria etc. The bee venom obtained from bees is used as a cure for Rheumatism and Arthritis and is available for hypodermic injection.

d) **Other useful products.** Shiny, metallic, beautiful elytra of the Bupestrid beetles are used as ornamentation. The fat matter obtained from silk worm chrysalid is used in the manufacture of soap. Silk worm chrysalid and wastes from lac dye etc. are used as manure. The excreta and dead bodies of insects serve as manure. Larvae of certain aquatic Neuroptera are used fish bait.

Helpful Insects

a) **Parasitoids.** An **'entomophagous parasite'** or **'parasitoid'** is an insect living on or in the body of another insect, whioch is called the host. The parasitoid lays eggs on or in the host at any particular growth stage viz., egg, larva or pupa. The larvae of the parasitoid on hatching from the eggs, feed on the body contents of the host, pupate either inside or outside on the host body and finally emerge as adults. The host is not killed immediately, but continue to live, mostly till the life cycle of the parasitoid is completed.

b) **Predators.** Predatory insects are those that catch other insects, kill them immediately and feed on them. The predators by virtue of their structural adaptations, well developed sense organs and ability to move swiftly are capable of catching their prey easily. A predator may feed upon a large number of small insects every day.

c) **Pollinators.** Insects are one of the most important agents responsible for effecting pollination, especially in cross pollinated crops. Pollination in more than 350 species of higher plants belonging to different orders is known to be assisted by various insects. Honey bees, bumble bees, wasps, black ants, butterflies and some moths, some flies, few beetles, few bugs and hoppers and some thrips are important pollinators.

d) **Weed killers.** Some insects attack and destroy unwanted weed and are called **'weed killers'**. The prickly pear *(Opuntia dillenii)* is a noxious weed and it has been practically eradicated by the cochineal insect - *Dactylopius tomentosus*, which was introduced to India from Mexico in 1925 and 5 years after its introduction, the weed was almost terminated by the insect in India. Lantana, an important and serious weed in the hilly regions is kept in check by the coccid beetle - *Orthezia insignis*.

e) **Scavengers.** Several insects, especially those living in soil feed upon dead and decaying plant and animal residues and thus help to remove such matter from the soil, which otherwise may cause health hazards. Further, these insects help to break complex organic matter of plant and animal origin into simpler substances, which are easily available as food for plants. Termites, maggots of many flies such as, sand flies, midges, fungus knats, hover flies, house flies etc., some beetles and their grubs such as, dung beetles, chafer beetles, skin beetles, powder post beetles, jewel beetles, water scavenger beetles etc. serve as scavengers.

f) **For research and scientific investigations.** To assess the degree of pollution in natural streams, canals, rivers, ponds, lakes etc. due to addition of factory effluents, raw sewage etc., some specific water inhabiting insects are used.

   To evaluate the toxicity of insecticides and other poisonous chemical substances some insects such as aphids, hoppers etc. are being used. Mosquito larvae and *Drosophila melanogaster* flies are extensively used for studies on bioassay of insecticide residues.

   Insects are extensively used in many in many biological studies, especially in the field of cytology and genetics. *Drosophila melanogaster* is chiefly used in many genetical studies because of their rapid rate of multiplication.

g) **Soil ameliorators.** Many insects such as ants, fly maggots, mud wasps, beetles and their grubs, cut worms, mole crickets, which live in soil burrow into the soil, disturb, turn and disintegrate the soil particles and help in aerating the soil Further the soil is enriched by the addition of insect saliva, excreta, exuviae and their bodies after death.

h) **For recreation and aesthetic value.** Collection and displaying insects is an art and a hobby for many people. Artists, designers and internal decorators have copied extensively from the beautiful color patterns of butterflies, beetles etc.

Insect Ecology

'Ecology' is defined as 'a study of the relations of organisms or group of organisms to their environment' or 'the science of inter relations between organisms and their environment'. The term **'environment'** is defined as 'the whole complex of climatic, edaphic and biotic factors that act upon an organism or an ecological community that ultimately determine its form and structure'. It includes the **'abiotic'** or **'non-living factors'** such as, temperature moisture, relative humidity, light, pressure, gravitational forces, soil, nutrients, chemicals etc., as well as the **'biotic factors'** or living organisms such as predators, parasites, pathogens etc.

Every living organism has certain level of tolerance to each and every factor of the environment. According to the **'law of tolerance'** organisms have an ecological minimum and maximum tolerance level to the different factors of environment, with a range in- between, which represents the limit of tolerance. Organisms may have a wide range of tolerance for one factor and narrow range for another. Generally organisms with a wider range of tolerance for all factors are likely to be most widely distributed. The limits of tolerance to reproductives such as, seeds, eggs and embryos are usually narrower than the adult forms.

Environmental Factors Affecting Insects

i) Abiotic factors

a) **Temperature.** Temperature, especially the intensity of heat energy is one of the most important abiotic factors of the environment that influences the life activities of insects. Insects differ widely in their tolerance limit to temperature. The term **'stenothermal'** is used to indicate species with narrow range of tolerance for temperature (e-g) Termites. On the other hand, the term **'eurythermal'** is used to indicate species with wider range of tolerance for temperature (e-g) Beetles. At low temperatures below the tolerance limit, many insects go into **'hibernation'** (e-g) Cotton pink boll worm. Similarly at high temperatures beyond the tolerance limit, many insects **'aestivate'** (e-g) Red hairy caterpillar *(Amsacta albistriga)*. If the temperature goes below the freezing temperature of the insect's body fluids, it leads to the death of the insect. At reasonably higher temperatures, the growth of the insect is encouraged to a certain extent. Further, at higher temperatures, there are more number of generations, the life cycle becomes shorter to a certain extent, the fecundity is more and the insects become more darker in color. At an optimum temperature range of 20° - 35° C the activity of most of the insects remain quite normal.

b) **Moisture or Water.** Water is required for the normal physiological functioning of insects. Most of the insects obtain sufficient quantity of water with their food, which mostly consists of succulent plant tissues. Many insects drink water, if they are dehydrated. Locusts, if they are starved for 24 hours or fed with dry food, drink water when they come across free water, but turn away if they find moist food. Further, during the process of digestion, the end product of oxidative metabolism is water. This water is not excreted, but reabsorbed into the body through the rectum. Terrestrial insects are solely dependent on this water for survival.

Moisture and rainfall influence the distribution and development of insects to a large extent. The period of development of most of the insects is rather slow when atmospheric moisture is high and quite rapid when the atmospheric moisture level is low. Heavy rains wash down and kill small, soft-bodied insects such as, thrips, aphids and hoppers. Low moisture in the soil adversely affects the larvae of insects, which burrow into the soil for pupation. If the soil is not sufficiently moist, red hairy caterpillars cannot burrow into the soil for pupation and many of them may perish. Continuous dry weather may adversely affect soil-inhabiting insects.

c) **Relative humidity.** This is another important environmental factor that influences the life activities of insects. Relative humidity affects the growth, development and reproduction of insects to a large extent. For most of the insects, a range of 70 - 90 per cent is optimum for the physiological functioning of the various organs of the insect body, which promotes their normal development and reproduction.

d) **Light.** Light is an important abiotic factor, which influences the activities of insects directly and indirectly because of its association with photosynthesis of plants that serve as food for most of the insects. The sun, moon and the stars are the main natural sources of light.

Insects develop various color patterns based on their exposure to natural light. The bright color patterns of insects, generally called as **'warning coloration'** provide considerable protection from their enemies. Some insects such as, stick insects, leaf insects etc., change their color and merge with the background, thus protect themselves from their enemies.

Generally light induces development of insects. The behavior of many insects is also greatly influenced by light. Bright sunlight induces honeybees to go out of their hives for foraging. The diurnal insects are very active only when there is light. Light also plays a significant role in the orientation and flight of insects.

e) **Atmospheric pressure.** It is another environmental factor that affects the presence of insects. Wind and water vapor movements play an important role in the mobility of insects, as they are the major agents that carry insects over long distances.

f) **Food.** Availability of food in the environment is very important for the survival and perpetuation of insects. Because of specificity of food habits, zones with different vegetation harbor different insect communities. Quantity of food available in an area determines the population of an insect species. Every insect species multiplies up to the limit of its food supply, when other environmental factors are favorable. The maximum population density of a certain insect species that can be supported in a given environment, including food for a sustained period of time is termed **'carrying capacity'** . One of the main adaptations for a particular insect species is that in case of non-availability of its primary host, it may survive on other alternate hosts and continue its life.

When there is a rapid increase in the number of a particular species, then there is an increase in the demand for food and other necessities of life, which results in competition for the common resources. This may ultimately lead to aggression or even cannibalism.

g) **Nutrients.** Insects require nutrients such as, carbohydrates, proteins, fats, minerals, vitamins etc. for their nourishment, normal growth, development, metabolic activities, distribution and reproduction. These are obtained from the biosphere through biogeochemical cycles. **'Biosphere'** is the largest biological system in which several ecosystems such as, **hydrosphere**, **lithosphere** and **atmosphere** operate together. Biogeochemical cycles consist of two basic types viz., gaseous type and sedimentary type. In the gaseous type, the atmosphere constitutes the major reservoir of the elements that exists in the gaseous phase. Carbon and nitrogen cycles are typical examples of this type. In the sedimentary type, the major reservoir in the lithosphere from which elements are released by weathering. Sulphur, phosphorus and iodine cycles are examples of this type. Hydrosphere constitutes the water cycle.

Rainfall, mist and dew are the main sources of water in the ecosystem. A large amount of water goes back into the atmosphere as water vapor as a result of evaporation from the sea, water reservoirs and from soil. Some water goes into the atmosphere as a result or transpiration by plants. In the water cycle, insects use water as a metabolic raw material obtained from the hydrosphere. They excrete some of it back into the hydrosphere while alive and after death the remaining water is also returned after decay.

The air in the atmosphere consists of oxygen (about 20%), carbon dioxide (about 0.03%), nitrogen (about 76%) , water vapor and minute traces of inert gases. Except the inert gases, all other components of air serve as metabolites and each circulates through a cycle in which organisms play an important role.

Oxygen is used by organisms for respiration. During this process, carbon dioxide and water are formed. During photosynthesis, plants liberate oxygen. Such free oxygen is may again be used in respiration or it may be returned to the environment as molecular oxygen, thus completing the oxygen cycle.

Atmospheric carbon dioxide is the only source of carbon, which is the basic constituent of all organic compounds. Carbon dioxide is used by all plants in photosynthesis and the end product (organic substances) of this complex process is used for the construction of more living matter. Some of the organic substances thus produced by plants is used as fuel in respiration by both plants and animals. During this process also carbon dioxide is released as a by-product, which in turn may be used by plants for photosynthesis or may be returned to the atmosphere, thus completing the carbon cycle.

Nitrogen is required by living organisms for the synthesis of proteins, nucleic acids and other nitrogenous compounds. Except nitrogen-fixing bacteria and nitrogen-fixing blue green algae, fungi and yeasts, atmospheric nitrogen cannot be used by other living organisms. Nitrogen is mostly taken by living organisms in the form of nitrates. In the nitrogen cycle, the organic nitrogen of plants and animals is broken down to ammonia by autotrophic bacteria as a result of **'ammonification'**. Some of this ammonia is oxidized to form nitrites and nitrates by nitrifying bacteria. This process is called **'nitrification'**. The denitrifying bacteria act on ammonia and liberate nitrogen into the atmosphere. This process is called **'denitrification'**. The atmospheric nitrogen may be absorbed by nitrogen-fixing bacteria such as, free-living *Azotobacter* or nodule-forming *Rhizobium* and converted into usable form of nitrogen.

Phosphorus, another important nutrient is required by living organisms as inorganic phosphates for the synthesis of nucleic acid, phospholipids and other phospho - related compounds. In nature, phosphatic compounds are obtained by weathering of phosphorus rocks. The phosphate is taken by animal and plant consumers and after their death decompose into organic phosphates and subsequently recycled into inorganic phosphate and is made available to consumers. The excreta of animals also contribute to certain amount of inorganic phosphate.

ii) Biotic Factors

Biotic factors involve the interaction of different living organisms in the environment. This may be interspecific interaction or intraspecific interaction. Interspecific interaction may be beneficial to either one or both the species or may have deleterious effect to any one of the interacting species. Intraspecific interaction is between individuals of the same species that are mostly social insects, which live in colonies such as, termites, ants and honeybees.

a) **Interspecific relationship** A large number of insects live at the expense of others and cause injury of varying degrees to the latter. The insects, which depend for their lodgings or sustenance or both on other animals are known as **'parasites'**. Numerous insects pass their early developmental stages inside the bodies of other insects as **'internal parasites'**, **'endoparasites'** or **'parasitoids'** and most of these are found in the groups of wasps and flies. The adult female parasitoid usually lays its eggs on or inside the body of its insect host. The egg hatches into as larva, which feeds on the internal contents of the host body, pupates and emerges as an adult parasitoid. In the meanwhile the host, though it survives the attack for a time, gradually dies. Frequently the parasitic habit of some insects becomes economically important, when the host happens to be a pest of economic importance. Several wasps and flies act as natural enemies in nature and control many caterpillar and beetle pests of crops such as, rice, millets, cotton, sugarcane, coconut etc. Parasites, which remain outside the body of the host and feed on the host are called **'external parasites'** or **'ectoparasites'**.

Some parasitoids are sometimes parasitized by some other insect species and these insects are called **'hyper parasites'**. When a host is parasitized by many parasites simultaneously, it is called **'superparatism'**.

When a parasite requires more than one host for the completion of its life cycle, it is called **'multiple paratism'**. A parasite, which requires a host compulsorily for completing its entire life cycle is called a **'obligate parasite'** or **'compulsory parasite'**. If a parasite, which in the absence of of its regular living host, is capable of surviving on other non-living food, it is called **'facultative parasite'** or **'optional parasite'**.

In contrast to this there are some insects, which act as hunters, capture and feed on other insect victims. These are known as **'predators'**. The only marked difference between parasitism and predatory habit is that in the former, the death of the host is a slow and gradual torture, while in the latter, the death of the prey follows almost immediately after its capture by the predator. The insect predators are usually larger in size, quite active, swift in action, have well developed sense organs and structural adaptations for catching the prey. Many predators remain still and suddenly seize the prey when it comes within the reach, while some others catch the prey in flight. Some of the predators have cryptic coloration and deceptive markings, as in Praying mantis, leaf and stick insects etc. They feed upon many small insects every day. Well known examples of predatory insects are beetles of different kinds, including the tiger, ground and ladybird beetles, the dragon flies, robber flies, ant lions, the mantis, some stink bugs, the giant water bug and a few aquatic beetles.

In addition to these important inter relations among insects as parasitoids and predators, there are instances of insects living as associates, mess mates, guests, slaves or friends with one another. These types of inter relationship between organisms of different species, which live in a more or less intimate association is termed **'symbiosis'**. Such association may be between two insect species or between an insect and an animal, microorganism or plant. One species may feed on the food collected or stored by another species, thus indirectly harming the latter. This type of symbiosis is common among white ants, wasps, bees and ants and is known as **'social symbiosis'**. When the association benefits both the symbionts, the phenomenon is termed **'mutualism'**. Termites and the flagellate protozoans inside their guts show mutualism. Termites feed on wood, which is practically cellulose, which can not be digested without the aid of flagellates present in their stomach. The waste product after the cellulose have been digested serve as food for the termites. Similarly some species of ants, white ants, wood-boring weevils and wasps cultivate certain types of fungi, which can grow only under the environment and care provided by the host. The fungi in return provide nutritious food to these insects. Ants, some aphids and coccids also exhibit this type of relationship. The ants provide shelter to these insects, which in turn supply honey dew to the ants. Some scarabeid beetles and stephylinids are accommodated in the nests of termites. In another type of social symbiosis known as **'commensalism'**, one of the two partners gets the benefit, while the other though not benefited directly, but is not harmed in any way. Insects use plants or trees as breeding sites and get the benefit. Honeybees visit flowers for obtaining nectar and pollen for their own use and in doing so, they incidentally bring about cross pollination. In yet another type of relationship known **as 'antagonism'**, one of the two partners get the benefit, while the other is harmed. There are two types of antagonistic relationship. One is **'antibiosis'**, in which the metabolic products produced by one of the partners is harmful to the other. Antibiotic substances such as, penicillin, streptomycin etc. produced by some of the fungi are harmful to others. The other one known as **'exploitation'** leads to direct harmful effect on the other. *Polyergus* ants make *Formica* ants as slaves and exploit their services.

b) **Intraspecific relationship**. Association of individuals of the same species that are mostly social insects, which live in colonies such as, termites, ants, wasps and honeybees is known as **'intraspecific relationship'**. A social insect may be defined as one in which the parents and the off springs live in mutual harmony in a common abode or nest. Many insects occur in groups and these groups differ in the factors responsible for bringing the individuals together, Some groups are simply the result of a positive reaction by many individuals to the same stimulus (e-g) Insects attracted to light or to a dead animal. Some may result from the simultaneous emergence of many adults as in may flies. The larvae of many Lepidoptera, which hatch from a given egg mass remain together during most of their larval stage as a result of their reaction to a common food supply. Ants, termites, honeybees and some wasps live in much more integrated colonies called **'societies'** and these societies have some special features. Among social insects, the intraspecific relations are for the good of the colony. Besides a large number of individuals living together with their mother in a common abode, polymorphism or caste differentiation of its members is evident. Besides polymorphism, division of labor is involved among the members for the welfare of the community as a whole. For example, in white ants, the queen, which is the sexually mature female entirely devotes her time to egg laying, often at the rate of one egg per second and is generally described as an egg-laying machine. When the egg-laying capacity of the queen decreases, her feeding is stopped and she dies of starvation. Her fatty remains are eaten by other castes of the colony. In her place supplementary reproductives are produced, which continue reproduction to a limited extent. The workers are very efficient laborers and attend to all sorts of duties including the care of the eggs, young ones and the royal pair, building and cleaning the termitarium, growing of fungus gardens and procuring food for all the individuals of the colony. The soldiers protect the colony from intruders. Similar organization exists among social ants also. Among social bees, there is further specialization and the sterile female workers according to their age perform all the duties of the colony except reproduction.

The workers in some ant colonies are of different sizes and the activities of these different sized workers may differ. The larger ones act as soldiers; the intermediate sized workers do most of the foraging and regurgitate food to other workers and the smaller ones are concerned chiefly with tending the brood, nest construction and other chores inside the nest.

Social insects maintain a means of communication between themselves. The chief communication in ants is by means of pheromones. A honeybee worker upon returning to the hive after finding a nectar source communicates information about this source to the other workers by means of characteristic dances.

Social insects have a home base, the nest and the individuals that forage away from the nest return back without any difficulty. Social insects form new colonies either by swarming as in the case of honeybees, where a queen along with a group of workers leave and establish a new colony and a new queen is developed in the group that did not leave or by the parental forms leaving and starting a new colony as in some ants.

Insect societies are often associated with various other organizations such as, other insects, mites and sometimes fungi. Some of these may be predators or parasites, while some are scavengers and the others guests, which are not only tolerated and actually fed by the host insects. Many social insects cultivate fungi inside their nests, which they use as food. Many ants tend aphids and feed on the honeydew secreted by them.

Another important aspect of the bond governing social behavior is the mutual exchange of nourishment between adult and young ones. Thus wasp larvae exude saliva in exchange for food supplied by the workers. This type of behavior is met with in termites also, but not in bees, perhaps because of the availability of pollen and nectar from external sources.

There are some differences between the societies of termites and those of Hymenoptera. Termites have a child labor society in that young insects from about the second instar onwards must take care of themselves and remain as workers for the rest of their lives, except those that become soldiers or reproductive forms. The Hymenoptera society is built on adult labor and the adult workers care for the helpless young ones.

The majority of insects are solitary and live for themselves in-between them and the social insects are those, which resemble or are gregarious during a part or whole of their lives. Desert locusts have gregarious and solitary phases. The Embidae live in silken tunnels in colonies of over 20 individuals in each. Living in such groups is probably for protection. Some Coccinellid beetles and fruit flies congregate for hibernation under the creepers or other host plants.

Intraspecific interrelationship has yet another aspect. Under normal conditions of stabilization there is enough food and space for a particular species in a particular locality where it multiplies. When the density of population becomes high, the overcrowding proves harmful to the species, as competition for food and space results, which may affect the individuals inhabiting the habitat. Too much overcrowding and exhaustion of food often results in cannibalism. This is particularly so among the predators such as, praying mantis. Even in polyphagous insects such as, red cotton bug and locust hoppers, cannibalism is noticed under such conditions. Insect pests of stored products such as, *Tribolium* beetles often start feeding on their eggs, when the population density becomes high. Larvae of *Spodoptera litura* also exhibit this tendency. Competition for space is important in some groups of insects as in the case of wood borers. In bark - borer beetles, mostly competition occurs mostly between the larvae and sometimes between the larvae and the adults.

c) **Community** A **'community'** or **'biota'** is a total association of several populations of different species that live, develop and reproduce in a stable manner within a specified habitat with mutual tolerance and beneficial interaction between them and ultimately die. Such a community life cycle may take hundreds or even thousands of years.

The existence and activities of each species of organisms making up the community is influenced not only by the physical and chemical environment, but also by other species of the community. Each and every species of the community exists with interdependence among the species. This is called **'communal interdependence'**. The interdependence among different species of a community is mainly for food, reproduction and protection. However, in every community some plants or organisms exert a dominant influence in determining the nature and function of the entire community because of their numbers, activities or other dominant characters and they are called **'ecological dominants'**.

The existence of any community is decided mostly by the availability of food for all the species comprising of the community. Transfer of food energy starts from plants through a series of species with different food habits such as, herbivores, omnivores and carnivores, which are called the consumers. After the death of the highest order of the consumer, the cycle is completed. Such transfer of food energy from the plants through a series of organisms with repeated eating and being eaten is referred to as a **'food chain'**. Two basic types of food chains are recognized. The first one, which is the **'grazing food chain'** starts from green plant base, goes to the grazing herbivores and finally to the carnivores. The second one, **'the detritus food chain'** starts from dead organic matter, goes to the microbes feeding on the dead organic matter and finally to detritus feeding organisms and their predators.

Habitats of Insects

i) **Terrestrial insects.** These insects live above ground in all their growth stages. They may live in forests, deserts, mountains, valleys and in such above ground parts. Most of them are polyphagous, feed on wild and cultivated crops and cause considerable economic loss, while some others are parasitic on livestock and humans.

ii) **Aquatic insects.** Aquatic insects spend either part or their entire life in water. Only about 3.0 per cent of the insects are aquatic in nature. They are found in the ocean, lakes, rivers, streams, ponds and in water reservoirs. Some of them prefer deep water, while others prefer shallow water. These insects have special adaptations for breathing under water and organs for locomotion in water.

iii) **Subterranean insects.** These insects spend either part of their lives or their entire lives under the soil. Several insects pupate under the soil, while the larval and adult stages are terrestrial. In some other cases, except the adult stage, the other stages are spent under the soil.

Types of reproduction in insects

Oviparity. Normally reproduction among insects is dependent upon the mating of the two sexes and the fertilization of the ovum by the spermatozoa. Most insects are oviparous in the sense they lay fertilized eggs, which hatch after deposition within a specific time frame. However, there are some exceptions to this generalization.

i) **Viviparity.** In this type of reproduction the embryonic development is completed within the body of the female parent and hence instead of laying eggs, the young larvae or nymphs are laid. There are five different types of viviparity.

a) **Ovo viviparity.** Many insect species retain the eggs in the genital tracts until they are ready to hatch and so hatching occurs just before or as the eggs are being laid. All the nourishment required for the development of the embryo is present in the eggs and no nourishment is supplied from the parent. It differs from normal oviparity only in the sense that he eggs are retained till the time of hatching (e-g) Some species belonging to Diptera, Coleoptera and Thysanoptera.

b) **True viviparity.** In some species, the eggs are retained after fertilization in the vaginal chamber of the female parent and the embryos receive nourishment directly from the parent instead of the yolk or in addition to the yolk. In such cases some anatomical adaptations are provided in the parent or egg, which facilitate the exchange of nutrients. Viviparous species commonly produce fewer number of offspring than related oviparous species and this may be due to the reduction in the number of ovarioles (e-g) Aphids and some flies.

c) **Pseudoplacental viviparity.** Some species produce eggs containing a little or no no yolk and the eggs are retained by the female. The eggs may receive nourishment through embryonic or maternal structures known as pseudoplacentae (e-g) Some species of Aphids.

d) **Adenotrophic viviparity.** In this type, eggs are ovulated singly and the embryo develops by utilizing nourishment from the yolk. However, after hatching, the young larva is retained in the uterus and nourished by special maternal glands. The larva moults twice within the uterus and then deposited outside (e-g) Some flies.

e) **Haemocaelus viviparity.** In this type, the development of the embryo takes place in the haemocoel of the female parent. In these cases there are no oviducts and the ovaries lie free among the fat body and when mature they break up and the eggs are dispersed in the haemocoel. The eggs surrounded only by a trophic membrane through which nourishment is supplied (e-g) Gall flies.

ii) **Paedogenesis.** In a few cases, immature insects possess functional ovaries and the eggs produced develop parthenoogenetically and this condition is known as **'paedogenesis'**. This is due to a hormonal imbalance and most insects reproducing paedogenetically are also parthenogenetic and viviparous (e-g) Some species belonging to Diptera and Coleoptera.

iii) **Parthenogenesis.** Sometimes eggs develop without being fertilized and this phenomenon is known as **'parthenogenesis'**. Failure of a female to find a male mate may result in parthenogenesis occasionally, but in a number of insects parthenogenesis is the normal means of reproduction and is common in a number of insect species. Except Odonata, Dermaptera and few others.

Parthenogenesis may be either facultative when it co-exists with bisexual reproduction or obligatory when males are absent or extremely rare and functionless. The unfertilized eggs may give rise to only females or only to males or to both the sexes. Moreover, parthenogenesis may be combined with paedogenesis and viviparity in some cases or may occur in alternation with a bisexual generation.

v) **Polyembryony.** In a few cases, an egg gives rise to two or more larvaeinstead of the normal single larva which is called **'polyembryony'**. In all these cases the eggs are relatively free from yolk and nourishment is derived from the host tissue in which they are present. (e-g) Some species belonging to Acrididae and some endoparasitic species belonging to Ichneumonidae, which are egg and larval parasites of Lepidoptera.

vi) **Hermaphroditism.** In this type of reproduction, in addition to the normal males, which occur very rarely there are males which are functionally similar to the females, where true females are absent. The reproductive system is somewhat complex and the gonad, which is formed by the fusion of two originally separate organs produces both spermatozoa and eggs (e-g) Phorids.

Insect Pest

An insect, which multiplies beyond a certain level and causes damage beyond a certain limit attains the status of an **'insect pest'**. In general, an insect, which infests a particular crop and causes an economic loss of 5.0 per cent or more is considered to be an insect pest.

Biotic or Natural Balance

An **'ecosystem'** is a self-sufficient and self-regulated habitat, where the biotic and abiotic components of the environment interact together for the exchange of energy in a continuing cycle. The structural components of an ecosystem consist of **'producers'**, **'consumers'**, **'decomposers'** and abiotic factors. The producers are autotrophic organisms that utilize sunlight as energy and convert atmospheric carbon dioxide into carbohydrates through the process of photosynthesis. The consumers viz., insects, animals etc. derive their food directly or indirectly from the producers. The decomposers such as, fungi, bacteria etc. break down complex compounds of dead producers and consumers and convert them into simpler substances to be used again by producers. The abiotic components consist of non-living materials such as, water, minerals, atmospheric gases, salts etc. Thus, in an ecosystem there is a perfect equilibrium among the living and non-living components and the living organisms live in harmony in a state of compromise. This self-determined natural course is known as the **'biotic balance'**. This equilibrium usually prevents the multiplication of any one particular species of insect beyond a certain level, which may upset the biotic balance. Such a regulative process in an undisturbed environment neither allow decline of any plant or animal population to extinction nor allow the population to increase beyond a particular limit. The population levels of insects may fluctuate from year to year, season to season and from place to place, but the fluctuations are always restricted to within a particular limit and as such the biotic balance is not upset. However, during certain natural calamities such as, floods, cyclones, storms, volcanoes, severe drought etc., the biotic balance may be upset leading to flare up of some of the insects

The chief factors responsible for maintaining the biotic balance are:

(i) Food, (ii) Climate (iii) Natural enemies and diseases.

Many forms of life, especially the lower forms often die of starvation if food is not available. If they happen to depend on particular plants for their food, the absence of such food arrests their growth and multiplication. Similarly extremes of climatic conditions also adversely affect the perpetuation of various animals and plants. The presence of various carnivores, which depend for their food on other forms of life are responsible for checking the multiplication of various smaller organisms. Besides such natural enemies, various diseases caused by fungal, bacterial, viral and other parasitic organisms also play a vital role in restricting the growth and multiplication of several organisms. When one or more of these checking factors are not prevented from playing their part unabated, then the biotic balance gets upset leading to flare up of some forms.

2

Morphology of Insects

The insect body is made up of multicellular tissues. The body is bilaterally symmetrical and is covered by a hard, chitinous exoskeleton. The chitin at the junction of the segments is thinner and pliable and hence insects can bend their bodies. The exoskeleton is made up of several layers of non-living cells produced from the epidermis. Because of the hard chitin us exoskeleton of the body, the growth of the insect is affected beyond a certain extent. So, to facilitate further growth,, this exoskeleton is cast off a few times and this is known as **'molting'**. The number of moultings in a particular species is more or less constant.

Insects are Arthropods in which the body is divided into three regions viz., head, thorax and abdomen **(Fig. 2, 3)**.

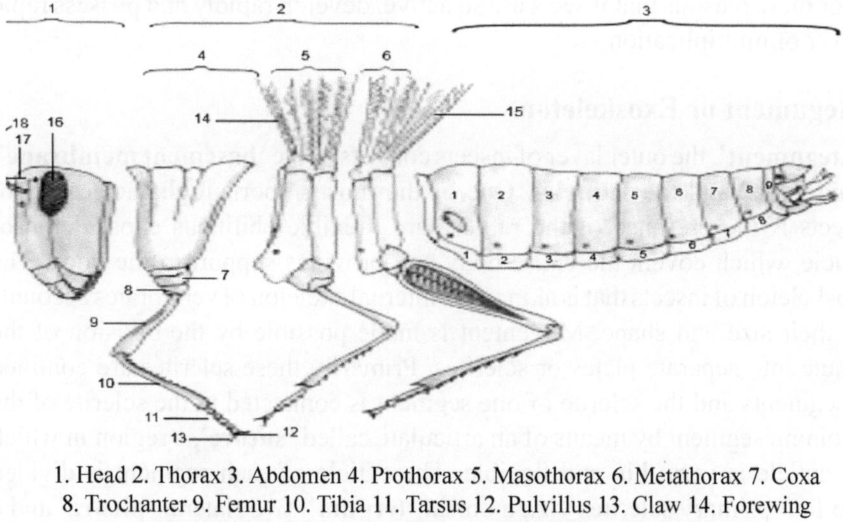

1. Head 2. Thorax 3. Abdomen 4. Prothorax 5. Masothorax 6. Metathorax 7. Coxa 8. Trochanter 9. Femur 10. Tibia 11. Tarsus 12. Pulvillus 13. Claw 14. Forewing 15. Hindwing 16. Compound eye 17. Simple eye 18. Antenna

Fig. 2. Body regions of a typical insect (e-g) Grasshopper (Lateral view)

The **'head'**, which consists of seven segments or **'sclerites'**, bears one pair of antennae, a pair of compound eyes and the mouthparts. Some wingless forms of insects lack antennae. Mouthparts are composed of a pair of mandibles, a pair of maxillae, each with maxillary palpi, small feeler-like sensory organs, a hypopharynx, which is a tongue-like organ and a labium usually with palpi. Modifications of these basic parts account for the variety of mouthparts among insects. The neck connects the head with the thorax. The **'thorax'** consists of three segments viz., pro, meso- and meta thorax. and this region bears three pairs of legs, one pair in each segment. The thorax also has two pairs of wings, usually one pair each on the meso and meta thoracic segments or one pair of wings and a pair of halteres, which are small, modified, club-shaped structures in place of the second pair of wings. In the immature stages and in certain primitive groups of insects wings are absent. Most ectoparasitic insects lack wings. The **'abdomen'** bears eleven or fewer segments and also the genital apertures, which are situated near the anus at the posterior end of the abdomen. The exterior genitalia are well developed and distinct. Excretion is performed by fine malpighian tubules attached to the anterior end of the hindgut. Respiration is by branched, cuticle-lined tracheae that carry oxygen directly to the tissues from paired apertures viz., the spiracles or stigmata, which are located on the lateral sides of the thorax and abdomen. During the process of respiration, they take in oxygen from the air and exhale carbon dioxide. The respiratory and circulatory systems of insects are so efficient, that the supply of oxygen and nutrition to different organs is rapid and complete. It is for these reasons that insects are so active, develop rapidly and possess rapid power of multiplication.

Integument or Exoskeleton

'Integument', the outer layer of insects consists of the **'basement membrane'**, **'epidermis'** and the **'cuticle'**. One of the most important characteristics of insects is the presence of the rather hard, flexible chitinous exoskeleton or cuticle, which covers the entire body and provides support to the body. The exoskeleton of insects that is akin to the internal skeleton of vertebrates accounts for their size and shape. Movement is made possible by the division of the cuticle into separate plates or sclerites. Primarily these sclerites are confined to segments and the sclerite of one segment is connected to the sclerite of the adjoining segment by means of an articulate called **'suture'**, a region in which the cuticle is very thin and flexible. The cuticle of each segment is divided into four primary sclerites viz., a dorsal **'tergum'**, two 'lateral **'pleura'** and a ventral **'sternum'**. The dermal glands present in the epidermal layer secrete the entire cuticle of the insect.

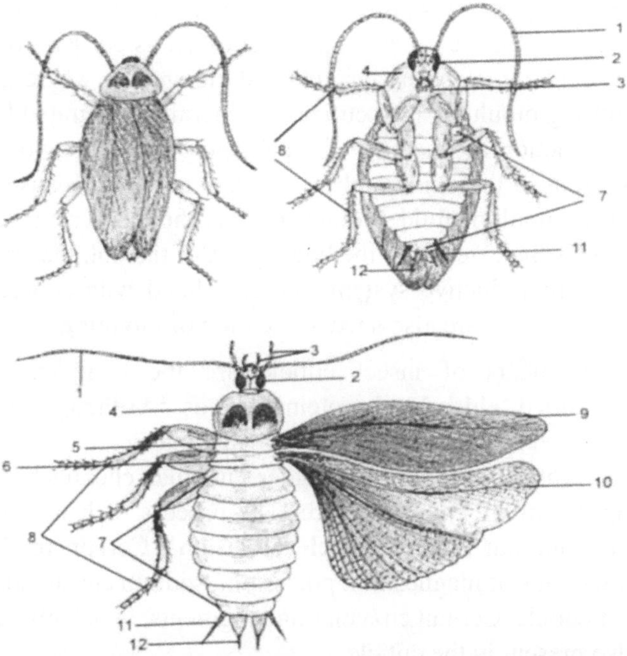

1. Antenna 2. Compound eye 3. Mouthparts 4. Prothorax 5. Mesothorax 6. Metathorax 7. Abdomen 8. Legs 9. Legs 10. Forewings 11. Hindwings 12. Anal cerci 13. Dorsal valves

Fig. 3. Body regions of a typical insect (e-g) Cockroach

Like that of the body, the cuticular skeleton of the appendages is divided into tube-like segments connected to one another by articular membranes, thus creating a joint at each junction. Such joints enable the segments of the appendages, as well as those of the body to move freely.

In addition to the exoskeleton, an **'endoskeleton'** is also developed. This may be an infolding of the procuticle that produces inner projections or **'apodemes'** akin to bones in vertebrates on which the muscles are inserted.

'Cuticle' is a complex, non-cellular layer secreted largely by the epidermis and forms the outermost investment of the body and its appendages. It is composed of a thin outer **'epicuticle'** and a much thicker **'procuticle'**. The epicuticle consists of an outermost **'cement layer'**, followed by a **'wax layer'** and then the **'cuticuline layer'** The cuticuline layer is further differentiated into an **'outer epicuticle'** and an **'inner epicuticle'** .The epicuticle does not contain chitin. The cement layer is a barrier to ions, while the wax layer is a barrier to water and helps in the conservation of water. The cuticuline layer is a barrier to oxygen and other gas molecules. The fully developed procuticle consists of an outer **'exocuticle'**, which is hard and dark and an inner **'endocuticle'** which is soft elastic and rubber-like. Both the layers are composed of chitin and protein in the form of a complex glycoprotein. Certain enzymes and pigments are also found in the cuticle. Both exo- and endocuticles act as barriers for lipophylic substances. Exocuticle is absent at joints and along lines where the skeleton will rupture during moulting. The cuticle is generally penetrated by fine pore canals, which function as ducts for the passage of secretions of underlying gland cells present in the epidermal layer. They act as a means to transport materials such as, moulting fluid, protein and polyphenols during the formation of cuticle. Besides the exterior of the body, parts of the gut, the tracheal tubes and parts of the reproductive systems are also lined with cuticle. All these internal cuticular linings are also shed at the time of moulting.

The main constituents of insect cuticle are the complex nitrogenous polysaccharide called **'chitin'** and proteins such as, **'Anthropodin'**, **'Resilin'** etc., generally called **'Sclerotin'**. Resilin provides elastic, rubber-like structure to the skin. Sclerotin mixes with the other cuticular chemicals to form the relatively impervious, flexible structure. The process of hardening of these substances after mixing is termed **'sclerotization'**. Certain lipids, phenols, inorganic compounds of magnesium, potassium, sodium etc. are also found as constituents of cuticle. Certain enzymes and pigments, which give color to the insects are also present in the cuticle.

Thus, the integument helps to prevent water loss from the body of insects, protects insects from predators, pathogens and parasites, lines and protects the living ectodermal invaginations such as, the stomodaeum, proctodaeum and tracheae and also provides a site for muscle attachment .A more or less distinct membrane called the **'basal membrane'** binds the inner ends of the epidermal cells. This is a continuous sheet, which is formed from blood cells and limits what enters the epidermal cells from the blood **(Fig. 4)**.

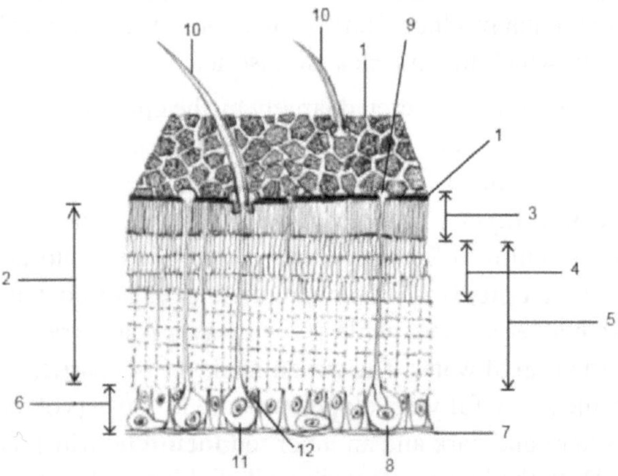

1. Epicuticle
2. Procuticle
3. Exocuticle
4. Mesocuticle
5. Endocuticle
6. Epidermis
7. Basement membrane
8. Dermal gland
9. Opening of duct
10. Seta
11. Seta forming cell
12. Socket forming cell (Tomogen cell)

Fig. 4. Structure of the integument of an insect

Stages in the Development of Cuticle

Despite the locomotor and the supporting advantages of an external skeleton, it poses problems for a growing insect such as, a nymph or larva from the time it hatches out from the egg. To overcome this problem, insects shed their skeleton periodically. This process of casting off the old exoskeleton and replacing it with a new, larger exoskeleton to accommodate its growing body is called **'moulting'** or **'ecdysis'**. Thus, the insect passes from one instar to the next instar after each moulting. Moulting involves a sequence of events beginning w ith the separation of the old cuticle from the epidermis dermal gland to the shedding of the old cuticle. This separation of the old cuticle from the underlying epidermis is called **'apolysis'**. Moulting is under hormonal control. **'Ecdysone'** secreted by certain prothoracic glands is circulated by the blood stream and acts directly on the epidermal cells. The production of ecdysone is in turn regulated by some other hormones. Secretion of ecdysone triggers the process of moulting. The glandular cells in the epidermis become enlarged and there is accumulation of lot of nuclear materials in the nucleus of the cells indicating the advent of ecdysis. A a result of such changes in the epidermal layer, a tension is generated at the surface of the epidermal cells, which is responsible for separating the cuticle from the epidermis. Meanwhile the epidermal cells secrete a new epicuticle. The first layer of epicuticle that is secreted glands of the epidermal cells is the cuticulin layer. The epidermis now secretes moulting fluid, which contains the enzymes chitinase and proteinase. The moulting fluid passes through the new epicuticle and erodes away the old endocuticle, which is not sclerotized. On the other hand, the exocuticle, which is sclerotized, is not digested. The muscle attachments and nerve connections are unaffected by the enzymes and he insect can continue to move. Following the digestion of the old endocuticle, the epidermis secretes a new procuticle. The exocuticle gets thickened or darkened as a result of scleotization and gets differentiated from the endocuticle, which is not sclerotized. Maximum sclerotization is found in the cuticle of the mesonotum of adult flying insects and mandibles of both adults and young ones of mandibulates. At this point the insect is enclosed within both an old and a new skeleton. The next step in cuticle formation is the secretion of the wax layer over the cuticulin layer of the epicuticle by the waxy material produced by the epidermal cells. Finally the cement layer is formed over the wax layer.

The materials required to form these layers are carried to the top through the ductsof the dermal glands in the epidermis. During the later stages of ecdysis, the insect swallows air or water in the case of aquatic insect, as a result the alimentary canal expands and exerts pressure on the haemolymph, which in turn exert pressure on the old skeleton. The old skeleton now splits along certain predetermined lines and the insect pulls out of the old encasement. The cast out or the moulted old skin is called **'exuviae'**. Besides ecdysone, other hormones such as, **'juvenile hormone'**, **'eclosion'** and **'bursicon'** are also involved in cuticle formation.

Immediately after moulting, the new integument is soft and pliable and is stretched to accommodate the increased size of the insect. Hardening of the cuticle results from tanning of the protein. Additional procuticle and epicuticle may be added following actual ecdysis by secretions through he pore canals. A cement layer often forms the final surface of the epicuticle. The stages between moults are known as **'instars'** and the interval between two moults of nymphs or larvae is called **'stadium'**. The duration of the stadium becomes longer as the insect becomes older. Insects have more or less fixed number of instars. Besides secreting the greater part of the cuticle, the epidermis produces moulting fluids, absorbs the digestion products of the old cuticle and repairs the wounds **(Fig. 5)**.

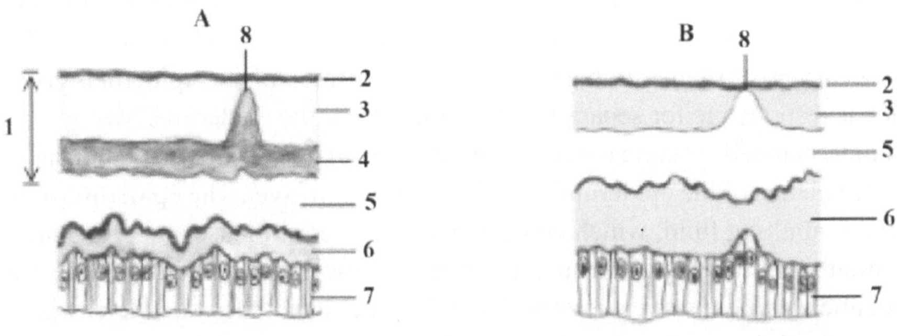

A. Digestion of the old endocuticle and secretion of new procuticle within both new and old skeleton

B. The insect before moulting encased within both new and old skeleton

1. Detached hypodermis 2. Epicuticle 3. Exocuticle 4. Endocuticle 5. Moulting fluid 6. New epicuticle and Procuticle 7. Hypodermis 8. Line of rupture of old skeleton

Fig. 5. Moulting in an insect

Cuticular Processes, Appendages or Outgrowths

1. **External processes.** The appendages or outgrowths predominantly present on the cuticle are the hairs found all over the body surface. These hairs are sensory hairs. Besides these hairs some insects have spines on the cuticle, which are used for defensive purposes. Spines are found in grasshopper and allied insects. Some insects have spurs, which are fixed spines found on the legs. The cuticle of many insects, especially the Lepidopterans is clothed with fine, soft, velvety scales, which give a specific color pattern to the insects. The mealy bugs have waxy, filamentous coating on the surface

of the body. The white flies also have white, waxy coating on the body and wings. In some insect species glandular bristles are present, which have specific functions such as, secretion of sex hormones or pheromones.

2. **Internal processes**. The internal processes are formed by the invagination of the body wall, known as **'apodemes'**, which provide the required provision for muscle attachment.

Body Regions of Insects

The adult insect body is divided into three regions viz. , The head, thorax and abdomen **(Fig. 2, 3)**.

The head

Of the three body regions viz. , head, thorax and abdomen, the head is generally the smallest. It is the anteriormost part and is connected behind with the thorax by the neck region. In general form and shape, though the head appears to be a single compact structure, it is composed of several sclerites, which have gradually become fixed together in the course of development of the insect to form the **'head** capsule' or **'cranium'** in the adult stage **(Fig. 6)**.

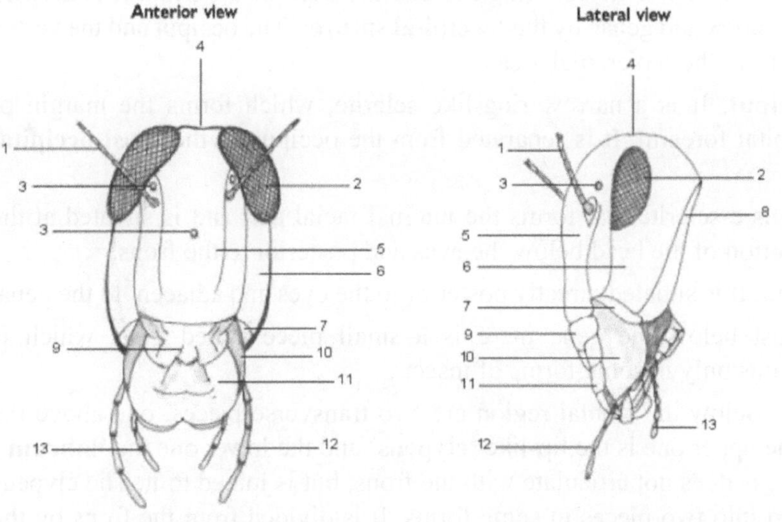

1. Antenna 2. Compound eye 3. Simple eye 4. Vertex 5. Frons 6. Gena 7. Gula 8. Post gena 9. Clypeus 10. Mandible 11. Labium 12. Maxilla 13. Labial palp

Fig. 6. Morphology of the head of an insect (e-g) Grasshopper

The highly sclerotized internal skeleton of the head supporting the various sclerites that form the head capsule or skull of insects is called **'tentorium'**. The principal sclerites of the tentorium are the **'vertex'**, **'frons'**, **'gena'**,

'post-gena', **'occiput'**, **clypeus'** and **'labrum'**. These sclerites are connected by means of sutures to form the head capsule. The epistomal, sub-cuticular, occipital and Sub-genal sutures are the prominent ones in the head. The sutures bind and retain the sclerites in position and maintain the shape of the tentorium. The capsule with a sclerotized upper portion contains the brain and a membranous floor, in which is situated the oral openings or mouth.

Vertex. In a typical insect's head such as, the head of a grasshopper, the sclerite forming the dorsal region or top of the head between the two compound eyes is the **'vertex'**.

Epicranial suture. It is an inverted 'Y'- shaped suture, the stem of which begins on the back of the head, crosses the vertex and forks on the face. The stem is called the **'epicranial stem'** and the two arms of the forked portion are the 'epicranial arms'.

Frons or **Front**. It is a sclerite, which forms the anterior face and lies between or below the epicranial suture. The **'median ocellus'** is situated on this sclerite. Ventrally it is bounded by the frontoclypeal suture. The pharyngeal muscles are attached to the frons.

Occiput. The area behind the vertex that connects the head with the neck is the **'occiput'**. It is a sclerite covering most of the back of the head. It is divided from the vertex and genae by the **'occipital suture'**. The occiput and the vertex together form the 'epicranial area'.

Post-occiput. It is a narrow ring-like sclerite, which forms the margin of the occipital forearm. It is separated from the occiput by the **'post-occipital suture'**.

Gena. It is a sclerite that forms the internal facial part and is situated at the lower portion of the head below the eyes and posterior to the frons.

Post-gena. It is situated directly posterior to the eyes and adjacent to the gena.

Gula. Just below the gena there is a small piece called gula, which is conspicuous only in some forms of insects.

Clypeus. Below the frontal region are two transverse pieces, one above the other. The upper one is the lip-like **'clypeus'** and the lower one the **'labrum'**. The clypeus does not articulate with the frons, but is joined to it. The clypeus is divided into two pieces in some forms. It is divided from the frons by the **'frontoclypeal suture'**. The dilators of the cibarium are attached to the clypeus.

Labrum. The labrum hangs below the clypeus and articulates by means of membranous connections with the clypeus.

The tentorium or skull is formed by two pairs of cuticular invaginations of the body wall that unite within the head to make a framework arching over the ventral nerve cord. It provides a rigid support for the attachment of muscles connected with the mouthparts. Tentorium is composed of four primary parts viz. , the paired anterior arms, the paired posterior arms, the corporotentorium or central mass and the paired dorsal arms. The anterior arms are invaginated from the anterior tentorial pits, while the posterior arms are invaginated from the posterior tentorial pits.

The **'corporotentorium '** represents the inward extension meeting and fusing with the anterior and posterior arms. The tentorial arms arise from some part of ridges and their points of origin appear externally as depressions and are called **'tentorial pits'**. The anterior tentorial pits lie in the epicranial suture and the posterior tentorial pits lie in the post-occipital suture. The dorsal arms are secondary outgrowths of the anterior arms.

The head of insects may assume various positions in relation to the axis of the body. The three most important positions are :

i) **Hypognathous type.** In this type, the long axis of the head is vertical and the mouthparts are ventral. It is found in most of the phytophagous insects, which have biting and chewing type of mouthparts (e-g) grasshopper, locusts, cockroaches etc.

ii) **Prognathous type.** In this type, the long axis of the head is horizontal and the mouthparts are anterior or so. It is found in of the Coleopteran insects, which have biting and chewing type of mouthparts (e-g) beetles, weevils etc.

iii) **Opisthognathous type.** In this type, the head is directed backwards and the mouthparts are posterio-ventral. This type of mouthparts are found in some carnivorous insects and Hemipteran bugs, which have piercing and sucking type of mouthparts (e-g) bugs, fleas etc.

Important Structures of the Head

The eyes, antennae or feelers and the group of appendages around the mouthparts are the important structures found on the head of insects.

Eyes

Insects have two types of eyes viz. , the **'compound eyes'** or **'faceted eyes'** and the **'simple eyes'** or **'Ocelli'**. In almost all adult insects one pair of compound eyes and one to three ocelli are present. The compound eyes, which are usually large in size and quite conspicuous are situated on the dorso- ventral portion of the head capsule. They are surrounded by a narrow ring-like ocular sclerite. Each compound eye is practically a group of minute faceted structures or **'ommatida'**, crowded together and functioning as a combined organ of vision. Though they are not homologous to the eyes of higher animals as organs of vision, they are remarkably well constructed. In shape, position in the head, size and structural peculiarities, the paired compound eyes exhibit remarkable variations and in some insects these organs have reached an extraordinary degree of perfection as organs of sight. The compound eyes can detect color, especially those of the bees and other pollinators, movement and to some extent, distances. In the males of some insects such as, house flies the two compound eyes are very large, practically occupying the greater portion of the head and almost touching each other at the median line.

'Ocelli' or simple eyes are found both in the adult and many larval forms, while typical compound eyes are generally found only in adult insects. In the case of adult insects, one to three ocelli are found, of which two are situated on the dorsal surface of the head, generally between the two compound eyes and the median ocillus is situated anteriorly near the centre of the frons. Though many larval forms have no visual organs, the ocelli if present are generally arranged along the lateral margins of the head and these lateral ocelli are in many cases different in some features from the dorsal ocelli found in adult insects. The ocelli record daily changes in light intensity useful to insects strongly diurnal or nocturnal.

Antennae

The insect head bears a pair of antennae or **'feelers'**. They are very motile, jointed appendages, which articulate in the head in front or between the eyes. They articulate in the antennal socket called **'antennifer'**, which is surrounded by a narrow ring-like antennal sclerite and can move in all directions. They are never more than two in number and all typical insects possess them. Each antenna is generally a slender, movable, whip-like structure made up of a number of pieces or joints showing in some insects differences in structure between the different joints also. The antennae are primarily sense organs and most of the **'sensilla'** are found in the annuli in the middle of the flagellum. Besides functioning chiefly as tactile and olfactory sense organs, in exceptional cases they are modified to perform other functions and serve as odor receptors, chemo receptors, hygro receptors and thermo receptors. They afford important secondary sexual characters. In some larvae, they are adapted for seizing the prey, while in other insects such as, male fleas the antennae are used for holding the females while mating. In *Hydrophilus*, the antennae are enlarged and are clothed with hydrofuge hairs to facilitate the formation of air tunnel for breathing. The antennae of butterflies and moths are provided with some transmitting and receiving organs. The female butterflies transmit certain waves, which are said to be received by males even at a distance of a few kilometers. The form, shape and other peculiarities of the antennae are very varied and many different types of these organs are recognized among insects. In some forms even the males and females show differences in antennal structure. These structural differences of antennae are of great taxonomic importance. In the more specialized insects, the antenna is divided into three parts viz., the **'scape'**, the **'pedicel'** and the **'flagellum'**. The scape is the basal segment of the antenna and is often larger than any of the succeeding segments. The pedicel is the segment immediately following the scape and it acts as a pivot between the scape and flagellum. The flagellum forms the remainder of the antenna. The flagellum varies greatly in the form and structure in adaptation to environments and habits. Some of the most common forms of antennae are described below **(Fig. 7)**.

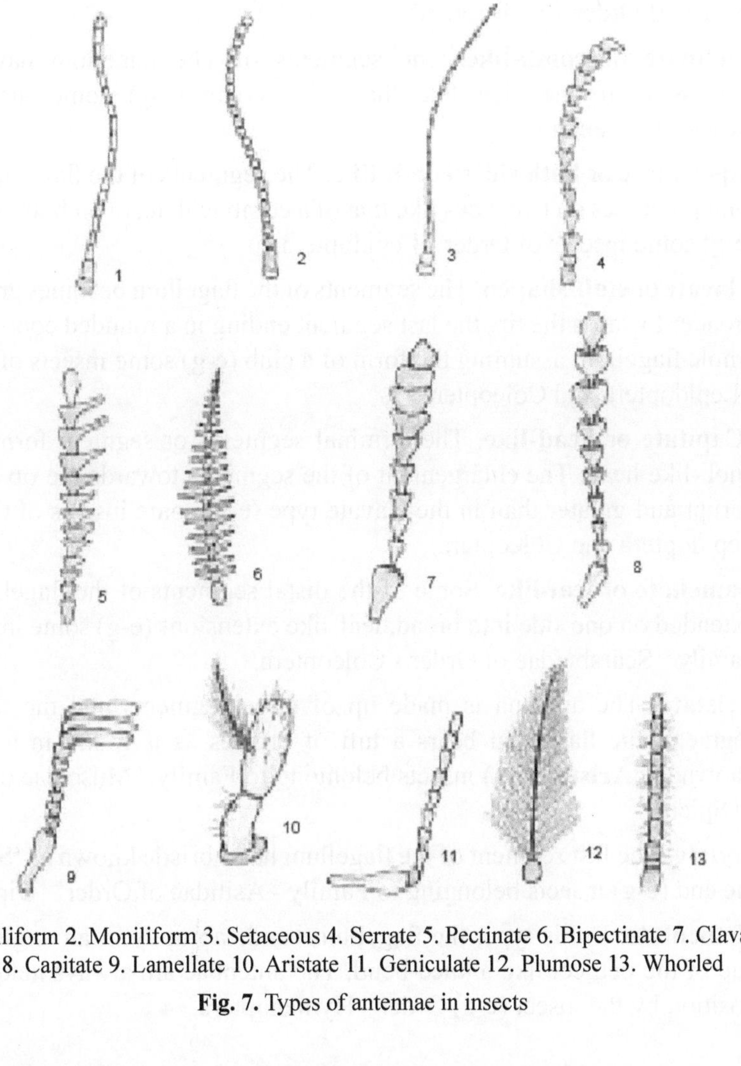

1. Filiform 2. Moniliform 3. Setaceous 4. Serrate 5. Pectinate 6. Bipectinate 7. Clavate 8. Capitate 9. Lamellate 10. Aristate 11. Geniculate 12. Plumose 13. Whorled

Fig. 7. Types of antennae in insects

i) **Filiform** or **thread-like**. In this type of antennae, all the segments of the flagellum are nearly uniform in thickness and imperceptibly reduced reduced towards the distal end and there are no distinct constrictions at the joints (e-g) Order - Orthoptera.

ii) **Moniliform** or **bead-like**. The segments of the flagellum are more or less globose, like a string of beads (e-g) Order – Isoptera.

iii) **Setaceous** or **bristle-like**. The segments of the flagellum become smaller and smaller and tapers almost to a point (e-g) some insects of the Orders - Lepidoptera and Coleoptera.

iv) **Serrate** or **saw-like**. The segments of the flagellum are more or less irregular, projecting in one direction like the teeth of a saw (e-g) some insects of Order – Coleoptera'.

v) **Pectinate** or **comb-like**. The segments of The flagellum have long processes on one side like that of a comb (e-g) some insects of Order – Coleoptera.

vi) **Bipectinate** or **both sides comb-like**. The segments of the flagellum have long processes on two sides like that of a comb with teeth on both the sides (e-g) some insects of Order - Lepidoptera.

vii) **Clavate** or **club-shaped**. The segments of the flagellum becomes gradually broader towards the tip, the last segment ending in a rounded cone and the whole flagellum assuming the form of a club (e-g) some insects of Orders - Lepidoptera and Coleoptera.

viii) **Capitate** or **head-like**. The terminal segments or segment forms large knob-like head. The enlargement of the segments towards the tip is more abrupt and greater than in the clavate type (e-g) some insects of Orders - Lepidoptera and Coleoptera.

ix) **Lamellate** or **leaf-like**. Some of the distal segments of the flagellum are extended on one side into broad, leaf-like extensions (e-g) some insects of Family - Scarabeidae of Order - Coleoptera.

x) **Aristate**. The antenna is made up of three segments and the terminal segment, the flagellum bears a tuft of bristles as is found in a crown, known as **'Arista'** (e-g) insects belonging to Family - Muscidae of Order - Diptera.

xi) **Stylate**. The last segment of the flagellum has a bristle known as **'Style'** at the end (e-g) insects belonging to Family - Asilidae of Order – Diptera.

xii) **Geniculate**. In his type, the flagellum bends abruptly at an angle at the end of the pedicel like a knee-bend. The antennae are always held in this position by the insect (e-g) Order - Hymenoptera.

xiii) **Plumose**. In this type of antennae, whorls of long hairs are seen from each joint of the flagellum and the antenna appears like a plume (e-g) some insects of Order - Diptera.

xiv) **Whorled**. In this type, whorls of bristles are seen at every junction of the segments of the flagellum (e-g) male mango mealy bug of Order - Hemiptera.

xv) **Flabellate**. In this type, there is a special attachment known as **'flabella'**, in which the flagellum appears to be lodged (e-g) insects of the Order - Strepsiptera.

Mouthparts

The mouthparts of insects have become modified in several ways for the purpose of ingestion of different types of food in different methods. The various types of mouthparts can be categorized into two groups viz., **'biting and chewing'** or **'cutting and chewing'** or **'mandibulate type'** of mouthparts and **'piercing and sucking'** or **'haustellate type'** of mouthparts. The former is found among those insects, which feed by mastication or chewing of their food that mostly contains solid stuff. This type of mouthparts is met with in the members of the orders - Orthoptera and Coleoptera. Grasshoppers and Lepidopterous larvae are the most common examples. The sucking type on the other hand is characteristic of those, which generally ingest liquid food by a process of sucking. The mouthparts have become greatly modified for piercing the epidermis of plants or the skin of animals and sucking up the sap or blood. Common examples are the plant bugs, mosquitoes, bed bugs, butterflies and blood-sucking flies.

1. Biting and chewing type or Mandibulate type of mouthparts

In a typical biting and chewing type of mouthparts, the following parts viz., **'labrum'**, **'mandibles'**, **'maxillae'**, **'labium'** and **'hypopharynx'** are present.

a) **The labrum** or **upper lip**. This is an unpaired, flap-like median plate attached to the clypeus and capable of movement or articulation. The main functions of the labrum are to close the front of the mouth cavity, protect the mandibles and guide the food into the mouth cavity.

b) **The mandibles**. One pair of mandibles or jaws is found, one on each lateral side of the head directly behind the labrum. Each mandible is generally a highly sclerotized, hard, thick, strong and solid appendage with a broad, triangular face and in some insects, it bears teeth. The mandibles are attached to the head by anterior and posterior articulations and they move in a transverse plane by two sets of muscles and are used to cut, chew and grind solid food.

c) **Maxillae** or **additional jaws.** The pair of maxillae or additional jaws lie directly behind the mandibles. They are more complex in structure than the mandibles and each maxilla is made up of four or five separate pieces, two basal pieces and three distal lobes, including a many-jointed palpus. The basal portion comprises of the **'cardo'** and **'stipes'**.

   'Cardo'. It is a triangular basal sclerite attached to the head and serves as a hinge for the movement of the remaining parts of the maxilla.

   'Stipes'. It is a more or less rectangular sclerite attached above the cardo, on which the distal parts of the maxilla are attached.

   Galea'. It is one of the distal lobes situated laterally on the outer side of the stipes. It is two segmented and the basal segment is called the **'sub galea'**.

   'Lacinia'. It is another distal lobe situated on the inner side of the stipes. This part also usually functions like the mandibles and it has a series of teeth along its inner edge.

   'Maxillary palp'. It is the third distal part of the stipes. On the outermost side of the stipes is a sclerite known as the **'palpifer'**, to which the five - segmented maxillary palp is attached. This antennae - like appendage functions as a sensory organ.

d) **Labium.** The labium or lower lip serves to close the mouth cavity from behind. Though the labium is found to be a single piece, it is actually composed of a pair of embryonic appendages, which have become fused along the median line and the labial appendages form the **'second maxillae'**. The labium is attached to the ventral wall of the hind part of the head (e-g) Grasshopper known as **'gula'**.

   The labium consists of the following parts :

   'Sub mentum'. It is a flat, leaf-like, basal plate attached to the head.

   'Mentum'. It is the central part of the labium, which bears a pair of three segmented **'labial palpi'**, one on each of the lateral sides. The lateral lobes of the mentum bearing the palpi are differentiated from the median portion and are called the **'palpigers'**.

   'Prementum'. It is the apical portion of the labium and is called **'ligula'**. This has four terminal lobes. The median pair is called **'glossae'** and is identical to the lacinia of the maxilla. The lateral pair is called **'paraglossae'** and is identical to galeae of the maxilla.

The three paired appendages viz., the mandibles, the maxillae and the labium (second maxillae) form the actual mouthparts.

In addition to the above parts, is an unpaired tongue-like median structure called the **'hypopharynx'** attached to the inner surface of the labium. The salivary glands open through this. Unlike the other mouthparts, the hypopharynx is not an appendage, but an unsegmented outgrowth of the body wall. Similarly a median structure called the **'epipharynx'** is found attached to the roof of the mouth below the labrum or clypeus in some forms of insects **(Fig. 8)**.

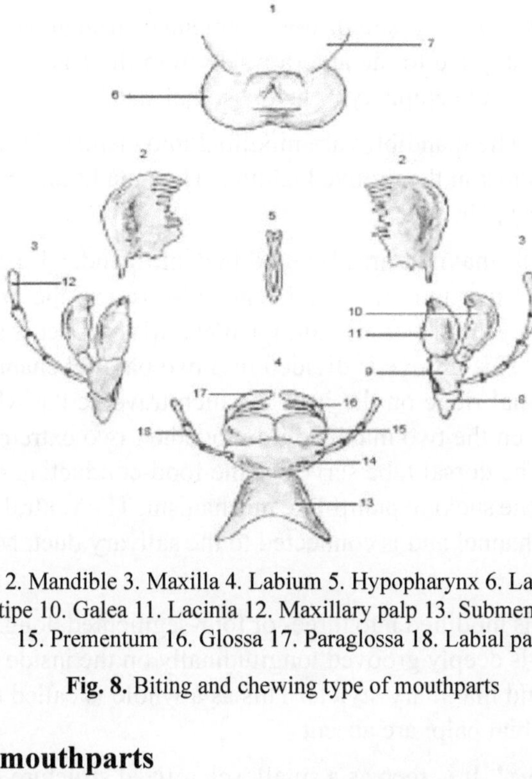

1. Upper jaw 2. Mandible 3. Maxilla 4. Labium 5. Hypopharynx 6. Labrum 7. Clypeus 8. Cardo 9. Stipe 10. Galea 11. Lacinia 12. Maxillary palp 13. Submentum 14. Mentum 15. Prementum 16. Glossa 17. Paraglossa 18. Labial palp

Fig. 8. Biting and chewing type of mouthparts

2. Suctorial mouthparts

Among insects with suctorial mouthparts, five different types are met with : They are: (i) Piercing and sucking type found in most of the plant bugs, biting flies, mosquitoes etc., where the organ is a long, curved or foldable beak, sharp at the tip (ii) Siphoning type found in butterflies and moths, where the sucking organ is a spiral tube, kept coiled up under the head (iii) Sponging type found in insects like the housefly, where the organ is distally flattened (iv) Chewing and lapping type found in the group of insects including the bees and wasps, where mandibulate and suctorial arrangements are more or less combined (v) Rasping and sucking type found in thrips, where the mouthparts are in-between chewing and piercing types.

i) **Piercing and sucking type of mouthparts.** These mouthparts are modified to a great extent for piercing the epidermis of plants or the skin of mammals and sucking up the sap or blood. In this type also there are two different types viz., (a) Bug type and (b) Mosquito type.

a) **Piercing and sucking type - Bug type.** In this type, the mouthparts are modified to pierce plant tissues and suck the cell sap from the cells. This type of mouthparts is found in insects belonging to the order – Hemiptera. The mouthparts comprise of the following parts :

'Labrum'. It is a very small, non-segmented, triangular, tapering lobe and is closely applied to the anterior side from the base of the beak. It is considered to be the epipharynx in a vestigial state.

'Mandibles'. The mandibles are modified into slender, bristle-like, sharp stylets, which rest in the grooved labium. The mandibular stylets form the anterior (outer) pair.

'Maxillae'. The maxillae are also modified into slender, bristle-like, sharp, stylets, which taper to a fine point. They also rest in the grooved labium and form the posterior (inner) pair. Each maxillary stylet is grooved along its inner side. This groove is divided into two parallel channels by means of a longitudinal ridge on the inside, which traverse the whole length of the stylet. When the two maxillae join together, two extremely fine tubes are formed. The dorsal tube serves as the food-conducting channel and is connected to the sucking pump-like mechanism. The ventral tube serves as the salivary channel and is connected to the salivary duct. Maxillary palpi are absent.

'Labium'. It is modified into three- or four-segmented, long, slender, rigid organ, which is deeply grooved longitudinally on the inside to enclose the mandibular and maxillary stylets. This as a whole is called the **'rostrum'** or **'beak'**. Labial palpi are absent.

'Hypopharynx'. It is seen as a small, sclerotized structure lying between the bases of the stylets and the salivary duct passes through it into the salivary tube of the maxillae.

'Sucking pump'. The pump chamber represents the pre-oral cavity. Its base is formed by the hypopharynx and its roof is formed by the wall of the epipharynx, below the clypeus and is provided with dilator muscles. This chamber acts as a force pump and opens into the oesophagus **(Fig. 9)**.

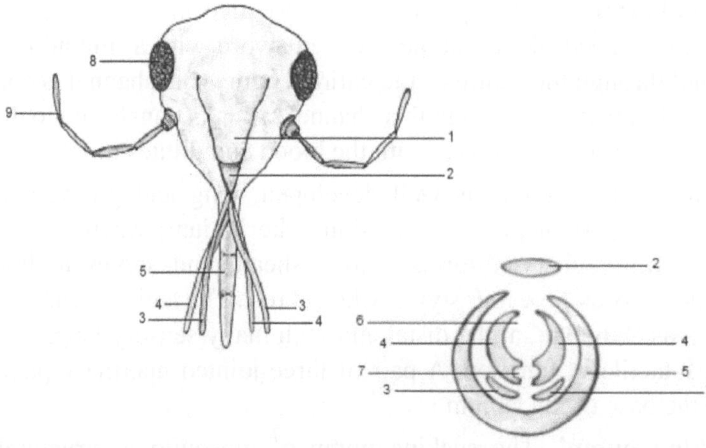

1. Clypeus 2. Labrum 3. Mandible 4. Maxilla 5. Labium 6. Food channel
7. Salivary duct 8. Compound eye 9. Antenna

Fig. 9. Piercing and sucking type of mouthparts (Bug type)

The protractor and retractor muscles attached to the mouthparts, by their rapid alternate actions enable the stylets to pierce the plant tissues deeper and deeper. All the four stylets are not protracted simultaneously, but only the mandibles penetrate the ues first, followed by the maxillae. After penetrating the plant tissues, saliva is injected through the salivary channel, which mixes with the cell sap and dilutes the sap. Then the dilator muscles contract, effecting an upward suction by which the liquid food is drawn through the food channel into the pump chamber, which is then drawn upward into the oesophagus.

b) **Piercing and sucking type - Mosquito type**. This type of mouthparts is adapted for blood sucking purposes and is commonly met with in the female mosquitoes, fleas and some blood-sucking flies. The following parts are found in this type of mouthparts :

'The labrum-epipharynx'. Both the labrum and epipharynx are fused together to form a long and slender inverted **'U'**- shaped structure, which along with the flattened hypopharynx forms a tubular channel through which blood is sucked in by the pumping action of the cibarium. This constitutes the food channel.

'Mandibles'. The mandibles are modified into extremely fine, needle-like stylets known as **'Mandibular stylets'**.

'Maxillae'. Like the mandibles, the maxillae are also modified into fine, needle - like stylets and are known as **'Maxillary stylets'**. The maxillae possess minute teeth distally. The maxillary stylets are responsible for piercing the skin and making a wound in the skin of the host.

'Hypopharynx'. The hypopharynx is modified into a long and slender structure, shaped like a double – edged sword with a minute transverse channel through the centre of the entire length. This channel is connected to the salivary duct. Through this channel saliva is transferred to the blood of the host. The saliva mixes with the blood and dilutes it.

'Labium'. The labium is well developed, long and grooved dorsally. It takes no part in piercing the skin. The labium, when covered with the labrum-epipharynx forms a long sheath and serves to house the hypopharynx and the four stylets when at rest. The labium bears two lobes known as **'labellae'** at the distal end with many sensory hairs, which are mainly tactile in function. A pair of three-jointed maxillary palpi arises from the base of the labium.

'Sucking pump'. The sucking pump of mosquito is structurally and functionally similar to that of the bugs. When the female mosquito starts feeding, the labellum at the tip of the labium is first brought against the skin and then the pointed maxillae and mandibles pierce and penetrate the skin. The labrum-epipharynx is also inserted into the puncture along with the hypopharynx. When the mouthparts are inserted to a sufficient depth, the labium is doubled back in the form of a loop. In this type of mouthparts, the puncture is always made by the maxillary stylets only **(Fig. 10)**.

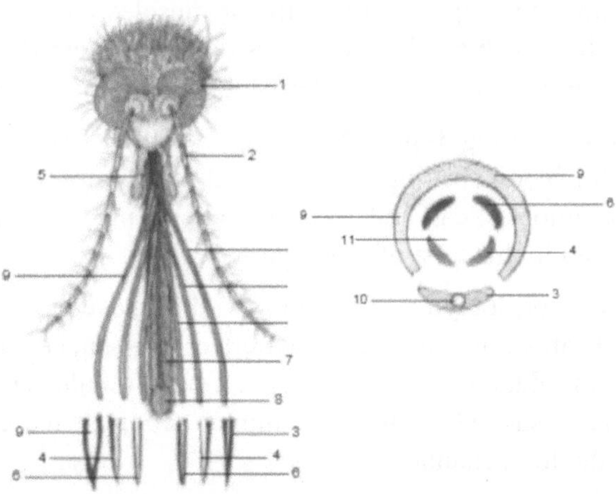

1. Compound eye 2. Antenna 3. Hypopharynx 4. Maxilla 5. Maxillary palp 6. Mandible 7. Labium 8. Labial palp 9. Epipharynx 10. Salivary duct 11. Food channel

Fig. 10. Piercing and sucking type (Mosquito type)

ii) **Siphoning type of mouthparts.** This type of mouthparts is met with in butterflies and moths, which feed on the nectar from flowers and other liquid food. The following parts are present in this type of mouthparts:

'Labrum'. It is a small, usually a narrow transverse band attached to the lower edge of the large clypeus, bearing a pair of small, hairy lobes known as **'pilifers'** on its lateral extremities.

'Mandibles'. The mandibles are entirely absent in most of the species however, in some macro Lepidopterous moths, vestigial mandibles are seen as small, immovable lobes.

'Maxillae'. This part is represented by Rudimentary maxillary palpi and the modified galeae of the maxillae. The galeae are greatly elongated and joined on their inner surface to form the external part of the sucking structure called the **'proboscis'**. The lacinia are reduced to insignificance. The proboscis is thus made up of the two lateral galeae, which are held together by interlocking devices. The opposed walls of the galeae are thickened and concave inside, forming in-between them a minute food channel, through which liquid food is sucked into the mouth. The basal part of each maxilla is differentiated into a small **'cardo'** and a slightly bigger **'stypes'**. The stypes bears a rudimentary palpus and the elongated galea. (e-g) Butterfly.

'Labium'. The labium is modified into a small lobe or a flat, but often reduced membranous sheath behind the base of the proboscis. Three-segmented labial palpi are usually well developed. They are covered with long hair-like scales forming two conspicuous brushes, projecting from the base of the proboscis.

'Sucking pump'. The sucking pump is well developed and functions as in the piercing and sucking type of mouthparts **(Fig. 11)**.

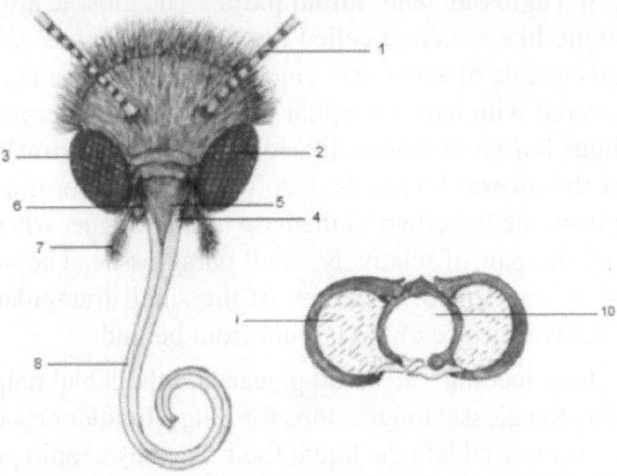

1. Antenna 2. Compound eye 3. Labrum 4. Pilifers 5. Epipharynx 6. Maxillary palp 7. Labial palp 8. Proboscis 9. Cross section of the proboscis 10. Food channel

Fig. 11. Siphoning type of mouthparts

When the proboscis is not in use, it is coiled like a spring of a watch underneath the head. However, in response to food stimulus, it is extended fully. The uncoiling of the proboscis may be due to blood pressure or due to the action of some muscles in the head along with blood pressure. The outer wall of each half of the proboscis shows a closely ringed structure, which may allow the coiling of the proboscis.

iii) **Chewing and lapping type of mouthparts.** This type of mouthparts is invariably a combination of mandibulate and suctorial types. This is met with in bees and wasps, which take liquid food such as, nectar from flowers by licking and collect pollen from flowers. The mouthparts are also used for moulding wax or other materials for building nests. The labrum and mandibles are of the chewing type, while the maxillae and labium are modified to form a **'proboscis'**. The following parts are present in this type of mouthparts :

'Labrum'. It is a small, semi circular lobe, which is almost vestigial.

'Mandibles'. Unlike the typical mandibles of the chewing type, here they are modified into rather flattened or spoon-shaped organs to serve as tools for performing various functions such as, moulding wax for building combs, collecting pollen from flowers etc.

'Maxillae'. In the maxillae, the inner lacinae are rudiment or absent. The galeae are modified into large, flat blades, much longer than the stipes. The maxillary palpi are reduced to small pegs. The cardo on which the galea articulates is also developed into somewhat long structure.

'**Labium'**. The labium comprises of the **'mentum'**, **'prementum'**, **'glossae'**, **'paraglossae'** and **'labial palpi'**. The glossae are modified into a long, tongue-like structure called the **'glossal tongue'** which is highly flexible and capable of active movement of protraction and retraction. It is densely covered with hairs except at the base . At the terminal end of the glossal tongue is a small spoon – like lobe called **'flabellum'**. The posterior s urface of the glossal tongue is deeply grooved to form a deep channel extending from the flabellum to the base of the tongue, where it is closely encircled by the pair of relatively small paraglossae. The labial palpi are long and four-segmented. The apex of the small triangular mentum fits with the concave surface of the labrum from behind.

When the bee starts feeding, the broad galeae and the labial palpi are brought together encasing the glossal tongue, thus forming a tubular proboscis. The end of the proboscis is inserted into the liquid food. Then by a rapid protractory and retractory movement of the tongue, the liquid food is drawn up into the food channel of the proboscis and then sucked up into the mouth by the sucking pump **(Fig. 12)**.

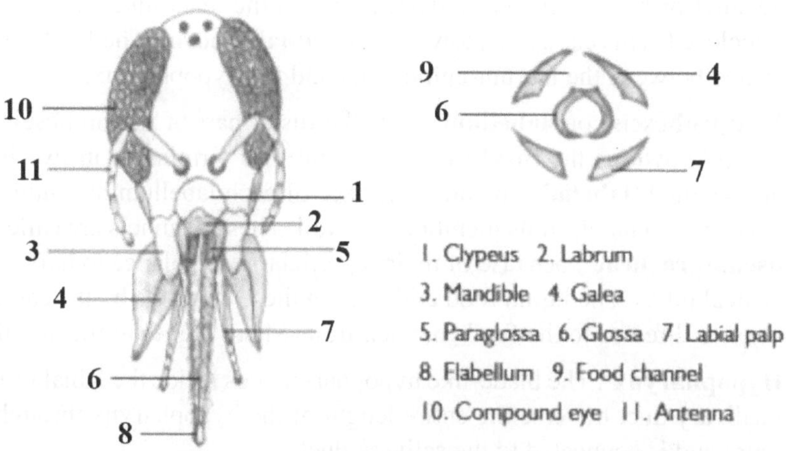

Fig. 12. Chewing and lapping type of mouthparts (e-g) Honeybee

iv) **Sponging type of mouthparts.** This type of mouthparts is also known as **'muscoid type'** and is seen in most of the flies such as, the common housefly. This type is adapted to consume either liquid food or food materials, which are readily soluble in saliva. The mouthparts comprises of the following parts :

'Labrum'. The labrum is a long tapering, strongly sclerotized, flap-like structure arising from the distal margin of the rostrum and covers the anterior portion of the haustellum. It is fused with the epipharynx and is termed the **'labrum-epipharynx'**.

'Mandibles'. The mandibles are completely absent.

'Maxillae'. The maxillae are also absent. Only one pair of single-segmented maxillary palpi is present.

'Labium' The labium is well developed and modified into a long proboscis. The proboscis comprises of the following parts :

a) **'Basiproboscis or 'rostrum'.** It is the basal part of the proboscis and is like a broad, inverted cone with flexible, membranous walls. On the lower part of the anterior portion of the rostrum, there are two lateral sclerites, which support the pair of maxillary palpi.

b) **'Mediproboscis' or 'haustellum'**. It is the middle part of the proboscis. It is a cylindrical structure, which projects downward and somewhat forward from the end of the rostrum when protruded. The posterior wall of the haustellum is made up of a prominent plate, which is the mentum of the labium/ The anterior portion is covered by the labrum-epipharynx. When the distal end is lifted, a deep, lengthwise groove is discernible in the anterior part of the haustellem, in which the blade-like hypopharynx is enclosed. This cavity is known as the **'labial groove'**. The food canal is formed between the labrum-epipharynx and the hypopharynx.

c) **'Distiproboscis'** or **'labellum'**. It is the distal part of the proboscis and is also known as the **'oral disc'**. The oral disc terminates in two broad lobe s called **'labella'**. The under surface of each labellum has numerous transverse channels in its membranous wall. These channels are called the **'pseudotracheae'**, because of their superficial resemblance to half - open tracheal tubes. The liquid food collected in these channels by the capillary action is directed to the food channel, from where it goes to the mouth.

   'Hypopharynx'. The blade-like hypopharynx lies inside the labial groove. A salivary duct traverse the entire length of the hypopharynx through the centre and is connected to the salivary duct.

The proboscis when not in use, remains retracted under the head. But when the fly starts feeding, the proboscis is protruded and thrusted into the liquid food. The liquid food enters the pseudotracheae and then conveyed to the food channel. Certain easily soluble solid foods such as, sugar, jaggery etc. are also consumed by flies. For this, the fly first exudes a droplet of saliva on the food. The food gets dissolved in the saliva. Then the fly consumes the dissolved food in the liquid state **(Fig. 13)**.

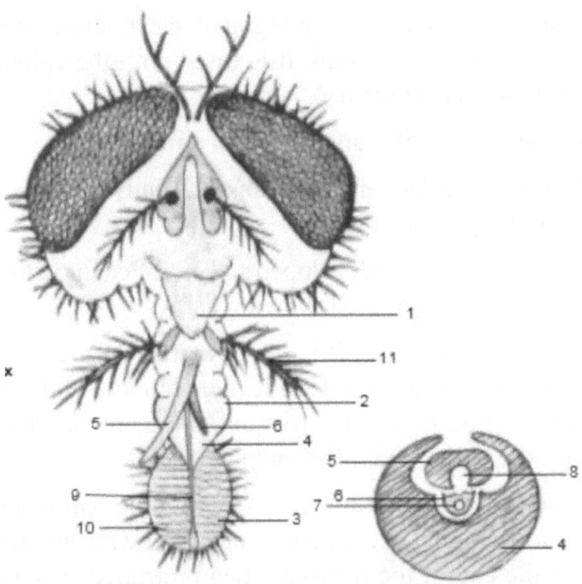

1. Proboscis 2. Haustellum 3. Labellum 4. Labium 5. Labium-epipharynx 6. Hypopharynx 7. Salivary duct 8. Food channel 9. Labial groove 10. Tracheal tubes 11. Maxillary palp

Fig. 13. Sponging type of mouthparts (e-g) Housefly

Thorax

The **'thorax'** is the second and middle region of the insect body. In the front, it is connected with the occipital region of the head by a flexible **'neck'** or **'cervical region'** and posteriorly it is attached to the base of the abdomen. This region is made up of three separate segments known as pro-, meso- and metathoracic segments. In some insects like grasshoppers, crickets etc., the prothorax is shield-like and is often much bigger than the hinder two segments. The meso- and metathoracic segments are generally fused together and in many insects, the chitinous exoskeleton of these segments is divided by sutures into different regions and structures in each. Each thoracic segment bears a pair of legs. The first pair of wings arises from the mesothotax and the second pair from the metathorax. The dorsal surface of the thorax is called **'tergum'** or **'notum'**, the sides are called **'pleura'** and the ventral surface is called **'sternum' (Fig. 1, 2)**.

In a typical insect such as, grasshopper the notum or tergal plate of the prothorax is shield-like and is known as **'pronotum'**. It completely covers the prothorax and extends over part of the mesothorax. The pleural plate is more or less a triangular structure just above the base of the prothoracic leg. The area underlying between the two prothoracic legs is known as **'prosternum'**. The meso- and metathorax are strongly united to form a compact body unit and accommodate the musculature necessary to combine both running and flying mechanisms as in one segment. Both these segments are collectively called the **'pterothorax'**, as it carries the wings. In Orthoptera and Lepidoptera, where both the pairs of wings are important in flight, the meso- and metathorax are well developed. On the other hand, in Diptera, where only the forewings are used in flight, only the mesonotum is well developed.

Each pleuron of the meso- and metathorax of the grasshopper is an enlarged, conspicuous lateral plate and is composed of two sclerites, which are separated by pleural sutures. The anterior sclerite of the mesothorax is termed as **'mesoepisternum'** and the posterior sclerite as **'mesoepimeron'**. Similarly the two sclerites of the metathorax are termed as **'metaepisternum'** and **'metaepimeron'**. Two pairs of spiracles are situated in these segments. The mesothoracic spiracle is located between the prothorax and mesothorax and metathoracic spiracle is located above the second pair of legs.

The sterna comprise of broad, flat plates between the meso- and metathoracic legs and are called **'mesosternum'** and **'metasternum'** respectively.

The basal portion of the leg known as the **'coxa'** reticulates with sclerites in the pleural area. With the notal and pleural processes of the meso- and metathorax, the two pairs of wings articulate.

Thoracic Appendages

The most important appendages of the thorax are the legs and wings.

Legs

The presence of three pairs of legs in an adult insect is such a fundamental character that an insect can be simply defined as a **'six-legged arthropod'** or **'Haxapod'**. These legs are present in almost all adult insects, but in rare cases legs may be reduced or absent in the adult forms. The reduction of legs is associated with a sedentary or parasitic habit. Some female psychids are completely apodous, because they never leave the bag constructed by their larvae. Legs are completely absent in female Strepsiptera, which are parasitic. In Nymphalidae of Order Lepidoptera, the forelegs are held permanently withdrawn against the thorax and hence these insects have only two pairs of functional legs. However, legs may be absent in some larval forms. The larvae of insects belonging to Order - Diptera, the parasitic larvae of the Orders – Hymenoptera and Strepsiptera are apodous. Each thoracic segment bears one pair of legs, each leg articulating within sclerites in each of the pleural region. The legs are typical jointed appendages and are made up of distinct pieces articulating with each other by movable joints.

Though the size, shape, armature and other peculiarities in the structure of legs show remarkable variation in different insects and are modified for various modes of life, in general each leg is made up of five distinct parts viz. , the proximal **'coxa'**, the second **'trochanter'**, the third **'femur'**, the fourth **'tibia'** and the fifth, distal **'tarsus'**. While the coxa, femur and tibia are always single pieces, the trochanter and tarsus are not always so. The trochanter is found divided into two in some insects of the Order - Odonata and some parasitic insects of the Order – Hymenoptera. The tarsus is made up of a number of joints in most of the insects, the number varying from one as in some primitive forms, to five pieces in others, which is more or less typical.

The coxa fits in a cup-like sclerite in the pleural region of the thorax and articulates at two points, thus limiting the movement. The trochanter is quite a small piece and serves as a freely movable joint between the coxa and the femur. The articulation of coxa with the trochanter further restricts the movement of the leg. The **'abductor muscles'** are the ones that extend the appendage away from the base, while the **'adductor muscles'** are the ones that pull or draw the appendage towards the base. The femur is the largest and most powerful part. The tibia is slender, often quite long and provided with downward projecting spines and spurs, which aid in climbing or for defensive or offensive purposes. The tarsus is usually subdivided into two to five segments. The pre-tarsus is provided with a set of claws and has one r more pads or cushions at their base or between the claws, which are useful for walking. A single pad between the claws is known as **'arolium'** and pads at the base are called **'pulvilli'**. When the arolium has a slender, bristle – like structure as is found in the tarsus of houseflies, it is termed as **'empodium'**. Though the legs of insects are primarily intended for locomotion, they are adapted to perform various other functions also in different insects such as, running, digging, swimming, jumping, foraging, clinging etc. The structural variations or modifications of legs in insects are important characteristics in the classification of insects **(Fig. 14)**.

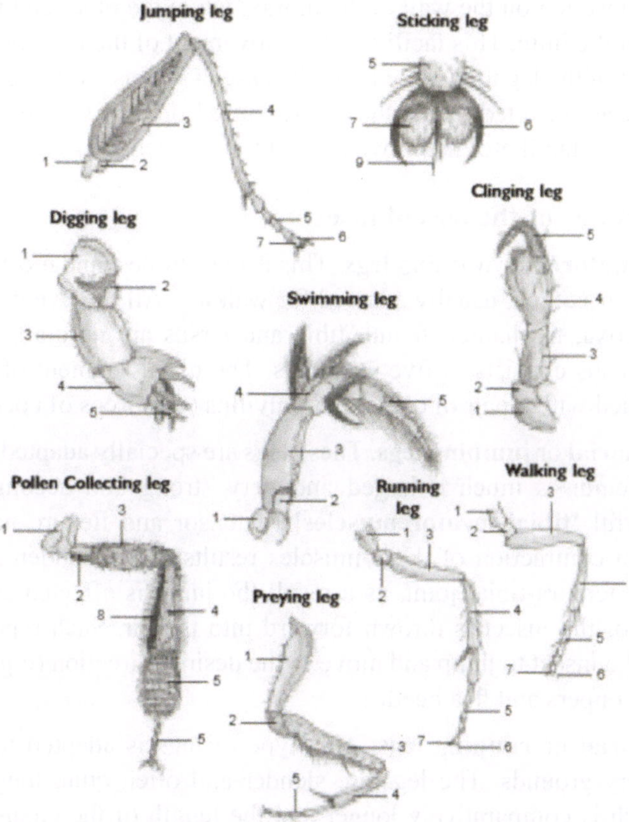

1. Coxa 2. Trochanter 3. Femur 4. Tibia 5. Tarsus 6. Claw 7. Pulvillus 8. Pollen basket 9. Empodium 10. Pollen basket

Fig. 14. Types of legs in insects

Muscles and Locomotion

The muscles of insects are classified into two categories viz., (i) **'Skeletal muscles'** and (ii) **'visceral muscles'** based on their functions. The skeletal muscles, otherwise called **'phasic muscles'** are involved in the movements of the appendages. The visceral muscles, otherwise called **'tonic muscles'**, which form layers enveloping the internal organs control the movements of the inner organs. In insects the muscles all striated. Each muscle is made up of a number of fibres, which are long, usually multinucleated cells, running the whole length of the muscle. Each fibre is is bounded by the plasma membrane of the cell and basement membrane, which jointly constitute the **'sarcolemma'**. The cytoplasm of the fibre is called the **'sarcoplasm'**.

The muscles involved in the movement of insects are (a) the **extrinsic muscles'** and (b) the **'intrinsic muscles'**. In the case of extrinsic muscle, one end of the muscle is inserted on the wall of the thorax, while the other end is attached to the coxa of the limb. This facilitates the movement of the leg from the point of attachment of the leg to the thorax. In the case of intrinsic muscle, both ends of the muscle are inserted within the leg, running from one segment of the leg to the next. This facilitates the movement of the segments within the leg.

Modifications of the legs of insects

i) **Ambulatorial** or **walking legs.** This is the simplest and most typical form of the insect leg, usually adapted for walking. All the five parts of the leg viz., coxa, trochanter, femur, tibia and tarsus are normally represented. The tarsus consists of five segments. The distal segment of the tarsus is provided with a pair of claws and pulvillus (e-g) Legs of cockroach.

ii) **Saltatorial** or **jumping legs**. These legs are specially adapted for jumping. The femur is much enlarged and very strong and accommodates the powerful **'tibial levator muscles'** (extensor and flexure muscles). The sudden contraction of these muscles results in the sudden straightening of the femora-tibial joint, as a result the jump is effected and the entire body of the insect is thrown forward into the air. Such repeated actions help the insect to jump and move in the desired direction (e-g) Hind leg of grasshoppers and flea beetles.

iii) **Cursorial** or **running legs.** This type of legs is adapted to run fast on slippery grounds. The legs are slender and often quite long. The tarsus of such is comparatively longer and the length of the tarsus touches the ground while running. Such a provision helps the insect to run fast on slippery grounds or on loose sand without slipping (e-g) The legs of wasps belonging to Ichneumonidae.

iv) **Raptorial** or **preying legs**. These legs are found in some of the predatory insects and are specially adapted for catching and killing the live prey. The coxa is fairly long and quickly movable. The femora are thickly slinose and have a groove along the lower side. The tibiae are also spiny and when bent they fit into the groove along the femora like pincers. By a quick movement of the coxae, the prey is caught in-between the opposing femora and tibiae. of both the legs and held tightly between the grooves and killed by the spines in the femora and tibiae. The tarsus is compossed of five segments and is much reduced (e-g) Forelegs of praying mantis.

v) **Fossorial** or **digging legs**. These legs are modified for digging purposes. The legs are very strong and powerful, broad and short in size. The coxa, trochanter and femur are short. The tibia, which is also short is provided with four, hard and strong scraper-like prolongations at its broad distal end. These are used for digging the ground and for cutting the roots of plants. The tarsus consists of two or three segments, each segment protruding like a tooth and assists the tibia while digging (e-g) Forelegs of mole cricket.

vi) **Foragial** or **pollen collecting legs**. Such legs are suitably modified for collecting and carrying pollen from flowers. The coxa, trochanter and femur are normal as in a typical leg. However, the tibia and tarsus are modified. The posterior region of the tibia is more or less dilated like a cavity and margined with long hairs. Thus the tibia is formed into a **'corbicula'** or **'pollen basket'**. The basi-tarsus is flattened on its side and provided with many rows of short, stiff bristles forming a brush. This brush helps the bees in gathering the pollen adhering to the hairs of its body and also from flowers and stuffing the same into the pollen baskets to be taken to the beehive. The other segments of the tarsus are very much reduced (e-g) Hind legs of worker honeybee.

vii) **Natatorial** or **swimming legs**. Insects living in aquatic habitats are provided with such modified type of legs. The femur, tibia and tarsus of the hind legs are flattened and broad antero-ventrally to form paddle-like structures and are provided with long rows of hairs. Such a modification helps the insect to use the legs as paddles for swimming in water (e-g) Hind legs of water beetles.

viii) **Scansorial** or **clinging legs**. This type of legs is specially adapted for clinging to the hairs of its host. The tibia and tarsus are very much modified. The tarsus is one or two segmented and it bears a single prominent claw at its end. The tip of the tibia is provided with a cushiony structure. Due to muscular action, the claw along with the tarsus is tightly approximated to the cushiony structure at the tibial end with the hairs of the host in-between, thereby allowing the insect to maintain a firm hold on to the hairs. Pulvillus is absent in this type of legs (e-g) Legs of louse.

ix) **Sticking legs**. This type of modified legs is met with in insects such as, common houseflies. The tarsus alone is modified. A pair of broad, pad-like structures, known as **'pulvilli'** are provided at the base, between the two incurved claws of the terminal segment of the tarsus. Numerous hollow, tubular hairs, which exude a sticky substance are attached to the pads. The sticky substance makes the fly to stick on to any polished, glossy or smooth surface such as, glass and can sit comfortably at ease even in an upside-down position on such surfaces. A bristle-like structure called **'empodium'** is present in-between the pulvilli, which aids the insect to life and release the insect (e-g) Legs of common housefly.

x) **Grooming legs**. In the distal end of the fore-tibia, several rows of setae-like hairs are provided to form a comb-like structure, which is used to brush and clean the body, (e-g) Legs of mosquitoes.

Prolegs

Most of the caterpillars have **'prolegs'** or **'abdominal legs'**. They are hollow, cylindrical outgrowths of the body wall, the lumen of which is continuous with the haemocoel. The apical portion of the proleg is less rigid than the sides and is called **'planta'**. The planta bears outwardly curved hook-like structures called **'crochets'**. The caterpillars move by a serial contraction of the longitudinal muscles along with leg movements. The contraction starts from the posterior end. Each segment of the abdomen is lifted as result of contraction of the longitudinal muscles in front, at the same time the prolegs are retracted. Following this, the ventral longitudinal muscles contract, which brings the segment down again as the prolegs are extended and get a fresh grip in front, thus completing the forward movement.

Wings

The wings of insects are borne on the meso- and metathorax. They are characteristic features of insects, but a wingless condition occurs in a number of different groups of insects. In some, the absence of wings is obviously secondary. For example, ants and termites have wings only in the males and at certain periods of the life cycle in the females. Workers always lack wings. Some parasitic insects such as, the lice and the fleas have lost the wings completely. On the other hand, insects belonging to some Orders viz., Protura, Thysanura and Collembola are primarily wingless. The group in which the wingless condition (apterous) is considered primary, is classified under the Subclass – **'Apterygota'** and is believed to represent the most primitive members of the Class – Insecta. Winged insects and those that are secondarily wingless are classified under the Subclass - **'Ptrygota'**. In Mayflies (Ephemeroptera), the last larval instars are winged and are capable of short flights.

Since wings are evaginations of folds of the integument, they are composed of two sheets or layers of cuticle. At a vein, the two cuticular membranes are thickened and separated, forming a tubular lumen surrounded by heavy cuticle. The veins thus formed are an effective supporting skeletal rod for the wing. Wing veins open into the body and contain circulating blood. The lumina of the main veins contain in addition to blood, tracheoles and sensory nerve branches.

Movement of Wings of Insects

Two sets of muscles called the direct and indirect flight muscles are associated with the movement of wings. **'Direct muscles'** are those that arise from the pleuron and inserted directly on to the sclerites of the wing base, thus securing the wings to the tergum. Each wing articulates with the edge of the tergum with its inner end resting on a dorsal pleural process, which acts as a fulcrum. The **'indirect muscles'** situated within the thorax viz., the **'dorsal longitudinal muscles'** and the **'vertical tergosternal muscles'**, though not directly associated with the wings are mainly responsible for the up and down movement of the wings. Upward movement of the wings results indirectly due to the contraction of the vertical muscles within the thorax, thereby depressing the tergum. Downward movement of the wings is effected either directly by contraction of the muscles attached to the wing base, as in dragon flies and roachesor indirectly by the contraction of the tranverse horizontal (longitudinal) muscles, thereby raising the tergum as in bees, wasps, butterflies and flies or by both direct and indirect muscles as in grasshoppers and beetles **(Fig. 15)**.

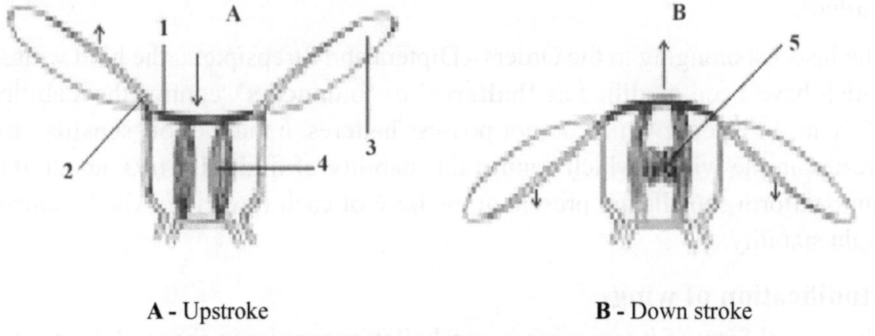

A - Upstroke **B** - Down stroke

1. Tergum 2. Pleural process 3. Wing 4. Verticl muscles contracted 5. Longitudinal muscles contracted

Fig. 15. Mechanism of the basic wing strokes in flight of an insect

Development of Wings

The insect wing is a flattened, double-layered expansion of the body wall and ita walls are composed of the same elements as the body wall viz., cuticula, epidermis and basal membrane and its lumen contains nerves, tracheae and blood.

The wings of insects with incomplete metamorphosis grow externally as the mouthparts and legs. The wing buds appear first as hollow, flattened outgrowth of the body wall along the lateral margins of the notum in the meso- and metathorax in the second or third instar of the nymph. At each moult, their size increase without much structural changes. After the last noult, the adult form emerges with full-fledged wings. In insects with complete metamorphosis, the wings start developing beneath the outer cuticle during the larval stage, usually within the epidermal pouches and develop fully in the pupal stage.

The epidermis of the very young wing pad consists of upper and lower layers, which correspond to the dorsal and ventral surface of he wing pad, but very soon the bases of the epidermal cells become stretched into long processes and the basement membrane of the lower and upper epidermal layers become opposed for most of their area forming the so called middle membrane of the wing. But the basement membrane does not come together along certain lines. The channels thus left open contain wing nerves, tracheae and blood. These channels determine the courses of the future veins of the wings. The epidermal cells condense along the vein channels in the later stages and form a thick cuticular layer, which constitute the walls of the wing veins. When the development is complete, the epidermis has almost disappeared and he matured wing is almost entirely a cuticular structure. On the emergence of the adult, the blood pressure inflates the wings to he full size and the cuticle hardens.

The insects belonging to the Orders - Diptera and Strepsiptera, the hind wings, which have been modified as **'halteres'** or **'balancers'**, control the stability of flight. In insects, which do not possess halteres, a number of **'sensilla'** are present in the wings, which control the stability of flight. In *Apis,* about 700 campaniform sensilla are present at the base of each forewing, which control flight stability.

Modification of wings

The typical form of insect wing is nearly flat, triangular in shape, delicate and membranous appendages of the body, which is stiffened by the wing veins. But striking modifications of this form exist and specific names have been given to them.

'Elytra' - The forewings of beetles and earwigs are hardened and modified as veinless wing covers, which serve chiefly to protect the dorsal wall of the body and the membranous flying hind wings that fold beneath them when the insect is at rest. This type of forewings is called **'elytra'**.

'Tegmina' - The fore wings of Orthoptera are somewhat thickened and more leathery than those of he beetles, but are used in part atleast for flying. This type of fore wings is known as **'tegmina'**.

'Hemelytra' - The true bugs of the Order – Hemiptera (Suborder – Heteroptera) have wings that are hardened at the basal half and the outer or apical half is membranous, while the hind wings are completely membranous. This type of fore wings is known as **'hemelytra'**.

'Halteres' - The true flies of the Order- Diptera have one pair of flying wings (forewings) located on the mesothorax and the hind wings are modified into stubby appendages with a terminal knob. These balancing organs, which control the stability of flight, also serve as sensory structures and are called **'halteres'**.

'Pseudo-halteres' - In Strepsiptera, the forewings are very much reduced. Such reduced forewings are called **'Pseudo-halteres'**.

'Scaly wings' - Moths, butterflies and skippers have transparent wings, which are covered by fine, minute, soft scales that form various, but typical color patterns and are called **'scaly wings'**.

The Wing Coupling Mechanism

There are great variations in the wings of insects. Many of these variations represent modifications, which accommodate the demands of flight characteristic of the particular group of insects. Wing coupling indicates the coupling of the fore- and hind wings together by means of different interlocking devices or by simply overlapping, so that the wings operate together. In several groups of insects, the two-winged condition is found to be apparently more efficient than the four-winged condition. However, in some insects such as, damsel flies, the the two pairs of wings move independently. As the fore wing moves up, the hind wing moves down. The more important mechanical coupling devices are discussed below.

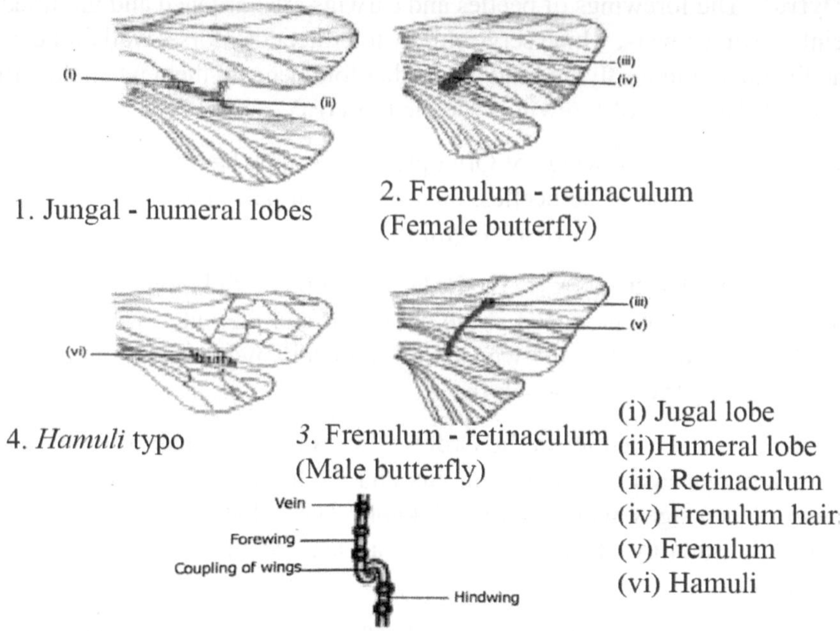

Fig. 16. Wing coupling structures

i) **Jugal and humeral coupling**. Among the Lepidoptera, Trichoptera and Mecoptera wing coupling depends on the modifications of the wing bases. In the forewing, the posterior margin is provided with a modified structure like a finger-like lobe, which is stiffened by a branch of the third anal vein and is called the **'jugal lobe'**. Similarly the anterior margin of the hind wing is also provided with a small lobe called **'humeral lobe'**. As the greater part of the posterior margin of the fore wing overlaps the anterior margin of he hind wing, it is held between the lobes, thereby the wings move together in flight **(Fig. 16)**.

In Mecoptera, the jugal and humeral lobes are provided with a few thick bristles. When the insect flies, the **'jugal bristles'** lie on top of the hind wing, while the **'humeral bristles'** press against the underside of the fore wing and thus coupling of the wings is effected.

ii) **Frenulum and retinaculum coupling**. In the higher forms of Lepidoptera, instead of jugal lobes, another structure called **'frenulum'** is provided. In female butterflies, it consists of a group of stout, stiff bristles arising from the hind wing and lying beneath the extended fore wing. In the fore wing another structure called the **'retinaculum'**, which consists of a patch of hairs near the cubitous vein is present. During flight, the frenulum engages with the retinaculum and thus coupling is effected **(Fig. 16)**.

However, in the male, the frenulum bristles of the hind wing are fused into a single, stout, cord-like structure and in the fore wing a curved process is provided from the sub-costal vein. While flying the **'frenulum cord'** is held by the curved retinaculumand thus coupling is effected.

iii) **Humulate coupling**. The insects belonging to Hymenoptera have a **'humulate'** type of wing coupling arrangement. The anterior margin of the hind wings is provided with a row of small, curved hook-like structures call **'humuli'**. The inner margin of the forewing is folded. While flying, the hooks fasten into the fold of the fore wing and thus coupling is effected **(Fig. 16)**.

iv) **Amplexiform coupling**. In some families of Lepidoptera viz., Bombicidae and Papilionidae, the posterior margin of the forewing and the anterior margin of the hind wing overlap and couple with each other by incurling of the margins **(Fig. 16)**.

In some other insects such as, Thysanoptera, the wings are provided with **'hooked spines'**, which couple with each other during flights.

Wing Venation

Wing shape, size, structure, clothing and especially the arrangement of wing veins are important characters used for identification purposes particularly in Diptera, Hymenoptera and Lepidoptera. An unique system developed to name the wing veins to identify and classify different insects. It has been shown that the position and number of veins have changed from the primitive orders to the more advanced orders as well as within each order. This along with the fossil remains of the wings of ancient insects has made possible to develop a classification of insects, which probably closely parallels the actual evolution of the Class - Insecta.

The wings of the more primitive insects are net-like, but has been a general tendency in the evolution of wings toward reduction in a few longitudinal veins and cross veins, thus giving a stronger support system to the wings. The arrangement of veins in a wing is very specific in certain genera and families of insects and provides an useful tool for systematics. The principal veins and their branches are all named and numbered. The areas or spaces of different sizes and shapes in-between the veins are called the **'wing cells'** and they contain blood vessels, tracheae and sensory nerves. The areas not delimited by veins are called **'open'**.

Most insect wings are more or less triangular in outline and therefore have three margins viz., the anterior margin or **'costa' (a-b)**, the outer **'apical margin' (b-c)** and the inner or **'anal margin' (c-d)**. Further, three well defined angles are distinguished. They are the **'humeral angle' (a)** at the base of costa, the **'apex angle' (b)** between he costal and outer margin and the **'anal angle' (c)** between the outer and inner margins **(Fig. 17).**

Wing veins appear as thickened lines and some of these lines run from the base of the wing towards the apex. The complete system of veins of a wing is termed as its **'venation'** or **'neuration'**. Extensive studies on tracheal proceedings have made it possible to construct a hypothetic type of wing venation by **Comstock** and **Needham** (1898) from which all other types have apparently been derived **(Fig. 18).**

There are eight principal longitudinal veins in a hypothetical wing type and of these four veins viz., the second to the fifth are branched or forked. The names of the principal veins are **'costa' (C)**, **'subcosta' (Sc)**, **'radius' (R)**, **'media' (M)**, **'cubitus' (Cu)**, **'first ansl' (1A)**, **'second anal' (2A)** and **'third anal'** (3A). Generally each vein is distally forked into an anterior convex branch and a posterior concave branch, which facilitates the folding of the wings when at repose.

Besides the longitudinal veins, there are cross veins between the longitudinal veins. The prominent cross veins are (i) **'Radial cross vein' - 'r'**, (ii) **'Radio-medial cross vein' - 'rm'**, (iii) **'Medial cross vein' - 'm'**, (iv) **'Medio-cubital cross vein' - 'mcu'** and (v) **'Anal cross vein' - 'a'**.

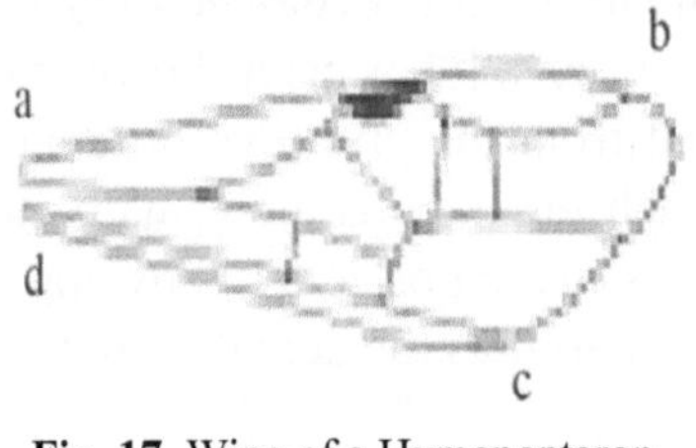

Fig. 17. Wing of a Hymenopteron showing the margins and angles

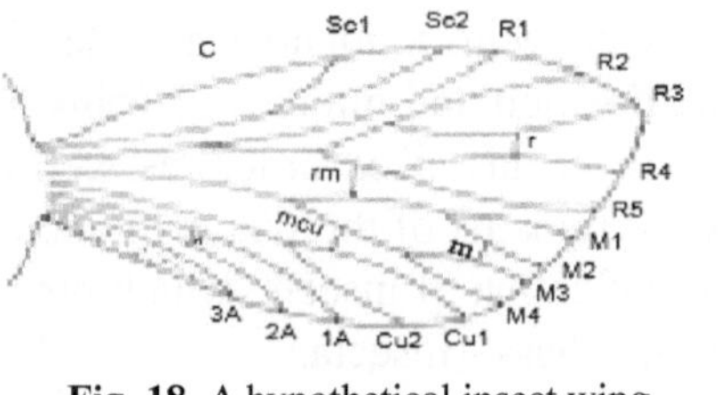

Fig. 18. A hypothetical insect wing showing the arrangement of veins

The insects of the lower orders have a tendency to fold the wings in a fan-like manner when at rest. In the folding arrangement, veins remain on crests and troughs of the folds. In the folds, the convex veins are on the crests and the concave veins are in the troughs. Costa, two anterior radial veins, two anterior median veins, anterior cubital vein and the anal veins are convex, while the subcostal veins, two posterior radial veins, two posterior median veins and the posterior cubital vein are concave veins. Folding is effected by the contraction of the flexor muscles arising from the pleuron and inserted into the third axillary sclerite, which pulls back the anterior portion of the wings to fold into position. In Odonata and Ephemeroptera, the attachment of the wings with the thorax is simple and direct, because axillary sclerites are absent. Hence, the wings of Odonata and Ephimeroptera can not be folded on their back when at rest.

Abdomen

The third and posterior-most of the three body segments of insects is the abdomen. Though it is almost impossible to demarcate more than eight or nine segments in the abdomen of most of the insects, this region is really composed of 10 or 11 ring-like segments. However, in primitive forms such as, Protura the abdomen has 12 segments. In the grasshopper, there are 10 abdominal segments and the super anal plate at the posterior end is considered to be the eleventh segment by some Entomologists. The reduction in the number of distinct segments o eight or nine is due to the modification of the terminal segments into accessory reproductory structures. The abdomen is generally the longest of the three body regions and the body surface is also smoother, pliable and not covered with hard chitin. The abdomen serves as a container for the principal viscera of the insect and is a chief part of the body involved in producing movements of respiration. In adult insects, a typical abdominal segment consists of a tergum, a sternum and a thin membranous connecting pleuron. The abdominal segments in adult insects do not generally bear any appendages su as, legs and wings except in a few exceptional cases, where paired appendages are present as in spring tails. In the embryonic stages of many insects however, vestiges of abdominal appendages are found and temporary ones such as, the prolegs of the larvae of moths, sawflies and butterflies also occur.

In Collembola, a special modified abdominal appendage called **'furca'** is present in the fourth abdominal segment. Only with the help of this appendage, the insect is able to jump. The legs are not useful for jumping.

Only 10 segments can be seen in both the sexes and the ninth and tenth are partly fused. The number of sternites is lesser than the tergites. The first eight sternites are distinct and the ninth is membranous, while the rest are modified to form the external organs of reproduction or **'genitalia'**, which surround the genital aperture. One of the external organs in some female insects is the **'egg depositor'** or **'ovipositor'**, which attains a length much longer than the rest of the insect's body, as is found some parasitic wasps, some grasshoppers and crickets. This organ is occasionally found in combination with a **'stinging apparatus'**, as in wasps and bees.

At the dorsal extremity of the last segment is found the large supra anal plate. Beneath this plate, the anus is situated in-between a pair of podical plates. The genital aperture is situated ventrally and a little anteriorly, somewhat below the ninth segment. In some forms, the posterior margin of the abdomen may be provided with processes known as **'anal cerci'** on either sides of the anus, as in cockroaches and crickets. In the earwigs, a forceps-like pair of anal cerci is provided.

In a line along each lateral side of the abdomen are situated the series of external respiratory openings known as **'stigmata'** or **'spiracles'**. These are minute apertures with horny rims situated in the pleural membrane, one on each side of the first eight abdominal segments. This structure besides the actual opening includes the vestibule to which the openings are connected and also the apparatus for closing and opening the hole. These eight pairs of spiracles are in addition to the two pairs in the thoracic segments, thus making a total of 10 pairs of spiracles. These spiracles are more clearly visible in the larval forms than in the adult insects.

A tympanal organ is provided on the first abdominal tergum. The thin pleural membrane, which stretches between the tergal and sternal sclerites, allows characteristic movements associated with the functions of breathing and further enables the abdomen for expansion and contraction, chiefly for the process of oviposition. The reproductive structures generally include appendages of the eighth and ninth segments in the female insect, but only the ninth segment in the males.

In general the abdomen is clearly demarcated from the thorax, but in higher Hymenoptera, the entire first abdominal segment is fused with the metathorax and is called **'propodium'**. The rest of the abdomen is called **'gaster'**. The portion between thorax and abdomen is called **'waist'**. But in Hymenoptera, the narrow portion between first and second abdominal segments is called 'waist', because the anterior end of the first abdominal segment is fused with the posterior end of the metathorax.

External Genitalia of Female Grasshopper

The external genitalia of a typical female insect such as, grasshopper consists of an egg-laying organ known as **'ovipositor'**, which is formed of the appendages of the eighth and ninth abdominal segments, besides the **'copulatory pouch'**. The ovipositor enables the female to deposit the eggs in the soil after digging a hole in the moist ground. In female grasshoppers and locusts, the intersegmental membrane between the abdominal segments 4 and 5, 5 and 6 and 6 and 7 is made up of elastic cuticle. The stretching of this intersegmental membrane increases the length of the abdomen and hence these insects can lay their eggs in holes in the ground at depths more than twice the normal length of the abdomen. But the males and the immature females do not have such stretching membrane. There are three pairs of valves in the female genitalia. The first pair of dorsal valves originates from the ninth sternum, which are triangular, pointed and hard, sclerotized structures. The inner second pair of valves are very small and lie in-between the dorsal valves. The third pair of ventral valves originates from the eighth sternum. They are similar to the dorsal valves in structure and are projected upwards. The inner valves are considered to be vestigial. The dorsal and ventral pairs of valves fit together to form the functional ovipositor. Besides these three pairs of valves, an **'egg guide'** is also present originating from the base of the eighth sternum. The egg guide arranges the eggs in a specified pattern **(Fig. 19)**.

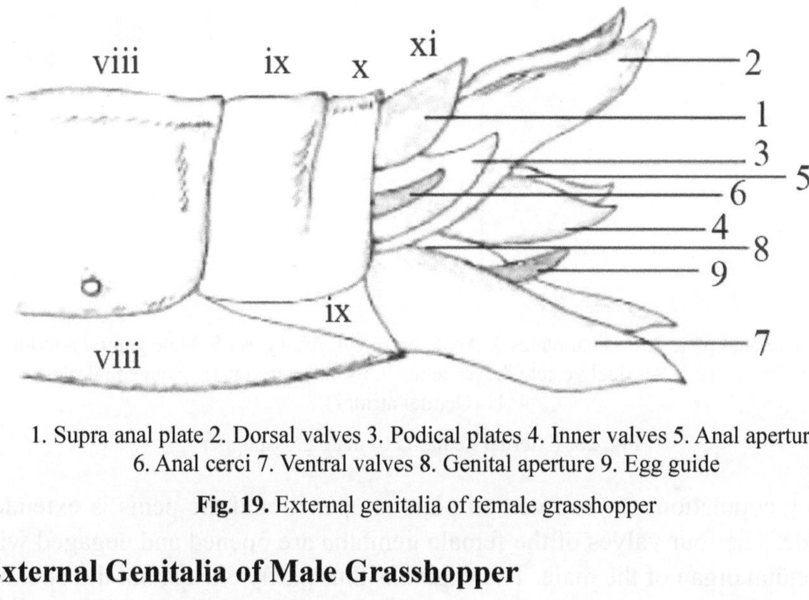

1. Supra anal plate 2. Dorsal valves 3. Podical plates 4. Inner valves 5. Anal aperture 6. Anal cerci 7. Ventral valves 8. Genital aperture 9. Egg guide

Fig. 19. External genitalia of female grasshopper

External Genitalia of Male Grasshopper

In the male genitalia also, the cuticular supra anal plate, the podical plates and the anal cerci are present. The external genital apparatus in formed from the ninth sternum alone, which breaks into two pieces, one proximal which is the true, normal sternite and the other distal, which serves as sub-genital plate, formed by the fusion of the coxites of this segment and is called **'gonocoxite'**. The abdominal end can be closed dorsally by the supra anal plate, laterally by the podical plates and ventrally by the sub genital plates. These plates collectively enclose a cavity known as **'genital atrium'** in which lies the actual copulatory organ, the **'phallus'** or the **'penis'**. The genital atrium consists of two parts, a proximal **'phallobase'**, which is the proximal part of the penis and a distal **'aedeagus'**, which is the male copulatory organ, including accessory structures and penis. The whole structure forms a sperm Ejection pump and the ejaculatory duct opens into it. The **'pseudosternite'** or **'epiphallus'**, a large and complex sclerite, which lies dorsally at the base of the phallobase, forms an important organ for copulation. Through the male **'genital aperture'**, two structures called **'parameres'** project laterally. The parameres also play a vital part in copulation. When the male is at rest, the sub genital plate presses against the supra anal plate and thus the end of the abdomen is closed. In this condition, the phallus is kept retracted in the genital atrium **(Fig. 20)**.

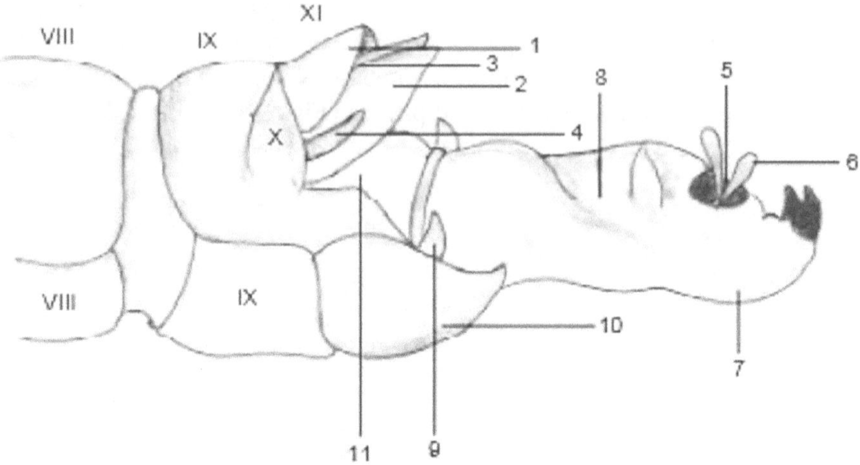

1. Supra anal plate 2. Podical plates 3. Anal aperture 4. Anal cerci 5. Male genital aperture 6. Paramere 7. Seminal vesicle 8. Aedaegus 9. Pseudosternum 10. Subgenital plate 11. Genital atrium

Fig. 20. External genitalia of male grasshopper

During copulation, the sub genital plate is opened and the penis is extended outside. The four valves of the female genitalia are opened and engaged with the genital organ of the male. The vagina is thus partly closed, but the opening of the spermathecal duct of the female nearly opposite to the aedeagus, thus accomplishing the reception of the spermatic fluid successfully.

3

Anatomy of Insects

Similar to animals of the higher order, insects also possess all important internal systems of organs necessary for the proper performance of the various vital functions. The most important of the systems are the alimentary or digestive system, the circulatory system, the respiratory system, the reproductive system and the excretory system. The fundamental arrangement of the various organs is more or less similar in all insects and does not exhibit much remarkable variations as seen in the external morphological features. The dorsal blood vessel or heart runs along the median line immediately below the dorsal wall. The digestive system including the alimentary canal and its appendages lie in the middle, below the heart. The nervous chain lies below the alimentary canal and immediately above the ventral wall. The organs connected with the reproductive system are found in the posterior region of the abdominal cavity on both sides of the alimantary canal. Though these arrangements are characteristic of all Arthropods, including insects, they are quite different in the higher animals **(Fig. 21)**.

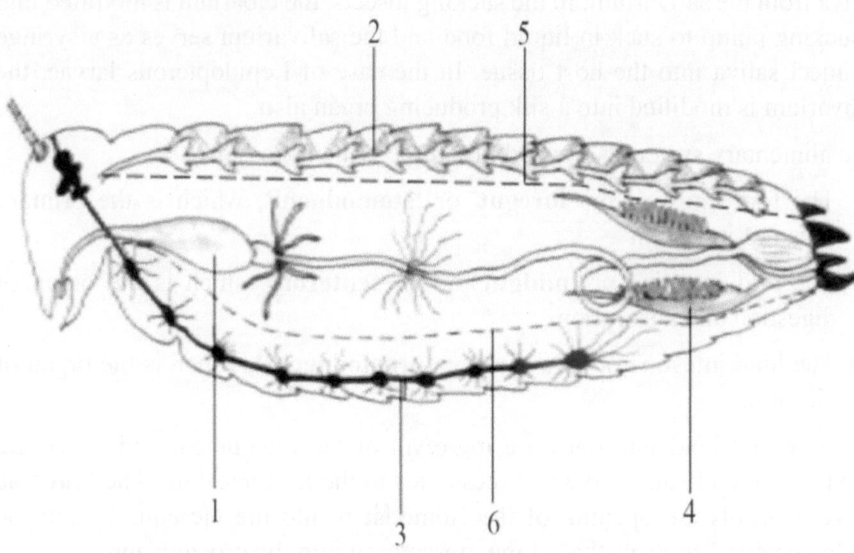

1. The alimentary system 2. The circulatory system 3. The nervous system 4. The reproductive 5. The dorsal diaphragm 6. The ventral diaphragm

Fig. 21. Arrangement of inner organs in insects (e-g) Grasshopper

The Alimentary System or Digestive System

The **'digestive system'** consists of organs of ingestion, digestion, absorption and ejestion. The digestive system of insects is usually complete, that is the tube extends from the mouth, through the body to the anal opening. The many modifications to the system directly relate to the type of food the insect ingest. Thus, plant sap and blood feeding insects have filtering mechanisms that remove water from the fluid food, which enable them to ingest large volumes of liquid food. Aphids for example do not have a complete digestive tract. Water is eliminated through a filtering device and most of the sugar, which is totally digestible is retained, while some sugar is excreted along with water. But the caterpillars feeding on the leaves excrete only dry pellets. According to the type of food they ingest, the mouthparts of insects are suitably modified.

The organs of ingestion are located in the head and they represent the mouthparts. A pre-oral cavity, which is not a true part of the intestine is enclosed within the mouthparts and are bounded by the epipharyngeal wall of the labrum in the front, the labium from behind and the mandibles and maxillae on the sides. In this cavity is located the hypopharynx, which bifurcates the oral cavity into two namely the anterior or dorsal, the **'cibarium'** and the posterior or ventral the **'salivarium'**. The true mouth opens into the cibarium, while the salivary duct opens into the salivarium. Solid food is cut, crushed and masticated in the cibarium, while the gnawing appendages are moistened by abundant supply of saliva from the salivarium. In the sucking insects, the cibarium is modified into a sucking pump to suck in liquid food and the salivarium serves as a syringe to inject saliva into the host tissue. In the case of Lepidopterous larvae, the salivarium is modified into a silk producing organ also.

The alimentary system is divided into three primary regions :

i) The fore intestine or **'foregut'** or **'stomodaem'**, which is the primary organ of ingestion

ii) The mid intestine or **'midgut'** or **'mesenteron'** which is the organ of digestion and absorption

iii) The hind intestine or **'hindgut'** or **'proctodaeum'**, which is the organ of ejestion.

The fore and hind intestines are ingrowth of the integument and hence are lined with cuticle and serve as accessories to the mid intestine. The **'cardiac valve'** controls the opening of the stomodaem into the mesenteron and the **'pyloric valve'** controls that of the mesenteron into the proctodaeum.

i) **The foregut** or **stomodaeum**. The **'stomodaeum'** arises as an invagination of the body wall and is therefore composed of the structure as that of the body wall viz., the cuticular innermost lining or **'intima'**, the epithelial layer continuous with the hypodermis and secreting intima, the basement membrane bounding the outermost layer of the epithelium, the longitudinal muscles, the circular muscles and the peritonial membrane consisting of connective tissues. Since the foregut is ectodermal in origin, it is shed at each moult in the same way as that of the rest of the cuticle.

The stomodaeum is differentiated into the buccal cavity, the pharynx, oesophagus often enlargesd into a crop and the gizzard or proventricule.

a) **The 'buccal cavity'**. The mouth or buccal cavity is situated at the base of the pre-oral cavity and is the oral and initial part of the stomodaeum on which are inserted a group of dilator muscles.

b) **The 'pharynx'**. The mouth leads into a tubular pharynx, which passes vertically upwards through the nerve collar into the oesophagus. It also possesses dilatory muscles and is well developed in sucking insects.

c) **The 'oesophagus'**. It is a simple, narrow straight tube of variable length. It passes through the neck into the thoracic region, where it dilates to form the crop.

d) **The 'crop'**. It is a large, thin-walled, pear-shaped, bladder-like lateral swelling, which extends through the thoracic region, far into the abdominal region and is by far the largest part of the alimentary canal.It is extremely variable in form and functions as a food reservoir. The walls of the crop, when not stretched with food content are thrown into numerous longitudinal folds, which allow its distension. Certain amount of digestion of food takes place in the crop, as it receives digestive fluids from the saliva **(Fig. 22)**.

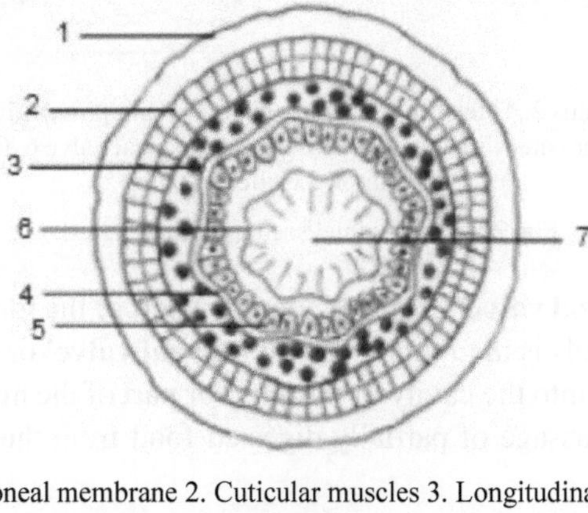

1. Peritoneal membrane 2. Cuticular muscles 3. Longitudinal muscles 4. Basement membrane 5. Epithelial layer 6. Intima (Cuticle) 7. Lumen

Fig. 22. T. S. of the crop

e) **The 'gizzard'**. At the posterior end of the crop lies the small, pear-shaped **'proventriculus'** or **'gizzard'**. It is a highly specialized structure. It has internally a circle of six highly cutinized thickenings called **'proventricular plates'** or **'teeth'**, which also form an efficient crushing apparatus. In the simplest forms, it Merely acts as a specialized structure between the crop and the stomach to regulate the passage of food material into the stomach. In the hindermost, tapering part of the proventriculus behind the thick proventricular plates, there is another circle of six lightly cutinized, thin proventricular plates, which are followed by six cushion-like pads covered with cuticular hairs, which are all directed backwards. Through these structures digestive fluids from the stomach may be permitted to flow forward into the gizzard and bring about partial digestion by mixing with the food mass **(Fig. 23)**.

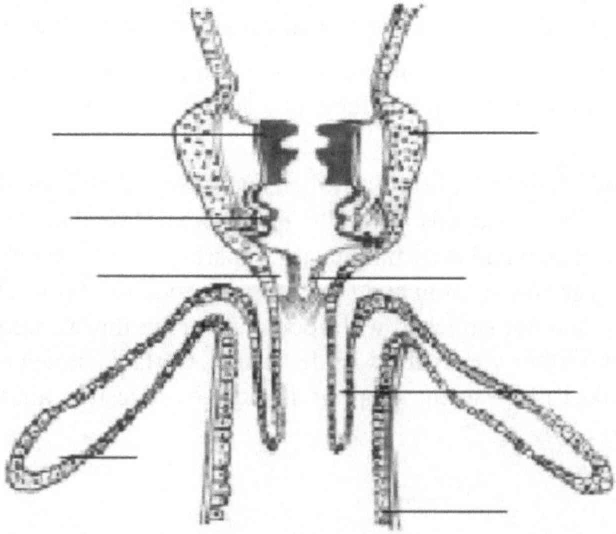

1. Proventriculus 2. Thick proventricular plates 3. Thin proventricular plates 4. Proventricular pads with cuticular hairs 5. Stomodaeal valve 6. Gastric caeca 7. Mesenteron 8.Cuticular hairs

Fig. 23. Longitudinal section of the gizzard

f) **The 'stomodaeal valve'**. At the hindermost part of the foregut, beyond the cushion-like pads is the region of **'stomodaeal valve'** or **'cardiac valve'**, which projects into the cavity of the anterior part of the midgut. This valve regulates the passage of partially digested food from the gizzard into the midgut.

ii) **The 'midgut'** or **'mesenteron'** or **'stomach'**. This is relatively a short tube or elongated sac of almost uniform diameter. Internally the stomach wall is lined by enteric epithelium and the outer ends of the epithelial cells rest on a basal membrane. An inner layer of circular muscles and an outer layer of longitudinal muscles follow this. The midgut differs from the stomodaeum in that, it has a thicker wall, spongy epithelium, absence of intima and the musculature is also weaker. The columnar cells of the stomach secrete enzymes and absorb the digested food.The regenerative cells renew the other epithelial cells destroyed through the various secretions or in the process of degeneration.Thin **'peritrophic membrane'** lines the inner surface of the stomach. This membrane forms a covering around the food mass moving through the midgut and protects the midgut epithelium, at the same time it is permeable to enzymes and digested foods. The peritrophic membrane also acts as a barrier to microorganisms .The insects, which live on liquid food, do not have a peritrophic membrane, From the anterior end of the stomach, arises seven or eight tubular mesenteric or **'gastric caecae'** with blind end. The caecae normally serve as a safe place for the multiplication of harmless bacteria in the alimentary canal **(Fig. 24)**.

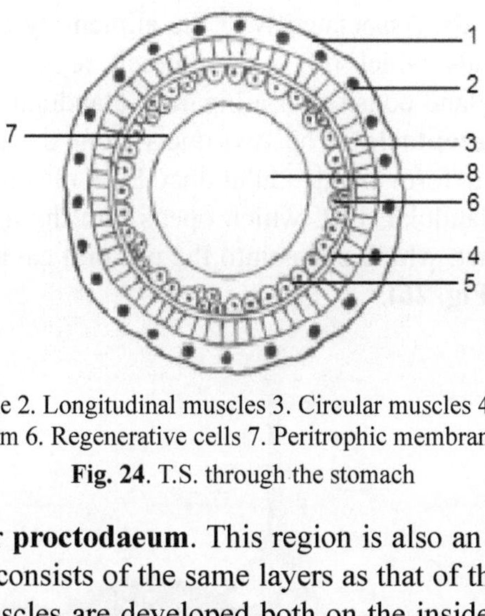

1. Peritoneal membrane 2. Longitudinal muscles 3. Circular muscles 4. Basement membrane 5. Epithelium 6. Regenerative cells 7. Peritrophic membrane 8. Lumen

Fig. 24. T.S. through the stomach

iii) **The hindgut** or **proctodaeum**. This region is also an invagination of the body wall and consists of the same layers as that of the stomodaeum, but the circular muscles are developed both on the inside and outside of the body layer of longitudinal muscles. The hind gut has a short tubular anterior portion, the **'small intestine'** or **'ilium'**, followed by a long and coiled **'large intestine'** or **'colon'** and a posterior broad **'rectum**. The anterior end of the proctodaeum is marked by the insertion papillae of yellow, silky, fine, blindly tubular **'malpighian tubules'**, which are arranged in six groups, each group consisting of about 15 tubules. At the entrance of the proctodaeum from the mesenteron is situated the **'pyloric valve'**. The inner wall of the rectum has six longitudinal ridges, projecting into the lumen. They are known as the **'rectal papillae'** and are believed to help in the absorption of surplus water and also in the distension of the rectum. The rectum opens to the exterior through the anus, which is situated at the caudal end of the abdomen **(Fig. 25)**.

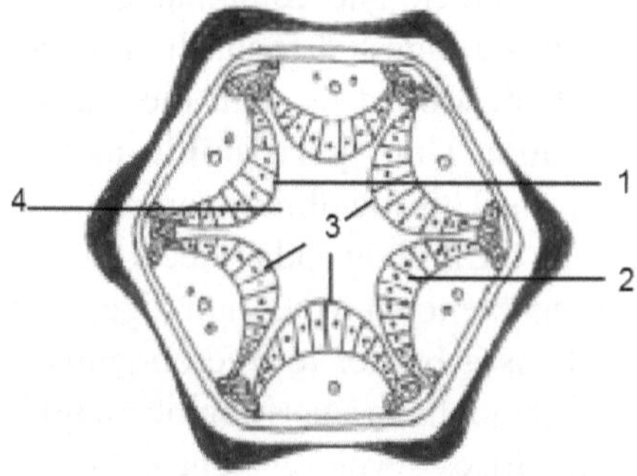

1. Chitinous lining 2. Epithelium 3. Rectal 4. Lumen

Fig. 25. T.S. of the rectum

iv) **Accessory glands**. Associated with the alimentary canal is found a pair of salivary glands, which lies in the thoracic region on either side of the foregut. Each gland consists of a bipartite glandular portion and sac-like reservoir or **'receptacle'**. The two ducts from the glandular portion of each side unite to form the glandular duct. Both the glandular unite to form the common glandular duct, which opens into the much larger common receptacular duct, which opens into the pre-oral cavity at the base of the hypopharynx **(Fig. 26)**.

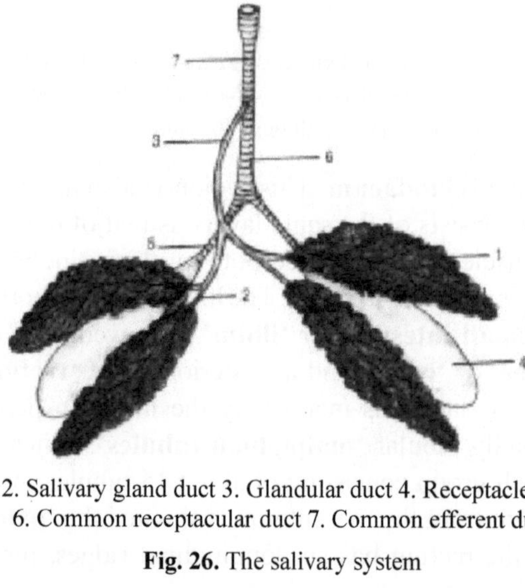

1. Salivary gland 2. Salivary gland duct 3. Glandular duct 4. Receptacle 5. Receptacular duct 6. Common receptacular duct 7. Common efferent duct

Fig. 26. The salivary system

Process of Digestion

While eating, the mandible are used for crushing and chewing the food, the lacinia, glossae and paraglosae are used for helping in holding and pushing the food into the mouth. The palpi are olfactory in function. During the masticatory process, the saliva from the common efferent salivary duct beneath the hypopharynx moistens the food. In this moistened condition, the masticated food is pushed down the oesophagus into the crop.

In the anterior part of the proventriculus, the food is further broken down by the proventricular teeth into small pieces, which are returned to the crop, where it is subjected to further action of the salivary secretion. The saliva contains amylase, invertase, maltase and lactase, with the help of which starch, sugar, maltose and lactose are digested to some extent. The digestive juices produced by the mesenteron and gastric caecae flow forward into the crop through the grooves between the teeth of the proventriculus and bring about partial digestion of the food stored in the crop before it is passed into the mesenteron. The six cushion-like pads covered with cuticular hairs directed backwards and situated in the posterior part of the proventriculus form a kind of strainer. Thus, only very fine particles are allowed to enter the mesenteron or midgut. Once the food particles have come into the midgut, it is not allowed to enter the mesenteron or midgut. Once the food particles have come into the midgut, it is not allowed to go back into the proventriculus by the stomodaeal valve.

The anterior part of the mesenteron is mainly secretory, while the posterior part is mainly absorptive. The digestive juice produced by the gastric caecae and the anterior part of the mesenteron contains amylase, invertase, lactase, maltase, tryptase and peptase and thus here carbohydrates, proteins and fats are all digested. The digested food passes through the permeable peritrophic membrane and absorption of this digested food is accomplished by the thick and spongy walls of the mesenteron and the mesenteric caecae. Some part of this absorbed food is utilized for the production of energy required for different life activities, while the rest is utilized for the formation of muscles and body tissues. If still some absorbed food is left, it is stored in the form of fat bodies to be used in emergencies. Fatty acids are accumulated in fat bodies as triglycerides, which provide reserve energy.

There are specific regulatory mechanisms, which trigger secretion of enzymes. In the **'secretagogue mechanism'**, enzyme secretion or production is induced by food directly stimulating the midgut cells. In the **'neural control mechanism'**, the impulse caused by the stomatogastric nervous system triggers enzyme secretion. In the **'hormonal control mechanism'**, the stretching of the wall of the foregut induces the production of appropriate enzyme.

The absorption of digested food takes place mainly in the midgut, specially the anterior portion including the caecae and to a lesser extent in the hindgut. Carbohydrates are absorbed as monosaccharides and proteins are absorbed after degradation to aminoacids. The undigested food passes through the proctodaeum and mainly in the rectum surplus water is absorbed back and finally the waste matter is expelled through the anus in the form of pellets **(Fig. 27)**.

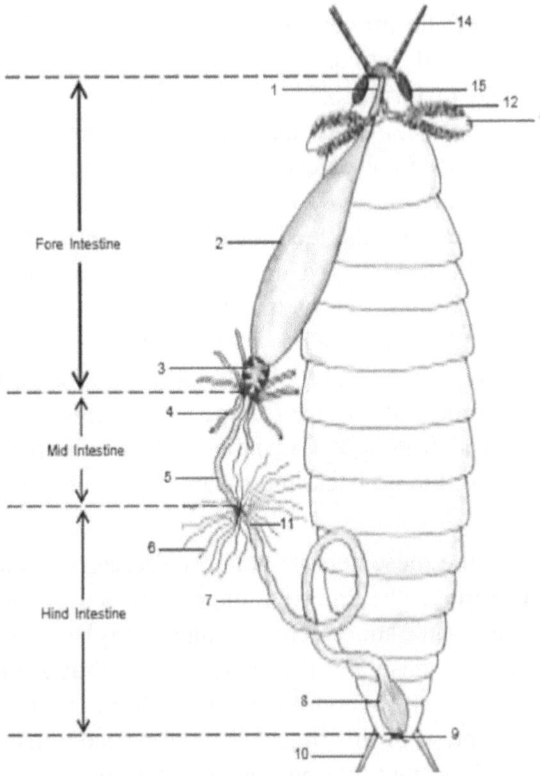

1. Oesophagus 2. Crop 3. Proventriculus 4. Gastric caeca 5. Mesenteron or mid gut 6. Malpighian tubules 7. Large intestine 8. Rectum 9. Anus 10. Anal cerci 11. Ilium (Small intestine) 12. Salivary glands 13. Salivary vesicle 14. Antenna 15. Compound eye

Fig. 27. The alimentary system (e-g) Grasshopper

In most of the insects belonging to Suborder - Homoptera in the Order - Hemiptera, due to an unusual modification of the alimentary canal, an organ is developed, which is known as **'filter chamber'**. In the filter chamber, two distinct parts of the digestive tube are closely associated with each other and are bound together by a connective tissue sheath. The parts involved in the filter chamber are the anterior ends of both mid- and hindgut. This chamber functions as a device that allows excess water and soluble carbohydrates in the food to be diffused through the anterior part of the midgut into the anterior part of the hindgut and are eliminated, while the proteins and fatty materials are retained in the midgut for further digestion and absorption.

The Excretory System

The process of elimination of harmful metabolic products is called excretion and the organs involved in this process constitute the **'excretory system'**. The principal excretory organs are the malpighian tubules. The primary function of the malpighian tubules is excretion and regulation of internal environment by controlling the ion and water balance and removing nitrogenous wastes. The tubules come in direct contact with blood withdraw water, dissolved salts, amino acids and nitrogenous wastes. In addition to the malpighian tubules, the integument, wall of the alimentary canal, tracheal system, nephrocytes, fat bodies and epidermis perform accessory excretory function.

The **'malpighian tubules'** were first discovered by the Italian Anatomist, **Malpighi** and were named after him. They are extremely fine, long, unbranched, hollow tubes, with blind end. They lie freely in the haemocoel, where they are continuously bathed in the blood or **'haemolymph'**. They are evaginations of the hindgut and each tubule is an outgrowth from the beginning of the proctodaeum. They open at their proximal extremities into the intestine near the junction of the hindgut with the midgut. The number of malpighian tubules is very variable, but usually occurs in twos or multiples of twos and their primitive number is six. In Orthoptera, the number of tubules varies from 30 - 200 and in Coleoptera and Hymenoptera it is 4 - 6 and 6 - 20 respectively. However, malpighian tubules re absent in Collembola, some Thysanura and aphids. Each tubule is lined with a glandular epithelium. In species with only a few tubules, they are usually long, slender and convoluted. If the number is large, then the tubules may be short and are often grouped in bundles.

Uric acid formed in the tissues and passes into the haemolymph is selectively absorbed by the malpighian tubule cells along with amino acids, water and dissolved salts. Together these substances are discharged into the hindgut. Reabsorption of water, some of the salts and other nutritive substances occurs in the course of elimination. The rectal papillae are responsible for reabsorption of maximum amount of water. However, reabsorption of some water and inorganic ions may take place in the proximal parts of the malpighian tubules themselves, the substance being returned to the haemolymph. But in most insects, the rectal epithelium transfers these substances back into the haemolymph, while the nitrogenous wastes pass out with the faeces.

Not all the waste products are removed by the malpighian tubules. A good amount of waste matter is also deposited beneath the cuticle and ultimately incorporated in the chitin. At each moult or casting, the cuticular covering is removed and along with that, much of the undesirable waste matter is got rid off. Further, the ectoderm acts as excretory tissue for the elimination of carbon dioxide during cutaneous respiration.

The fat body may also be utilized as a storage place for uric acid. Excretory materials in the form of urates are deposited in the fat body. Any excess amount of waste matter not dealt with by the usual excretory organs collects in the cells of the fat body. Although *Periplaneta* has malpighian tubules, they are not involved in the excretion of uric acid. Instead it is accumulated in the fat body when there is plenty of nitrogen in the diet. Storage excretion is also found in the larvae of *Apis*, where excretion would foul the cell in which they live. In such cases also uric acid is accumulated in the fat body.

In the walls of the midgut of the alimentary canal of various insects, there is an accumulation of crystalline bodies. These bodies are salts of calcium and uric acid, which are eliminated directly through the walls of the hindgut. In Collembola and aphids, where malpighian tubules are absent, the epithelial cells of the midgut, casted off with each moulting contain excretory materials.

Nephrocytes, which are groups of cells found in localized regions of the body also take part in excretion. Nephrocytes present in the surface of the heart called pericardial cells are also present in the pericardial septum. Nephrocytes take up foreign chemicals of relatively higher molecular weight, which can not be taken from the haemocoel by the malpighian tubules and excrete them.

Insects are best adapted for the prevention of water loss from their bodies. The epicuticle is impregnated with waxy compounds, which reduce surface evaporation. The excretion of uric acid reduces loss of water due to protein metabolism. The absorption of water by the rectum further conserves water that would otherwise be lost through excretion and ejestion.

The malpighian tubules perform a few other functions also. In Cercopidae (Hemiptera), the malpighian tubules secrete the **'spittle'**, in which the immature stages remain surrounded. Many immature Neuroptera and larvae of some Coleoptera produce silk from specialized cells in the malpighian tubules. In *Bolitophila* (Diptera), light is emitted by thoracic organs of the larvae formed by the free ends of the malpighian tubules lying just beneath the transparent cuticle of the thoracic region.

The Circulatory System

The dorsal vessel, the dorsal and ventral diaphragm and the accessory pulsatory organs are the main organs involved in the circulatory system. The blood cavity of the insect, which is filled with blood or **'haemolymph'**, is called **'haemocoel'** and the blood cells or **'haemocytes'** are found freely suspended in the blood. Besides haemocytes , the haemolymph contains a fluid component called **'plasma'**. Insect blood, which is devoid of haemoglobin, is not red, but greenish or yellowish in color or rather clear. Haemocytes are involved in wound healing and coagulation. Plasma serves in the transport of various substances in the body. Nutritive substances are carried from the alimentary canal and storage tissues to the sites where they are metabolized. Excretory products are also carried from their place of origin to the malpighian tubules. Further, hormones are transported to their site of action from the endocrine organs.

The blood (Haemolymph) has four known functions: (i) Digested food materials are absorbed from the digestive system and conveyed to the tissues and waste products are carried from the tissues to the excretory organs (ii) The cells, which are not provided with tracheoles obtain their supply of oxygen from the dissolved store in the blood (iii) The haemocytes in the blood eliminate certain bacteria and parasites (iv) The entire volume of blood enclosed within the cavities forms a closed hydraulic system capable of transmitting pressure from one part of the body to another.

Among insects, the circulatory system is not a closed one, but largely an open system, there being only a single closed organ or dorsal vessel. The greater part of the circulation of blood takes place in the cavities of the body and its appendages such as, legs, wings etc. The spaces not occupied by the internal organs are occupied with blood. These spaces may be enclosed by special membranes and form definite sinuses. With the exception of the aorta or the anterior prolongation of the dorsal vessel, there are as a rule no definite veins or arteries found in insects. In the appendages and wing veins, the blood flows through ingoing and outgoing streams along definite channels analogous to blood vessels.

The general body cavity or haemocoel is divided into three sinuses by means of two fibro-muscular partitions or diaphragms. The **'dorsal diaphragm'** is the principal septum and it extends across the abdominal cavity above the alimentary canal and the blood-space thus enclosed is called the **'dorsal sinus'** or the **'pericardial haemocoel'**. The **ventral diaphragm'** stretches across the abdominal cavity just above the ventral nerve cord and the space enclosed is called the **'ventral sinus'** or the **'sternal haemocoel' (Fig. 21)**. Between the dorsal and ventral sinuses is the large **'central sinus'** or the **'provisceral haemocoel'** containing the principal internal organs. The heart lies in the dorsally situated pericardial haemocoel, which is separated from the perivisceral haemocoel by the dorsal diaphragm. To the floor of the dorsal diaphragm are attached the inter segmental **'alary muscles'**. Each alary muscle is triangular or fan-like with its broad end near the middle of the pericardium and its pointed end attached laterally to the terga.

The main circulatory organ in insects is known as the **'dorsal vessel'**, It lies directly beneath the dorsal body wall and extends from near the caudal end of the body through the thorax into the head. It is a principal pulsating organ, which is responsible for the flow of blood. It is divided into two parts :- The posterior abdominal portion is called the **'heart'** and the anterior thoracic portion, the **'aorta'**.

The heart runs as a slender tube, more or less swollen in each segment and closely attached to the ventral side of the terga and held suspended to the ceiling of the cavity of the haemocoel. The tube is closed at the posterior end. The wall of the heart is composed of two distinct layers, an inner muscular layer and an outer connective tissue or **'peritonial layer'**. The series of fan-like triangular wing muscles or **'alary muscles'** arising from each side of the heart and extending from the heart to the lateral walls of the body constitutes the dorsal diaphragm. Both the dorsal and ventral diaphragms are perforated by small apertures.

Each swollen portion or chamber of the heart is called a **'ventricle'.** The chambered- heart usually starts in the second thoracic segment and ends with the abdomen. The number of chambers is variable and usually coincides with the number of the segments the heart occupies in the abdomen, the maximum being thirteen. The hind end of each ventricle is divided into three lobes, one dorsal and two lateral lobes just beneath the dorsal lobe. The dorsal lobes are attached below the interspaces of the terga. When the terga are contracted due to the contraction of the alary muscles, blood from the provisceral haemocoel passes into the pericardial haemocoel and due to the pressure exerted on each dorsal lobe, blood of each ventricle moves forward. In this way the dorsal lobes help in the blood circulation.

A slit-like opening is provided at the base of each lateral lobe of the of the ventricle, through which the blood from the haemocoel enters the heart. These openings are known as **'ostia'**. Each ostium is provided with an **'ostian valve'**, which admits the blood into the heart, but does not allow the blood to go back into the haemocoel. An **'auricular valve'** is also present at the anterior end of each ventricle.

The alary muscles cause the pulsation of the heart by their successive contraction and expansion. During this cycle of operation, two separate activities namely the **'systole'** and **'diastole'** are regularly performed. During the **'systolic act'**, the alary muscles attached to the particular ventricle contract and the paired ostial valves at the posterior end are closed and therefore the only course left for the blood in that ventricle is to go forward into the next ventricle above through the anterior auricular valve, which is opened. Systole is immediately followed by diastole in the ventricle, when the alary muscles expand. During the diastolic act, the anterior auricular valve remains opened, at the same time the ostial valves at the posterior end also open and blood enters the ventricle from the pericardial haemocoel. Thus the ventricle, which undergoes diastole, will receive blood supply from two sources, firstly from the pericardial haemocoel through the ostial valves and secondly through the auricular valve from the posterior ventricle, which had undergone systole.

Each ventricle contracts and expands one after the other starting from the caudal end of the heart forward due to the rhythmic contraction and expansion of the alary muscles in succession and thus a continuous circulation of blood is maintained. The prolongation of the heart at the anterior end from the first chamber is the **'aorta'**, which is a simple tube that extends through the thorax into the head, where it opens in the vicinity of the brain. The aorta functions as the principal artery of the body.

In the thorax, the blood collects in the sternal haemocoel, from which it is circulated through the legs and from the pericardial haemocoel blood is circulated through the wings. To further facilitate blood flow, especially through appendages **'accessory pulsating organs'** are present in the head, antennae, thorax, legs and wings, which drive blood into them **(Fig. 28)**.

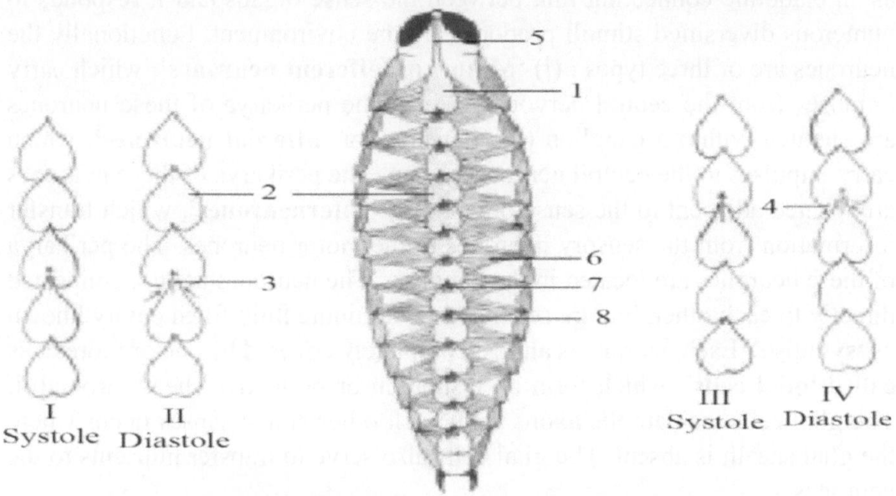

1. Heart 2. Ventricle 3. Auricular valva 4. Ostial valve 5. Aorta 6. Alary muscles 7. Upper chamber of the ventricle 8. Lateral chambers of the ventricle

Fig. 28. The circulatory system (e-g) Grasshopper

The general course of circulation of blood is represented in the line diagram given below :

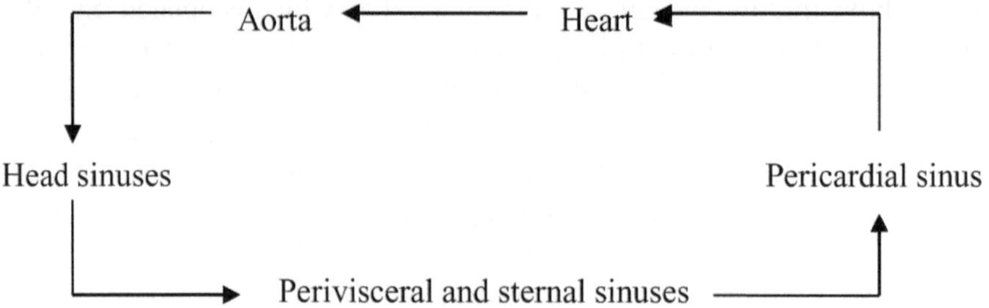

In insects there are no capillary functions and the blood simply flows through large sinuses and it is for this reason that the whole circulatory system is said to be of **'open type'** to distinguish it from the **'closed type'** as in vertebrates.

The Nervous System

The nervous system of insects is composed of excessively elongated cells, which transmit electrical disturbances or impulses from one part of the body to another. The structural unit of nerve impulse conduction is called **'neurone'**. The neurone comprises of a cell body - **'perikaryon'**, an **'axon'** and **'dendrites'**. The cell body has a **'nucleus'**, **'mitochondria'** and other organelles. The axon is an elongated cytoplasmic process, often branched distally. The dendrites, which are short, are the outgrowths of perikaryon. The nervous system serves as an elaborate connecting link between the sense organs and it responds to numerous diversified stimuli produced by the environment. Functionally the neurones are of three types : (i) **'Motor'** or **'efferent neurones'**, which carry impulses from the central nervous system. The perikarya of these neurones are situated within a ganglion (ii) **'Sensory'** or **'afferent neurones'**, which carry impulses to the central nervous system. The perikarya of these neurones are located adjacent to the sense organs (iii) **'Interneurones'**, which transfer information from the sensory neurones to the motor neurones. The perikarya of these neurones are located in the ganglion. The neurones are not connected directly to each other, but are separated by a minute fluid-filled cavity known as **'synapse'**. Each neurone is almost completely covered by one or more cells called **'glial cells'**, which form an insulation or protective sheath around it. The glial cells insulate the axons from each other and synapses occur where the glial sheath is absent. The glial cells also serve to transfer nutrients to the neurones.

Anything that produces a change in the metabolic activity of any living organism is called a **'stimulus'** and the response due to the stimulus is called **'sensitivity'**. Transmission of stimulus from molecule to molecule progressively is known as **'conductivity'**. The properties of conductivity and sensitivity are highly developed in the protoplasm of nerve tissues and as such, insects have a highly developed nervous system, which serves to coordinate their activities under varied conditions both inside and outside the body.

The nervous system of insects may be divided into three types :

(i) The central nervous system (ii) The visceral or sympathetic nervous system and (iii) The peripheral nervous system.

i) **The central nervous system.** The main organs of the central nervous system, which is the **'ventral nerve cord'** is located ventrally along the median axis of the body. Only the brain is located in the head dorsally above the oesophagus. It consists of the brain, sub-oesophageal ganglion and the nerve cord. The brain situated in the head capsule comprises of three ganglion and is connected to the sub-oesophageal ganglion situated very near to the mouthparts. Three thoracic and six abdominal ganglia follow it. The nerve cord runs in two parallel lines.

a) **The brain**. The brain or **'supra-oesophageal ganglion'** is a centre of association between major sense organs. The brain is composed of a **'protocerebrum'**, a **'deuterocerebrum'** and a **'tritocerebrum'**. The protocerebrum is the largest part of the cerebral mass, which contains both neural and endocrine elements consists of a fused pair of ganglia. This fused pair of ganglia innervates the compound eyes and ocelli. The deuterocerebrum represents the fused ganglia of the antennary segment and innervates the antennae. The tritocerebrum consists of a pair of small ganglia and controls the major sympathetic nervous system. The brain controls the reflex responses in accordance with the stimuli received from sense organs , thereby the brain is responsible for orientation and behavior of insects.

b) **The sub-oesophageal ganglion**. This is a large nerve centre located in the head, beneath the oesophagus. From the brain, a pair of large connectives known as the **sub-oesophageal connectives'** arise and circle the oesophagus, one on either side and join beneath the oesophagus, thus forming a loop around the oesophagus. This ganglion is the fused ganglion of the mandibles, maxillae and the labial segments and gives rise to nerve trunks servicing the mouthparts namely mandibles, maxillae, labium, hypopharynx, salivary ducts and some muscles of the neck. From this nerve centre or ganglion, a pair of connectives passes through the neck ino the thorax.

c) **The ventral nerve cord**. The ventral nerve cord consists of a series of ganglia in the ventral portion of each segment of the thorax and abdomen.

They are connected into a longitudinal chain by means of a pair of connectives. This chain forms the ventral cord. A pair of connectives passing through the neck joins it to the sub-oesophageal portion.

The thoracic ganglia control the organs of locomotion. Each ganglion gives off two pairs of principal nerves, one of which innervates the general musculature of the segment, while the other innervates the leg muscles. In the mesothorax and metathorax an additional pair of nerves is present, which controls the movement of wings. Similarly, there are five paired ganglia in the five abdominal segments. The sixth and terminal ganglion situated in the sixth abdominal segment is always composite and is the ganglionic centre formed by the fusion of at least three primitive paired ganglia of the posterior segments. Each abdominal ganglion gives off a pair of principal nerves to the muscles of the segment and the terminal ganglion also innervates the reproductive organs.

A segmental ganglion of the thorax or abdomen is usually an oval or polygonal mass of nerve tissues and the ganglia are continuous anteriorly or posteriorly with inter-ganglionic connectives. The transverse band of fibres that unite the two units of a pair of ganglia is termed as a **'commissure'**. The ganglia and the nerves of the ventral nerve cord are supplied with tracheae. Nutrition and excretion of the ganglia and nerves are provided by diffusion **(Fig. 29)**.

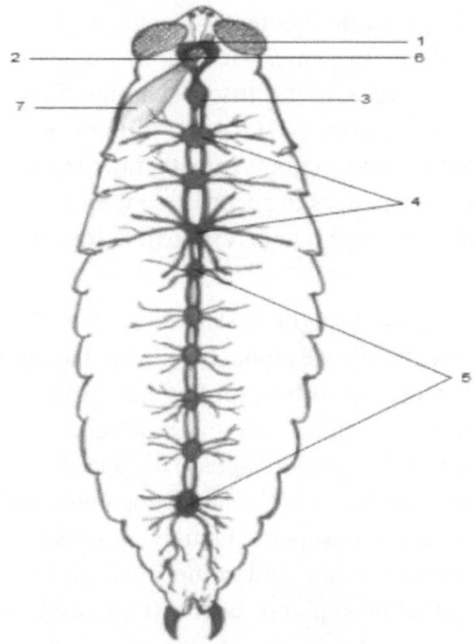

1. The brain 2. Nerve cord 3. Sub-oesophageal ganglion 4. The thoracic ganglia 5. Abdominal ganglia 6. Oesophagus 7. Crop

Fig. 29. The nervous system (e-g) Grasshopper

ii) **The visceral** or **sympathetic nervous system**. To control some of the involuntary motions of the anterior portions of the alimentary tract and dorsal blood vessels, insects a so=called sympathetic nervous system. The system is divided into (a) Oesophageal sympathetic (b) Ventral sympathetic and (c) Caudal sympathetic nervous systems.

a) **The oesophageal sympathetic nervous system**. This system consists of a pair of oesophageal or pharyngeal ganglia, which lies on the oesophagus just behind the brain and each ganglion is joined with the tritocerebrum by a pair of fibres. They are also connected to the protocerebrum. From this ganglion, a recurrent nerve runs back beneath the brain connecting with a system of small ganglia and nerves. This group innervates the stomodaeum, salivary ducts, aorta and certain muscles of the mouthparts.

b) **The ventral sympathetic nervou s system**. This system consists of a pair of transverse nerves associated with each ganglion of the ventral nerve cord and each pair is connected to the ganglion preceding it by a longitudinal median nerve. The transverse nerves pass to the spiracles of their respective segment.

c) **The caudal sympathetic nervous system**. This system arises from the compound posterior ganglion of the abdominal nerve cord and innervates the reproductive system and posterior part of the gut.

iii) **The peripheral nervous system**. Immediately below the hypodermal layer of the body wall, there are many bipolar and multipolar nerve cells, whose prolongations form a network of nerves. These include all the nerves radiating from the ganglia of the central and sympathetic nervous systems. The terminal prolongations of the bipolar nerve cells innervate the sense hairs of the body wall.

The Endocrine System

The **'endocrine system'** is an integrated part of the nervous system in which the impulses caused as a result of responses to stimuli in certain ganglia or neurones inactivate particular glands to secrete some chemical substances such as, hormones, which are vital for the metabolic functioning of the body or for the continuation of the life cycle. Hormones are internally secreted chemical messenger substances produced by endocrine tissues (glands) and are carried in the blood or lymph top various organs of the body. They coordinate the activities of these organs or physiologic processes within the same organism or activate other organs to to produce hormones. Besides regulating the metabolic functioning , hormones play a vital role in the physiological processes such as, moulting, metamorphosis, oocyte production, color change diurnal rhythm etc. The products of the neurosecretory cells of the brain pass along axons, usually to the corpora cardiaca or corpora allata. The secretions of the neurosecretory cells in the median part of the protocerebrum promotes the functioning of prothoracic glands, stimulate protein synthesis, control water loss, oocyte development ad activities of corpora allata. Thus the neurosecretory cells form a link between the endocrine system and the nervous system.

The endocrine system consists of (i) The **'neurosecretorycells'**, chiefly lying in the dorsal middle region of the brain (median part of the protocerebrum), lateral neeurosecretory cells, lying within the brain, but placed laterally and the neurosecretory cells sub-oesophageal ganglion and the ganglia of the ventral nerve cord. The neurosecretory substances produced by the neurosecretory cells are called **'brain hormones'**. The neurosecretory cells produce hormones other than **'ecdysteroids'** and **'juvenile hormones'**, which include **'eclosion hormone'**, **'diuretic'** and **'antidiuretic hormone'**, and **'pheromone biosynthesis stimulating hormone'**. (ii) A pair of glands **'corpora cardiaca'** and **corpora allata'** and (iii) A pair of ventral or thoracic or **'ecdysal glands'**. The neurosecretory substances produced by the neurosecretory cells are called **'brain hormones'**.

'Glands' are cellular organs or structures, which secrete certain characteristic products such as, wax, saliva, silk, hormone, pheromones etc. either within or outside the body. Endocrine glands are ductless glands, while exocrine glands have ducts.

'Corpora cardiaca'. They are a pair of endocrine structures on either side of the aorta behind the brain and forming a part of its wall, in which the neuro hormones accumulate. Each structure contains endings of axons from the brain cells and other axons leading to the corpora allata. In addition there are **'glial cells'**, which enclose the nerve cells or nerve fibres by cytoplasm processes or **'glia'** and insulate the nerve elements. They form an integral part of the stomatogastric system and have secretory cells with cytoplasmic projections, which serve to release neuro hormone secretions into the blood or lymph. Thus, corpora cardiaca serve to store and release hormones from neurosecretory cells to the target sites through lymph or directly through axons. These hormones are concerned with the regulation of heart beat and other physiological functioning.

'Corpora allata'. They are glandular bodies, usually one on either side of the oesophagus, but may be fused into a single structure . Each is connected to the corpus cardiacum of the same side by a nerve cord from the neurosecretory cells of the brain. A fine nerve cord connects each corpus allatum with the sub-oesophageal ganglion.

The corpora allata secretes **'juvenile hormones'**, which directly regulate metamorphosis, affect growth and reproduction. The hormones are essential for the development of eggs in females and deposition of yolk inside the eggs. In the males, the hormones are essential for the functioning of accessory glands. They also play a major role in polymorphism , regeneration of cells , metabolism of fats, water balance, color change and imaginal life process.

'Prothoracic glands or **ecdysal glands'**. They are a pair of diffused glands at the back of the head or prothorax that secretes a hormone called **'ecdysone'** or **'moulting hormones'**, which controls aspects of moulting, metamorphosis and maturation of insectsvia action of corpora cardiaca. Ecdysone is a chemical substance secreted by the prothoracic glands after each moult or at the termination of the larval diapause or pupal diapause, which stimulates growth and development of various somatic tissues and other physiological and morphological changes necessary for the continued repetition of the moulting cycle. The ecdysal fluid or moulting fluid secreted by glands in the epidermis of an insect as a result of the hormonal action dissolves away the endocuticle by enzyme action. **'Diapause'** is a state of suspended activity metabolism, which may occur at any stage in the life cycle of an insect as a means of surviving unfavorable conditions. Diapause is normally brought about by a change in the environmental conditions, especially reduced day length and requires a particular stimulus for its termination and for resuming normal activities and metabolism.

Insect endocrine system is a vital organ system and the hormones secreted by the system regulates the vital body functions such as, stimulating salivary secretion through enzyme action, cell regeneration, color change etc. Any blockage, disruption or inhibition in the functioning of the endocrine system may result in behavioral or physiological disturbance, which may prove fatal to insects. The hormones from the prothoracic glands and corpora allata are produced in glandular cells and are released directly into the blood or lymph. The juvenile hormones, moulting hormones, their chemical analogues and mimics have great value as pest control agents and are called as **'third generation pesticides'**.

'The ring gland or **'Weismann's ring'**. It is a gland surrounding the main dorsal blood vessel just above the cerebral ganglia in some insects belonging to Diptera such as, houseflies, blowflies etc. It is a gland formed from the corpora cardiaca, corpora allata and prothoracic glands, all fused together in the form of a ring. The ring gland is connected to the brain by a pair of nerve cord, besides having connection with the recurrent nerve. It secretes a hormone, which induces the formation of puparium of the Cyclorrhapha type, in which the adult emerges from the puparium through a circular opening at one end of the puparium.

The Exocrine System

The **'exocrine' system'** consists of exocrine glands, which discharge their secretions outside the body or lumen. Unlike endocrine glands, which have no ducts, these glands have well defined ducts and are called **'ducted glands'**. The exocrine glands secrete exocrine chemical secretions known as **'pheromones'**. Pheromones, which are also known as **'ectohormones'** are exocrine chemical messenger secretions produced by an organism, which influences or activate the behavior or physiology of other individuals of the same species. It is actually a messenger that gives information to other insects of the same species and causes a specific reaction in the receiving insects influencing inter-individual communication and behavior of other individuals in social insects. Such communications may pertain to availability of food, aggregation, alarm, sexual call etc. Besides such pheromones, some exocrine glands secrete certain specific substances such as, poisons, defensive secretions, dyes, repulsive secretions etc.

Types of Pheromones

'Sex pheromones'. The females, to attract the males for the purpose of mating generally secrete pheromones. Sex pheromones are produced from **'scent glands'**. These scent glands of the females are usually in the form of glandular epithelium near the tip of the abdomen or invaginated glands opening intersegmentally or on either side of the genital pore. In some forms pheromones are produced from eversible glands present at the terminal part of the abdomen. Sex pheromones are widely present in the females of many insect species and especially in Lepidopteran insects, secretion of sex pheromones is quite a common feature. Female sex pheromones are received by specialized sensilla of the male antennae and can be perceived from long distances.

Several male Lepidoptera also produce a sex pheromone known as **'aphrodisiac'** that stimulates sexual desire in females. In some male butterflies, the **'aphrodisiac pheromone'** or **'sex attractant'** is produced from scent glands called **'Androconia'**, which are found in the wing membrane and are connected to the base of the scales of the wings or produced from **'abdominal hair pencils'** or **'hair tufts'** or **'scent brush'**.

(e-g) In the honey comb moth - *Galleria mellonella,* male sex pheromones are produced in the scent wing glands; In the monarch butterflies - *Danaus plexippus,* the males produce sex pheromones from extrusible glands of the abdomen called **'hair pencils'**.

Some synthetic sex pheromones are being widely used as sex lure for the control of certain insect pests :

(e-g) **Gossyplure** - Sex lure for cotton pink boll worm *(Pectinophora gossypiella)*

Pherodine SL - Sex lure for American cotton boll worm – *(Spodoptera litura)*

Disparlure - Sex lure for gypsy moth *(Porthetria dispar)*

Bombycol - Sex lure for silkworm moth *(Bombyx mori)*

'Alarm pheromones'. Alarm pheromone substances are produced in social insects, in which an individual of a caste of the species warn other fellow individuals about impending danger from enemies that threaten them. These substances instigate either flight or aggression to escape from the enemies. They are emitted either by the by the mandibular glands, anal glands or sting apparatus. These substances are highly volatile and sometimes act as defensive substances. The response attracts other workers or soldiers for the purpose of defense or aggression.

(e-g) In the case of honeybees, the response to the pheromones are mass flight and direct attack of the enemies.

In the case of white ants, the soldiers aggregate around the damaged termitarium or termite colony and adopt an aggressive and threatening posture with the mandibles opened widely.

'Aggregation pheromone' Individuals of species of insects liberate pheromones that guide other insects of he same species to the host plant, In some cases, the pheromones are released by either one or both the sexes, which attract both the sexes to aggregate at a suitable food source for feeding, mating or for oviposition.

(e-g) **'Grandure'** – This aggregation pheromone produced by the males of the cotton boll worm *(Anthonomus grandis)* attracts both the sexes to the particular feeding site.

The aggregation pheromone consisting of a mixture of fatty acid esters and other substances attracts both the sexes to aggregate in a particular feeding site.

The aggregation pheromone secreted by the male nitidulid bug *(C. legubris)* attracts both the sexes at a suitable food source, where mating and oviposition takes place.

'Trail marking pheromones'. In the case of some insects such as, ants, white ants etc., one individual follows another in a distinct line following trail making pheromone. This interspecific chemical messenger pheromone produced by foraging ants and termites indicates the sources of requisites to other colony members. These pheromones usually have a relatively short transmission range (e-g) The ants *(Formica rufa)* use formic acid as trail marker.

In Dolichoderine ants, the trail the trail marking pheromone is secreted from the exocrine glands known as **'Pavan's gland'**, which opens at the ventral surface of the abdomen above the sixth abdominal sclerite.

The secretions of exocrine glands function in many other ways :

'Scent glands'. In the queen honeybee, a sac-like gland is present, which opens at the inner angle of each jaw. The secretion of these mandibular glands of the queen bee called **'queen substance'** controls queen rearing and ovary development in workers and attracts drones in mating flight and workers in swarming flight.

'Nasanoff gland'. This gland is present in-between the fifth and sixth abdominal tergites in worker honeybees. It is made up of a number of large cells, each of which is provided with a narrow duct leading to the exterior through the cuticle. The excretions from these glands are exposed while fanning their wings after locating a new food source and the volatile substances such as, **'geraniol'** and **'citrol'** emitted from these glands serve to orient workers to the newly found foraging area.

'Labial glands'. These are the usual salivary glands of insects, opening by a median duct between the base of the hypopharynx and the labium or on the hypopharynx , which secrete saliva.

'Wax glands'. In honeybees the glands under the abdomen produce bee's wax

Many insecs belonging to Homoptera posses wax glands and their secretion in the form of waxy powder or waxy threads.

(e-g) White flies *(Bemisia tabaci)*

'Poison glands'. Poison glands are present in a number of Hymenoptera, which have stings. In ants with sting, the poison gland consists of a pair of tubules, which unite to form a convoluted duct in the form of a seta opening into a reservoir above the sting through which the irritating venom from the poison gland is injected.

'Dufour's gland'. This gland is found in female wasps. It is an exocrine poison gland in Hymenoptera, which secretes the alkaline element of the poison carried by the sting. The gland is a small, simple globular sac, with thin walls and a delicate muscular sheath. The poison from the poison gland opens into the poison duct near the base of the sting of the females.

'Mandibular glands'. In Lepidopterous larvae, the saliva from the tubular salivary glands in the thorax on each side of the fore intestine, communicates with the mouth by a pore at the base of the mandibles. This saliva is useful in selecting the right type of food.

'Dye glands'. Some coccids such as, the cochineal insects *(Dactylopus coccus)* posess the dye glands and produce dye stuffs.

'Odoriferous glands'. A number of Heteropterans have characteristic dorsal abdominal odoriferous or repugnatorial glands. The glands usually open on the sides of the thorax in the form of ostioles.

(e-g) The painted bug of mustard *(Bagrada cruciferarum)*; Bed bug *(Cimex hemipterus)*.

The Respiratory System

'Respiration' is a process of interchange of gases between the environment and the blood or cellular tissues of the organism. The great majority of insects are terrestrial and breathe atmospheric air. In insects there is no lungs for respiration as in vertebrates. Most of the insects breathe by means of a network of breathing tubes named **'tracheae'**, which divide and subdivide as fine-walled, smaller vessels called **'tracheoles'** and ramify into every minute part of the body. The tracheoles floating in the blood of the haemocoel form a regular network and through their thin walls, necessary exchange of air takes place in the living cells of the different parts of the body. The tracheae are open tubes, which permit direct contact of air with the tissues through a number of openings called **'spiracles'**. The tracheae are lined with a thin layer of cuticle, which is thickened to form spiral rings that give them rigidity. In a relatively small number of insects, respiration is cutanous, while in the young ones of aquatic insects, the respiratory organs are gills, which arise as outgrowths **(Fig. 30A)**.

The tracheae receive air through ten pairs of **'spiracles'** or **'stigmata'**, which are located along the two sides on the pleura of the insect body. Two pairs are situated in the thorax, the first pair between prothorax and mesothorax and the second pair between the mesothorax and metathorax. The remaining eight pairs are are located in the abdomen on each side of the first eight segments. Tracheal spiracles in their simplest form are merely holes in the integument. However, in most insects, the spiracles lie in a pit or **'atrium'**, from which tracheal tubes arise. The opening and the atrium are collectively known as spiracle. The spiracle is generally provided with a closing mechanism and in most terrestrial insects, the atrium contains filtering devises such as, **'sieve plates'** and **'felt pads'**. The filtering structures prevent the entrance of dust and parasites into the tracheae and also help to reduce water loss. The spiracles are provided with some type of closing device. The device may be external, usually in the form of two, opposed lips or it may be internal, usually in the form of a clamp with pinches. The opening and closing of the spiracles are controlled in part by direct innervation and partly by the oxygen-carbon ioxide tension of the blood **(Fig. 30B)**.

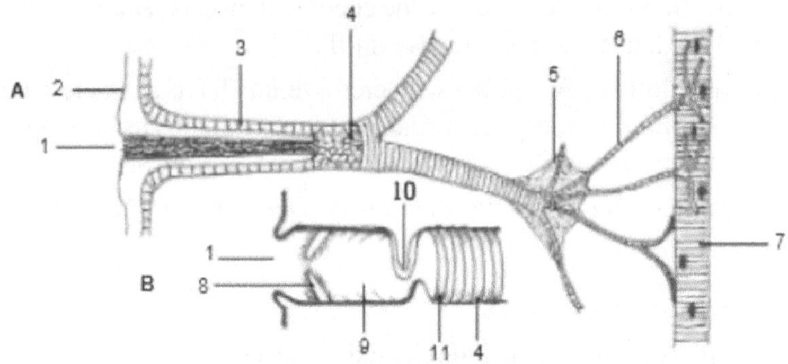

1. Spiracle 2. Body wall 3. Epithelial cells 4. Trachea 5. Tracheole cell 6. Tracheole 7. Muscle 8. Filtering apparatus 9. Atrium 10. Valve 11. Taenidia

Fig. 30. Spiracle, tracheae and tracheoles

The tracheae are elastic tubes, which arise as invaginations of the body wall. The three layers of the body wall are directly continuous with corresponding layers in the wall of the tracheae. The layers are the intima, the epithelium and the basement membrane.

The intima is the chitinous, inner layer of the tracheae, being directly continuous with the cuticula of the body walland is moulted at each ecdysis. A peculiar feature of the intima of the tracheae is the provision of spiral thickenings termed as **'taenidia'**, which prevents the tracheae from collapsing. The taenidia are absent in the tracheoles and in the air sacs. The epithelium of the tracheae is a cellular layer, which is a continuation of the epidermis of the body wall. The basement membrane is a delicate layer, which supports the epithelium.

In many winged insects, the tracheae are dilated in various parts of the body to form thin-walled vesicles or **'air sacs'**. They are extremely delicate and usually lack tacnidia. Air sacs are commonly met with among Diptera, Lepidoptera and Odonata.

The principal function of the air sacs is a respiratory one, as they serve to increase the volume of air in the tracheal system. An insect with air sacs has a lower specific gravity and they make light easier. In the head, there is a pair of large air sacs behind the eyes and many smaller ones in the region of the mouthparts. In the thorax, there are a pair of large air sacs and a pair of small ones connected to the metathoracic spiracle. In the abdomen, there are eight pairs of large air sacs. The air sacs on each side are connected by a longitudinal tube running along the alimentary canal. Besides these large ones, there are numerous smaller air sacs and inflated sections of the tracheae. In aquatic insects, the air sacs act as hydrostatic organs, which enable the insect to float at any level in the water **(Fig. 31)**.

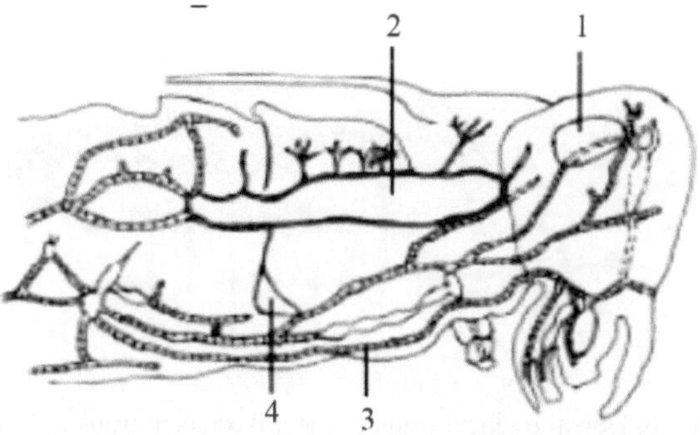

1. Cephalic air sac 2. Thoracic air sac 3. Ventral tracheal trunk 4. Mesothoracic air sac

Fig. 31. Vertical, longitudinal section of the head and thorax

The pattern of tracheal system is quite variable, but a pair of longitudinal trunk with cross connections form the main system of most insects.

The successive spiracular trunks join together by their anterior and posterior branches to form the lateral longitudinal trunk, thus joining all the spiracles of the same side. In the thorax, it opens in a large thoracic air sac and posteriorly it divides into two branches. The first branch, the dorsal trachea joins the dorsal tracheal trunk and the other, the ventral trachea joins the ventral tracheal trunk.

The dorsal segmental tracheaejoin on either side to form a pair of dorsal tracheal trunk. These trunks run parallel on both sides of theheart extending from the first to the eighth abdominal segments.The anterior and posterior ends ofeach trunk join with the lateral trunk of its side in the abdominal segments. A tranverse tracheal tube known as **'dorsal commissure'** in each segment connects both the dorsal tracheal trunks with each other.

The ventral tracheal trunks lie on the ventral side of the body and run parallel to the ventral nerve cord. Like the dorsal trunks, they are also connected with the lateral trunks of their corresponding sides by transverse tubes in each segment of the abdomen. Each trunk extends from the head to the eighth abdominal segment. Many of the branches form a network in the head and abdominal region. Both the ventral trunks are segmentally connected by a transverse ventral commissure **(Fig. 31, 32)**.

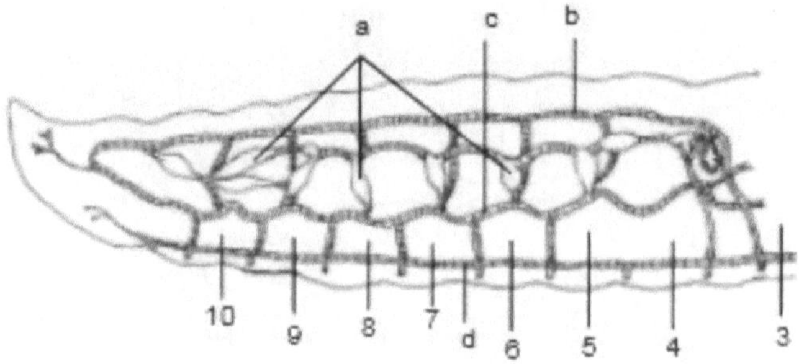

a. Abdominal air sacs b. Dorsal tracheal trunk c. Lateral tracheal trunk d. Ventral tracheal trunk

Fig. 32. Vertical, longitudinal section of abdomen

From the lateral tracheal trunks, branches lead to different parts of the body. One branch in each segment goes to the alimentary canal. This branch divides into two sub-branches and ramifies on the upper and lower sides of the alimentary canal. One branch goes to each air sac. Many branches in the thoracic region lead to the wings and legs. One branch in each of the eight abdominal segments goes to the ventral side and join the ventral tracheal trunk.

The tracheoles are exceedingly slender unicellular tubes, less than one micron in diameter and contain no tacnidia. These fine, cuticular tubes are often given off in clusters from the tracheae and then further branch into a fine network over the tissue cells **(Fig. 30A)**.

During respiration, all the spiracles do not function in the same way. While the two pairs of thoracic spiracles and the first two pairs of the abdominal spiracles serve only for inspiration, all the other six pairs of the abdominal spiracles are expiratory. When the abdomen and rarely the thorax expands, the first four pairs of spiracles open, while the the last six pairs remain closed. Air enters the tracheae through the spiracles and then into the tracheoles and ultimately end in the cells of the tissues. Through a process of diffusion, the cells take oxygen from the air and the carbon dioxide released by the cells enters the tracheae through the tracheoles. When the abdomen contracts, the first four pairs of spiracles are closed, while the last six pairs of abdominal spiracles open and the carbon dioxide, which enters the tracheae through the tracheoles are expelled out through the spiracles **(Fig. 33)**.

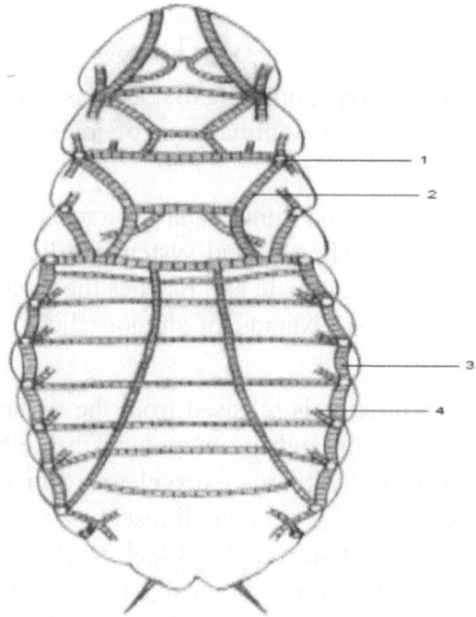

1. Spiracle 2. Trachea 3. Lateral tracheal trunks 4. Tracheoles

Fig. 33. The respiratory system (e-g) Cockroach

There are different types of respiratory systems in insects. This depends on the basis of functioning and number of spiracles :

i) **Polypneustic.** In this system there are at least eight functional spiracles on each side of the body. This is further subdivided into three types :

a) **Holopneustic.** This is the normal type and in this type, all the ten pairs of spiracles viz., one mesothoracic, one metathoracic and eight abdominal spiracles are functional (e-g) Nymphs and adults of many orders, which are terrestrial.

b) **Peripneustic.** In this type, all the spiracles except any one or two pairs are functional. Usually nine spiracles viz., one mesothoracic and eight abdominal spiracles are functional (e-g) Larvae of Lepidoptera, Neuroptera and Mecoptera.

c) **Hemipneustic.** In this type, eight spiracles viz., one mesothoracic and seven abdominal spiracles are functional (e-g) Mycotophilid larvae

ii) **Oligopneustic.** In this system, there are only one or functional spiracles on each side of the body. It is further divided into three types :

a) **Propneustic.** In this type, only the anterior-most pair of spiracles (Mesothoracic spiracles)is functional (e-g) Pupae of mosquitoes.

b) **Metapneustic.** In this type, only the last pair of the spiracles (Post abdominal spiracles) is functional (e-g) Larvae of mosquitoes.

c) **Amphipneustic.** In this type, only the first and last pairs of spiracles (one mesothoracic and one post-abdominal pairs of spiracles) are functional (e-g) Maggots of housefly.

iii) **Apneustic.** In this type, no functioanal spiracles are present and so respiration is cutaneous. The tracheal system is filled with fluid and the oxygen passes through the cuticle and reaches the tracheal system within which it is diffused (e-g) Naiads of dragon flies, larvae of parasitic Hymenoptera.

Some amount of carbon dioxide is released from the tissues directly into the haemolymph and diffuses out through the integument in the absence of spiracles. This is possible because the chitinous exoskeleton, which is impervious to water, is pervious to gases. Some very small insects such as' Collembola and Proturans, which live in moist surroundings lack tracheae and gas exchange occurs over the general body surface. Some aquatic, immature insects insects also lack tracheae, especially during the early stages. However, tracheae are usually present in aquatic immatures and always in adult insects that live in water. The adults rarely utilize air from air bubbles or films trapped against the body surface, but the nymphs and larvae of certain groups may posses special adaptations for gas exchange in water. Damsel fly and mayfly nymphs posses abdominal gills. The gils are provided with tracheae and gas exchange occurs across the gill surface between the water and the tracheae. Dragon fly nymphs pump water in and out of the rectum, which contains gills supplied with tracheae. Usually gas exchange in aquatic immature insects occur across the general integument between the tracheae and water. Some larvae such as, those of the mosquitoes have a few functional spiracles associated with one or more breathing tubes. The larva rises to the surface of water periodically and obtain air through the tubes.

The Reproductive System

Most insects are bisexual, with as many males as females. The chief organs of reproduction are situated in the abdominal cavity. The essential generative organs in the female are the paired **'ovaries'** and their **'oviducts'** and in the male they are the paired **'testes'** and their ducts. In many insects, these primary organs are supplemented by accessory structures of different kinds, more or less corresponding to each other in both sexes and most of the parts in both the sexes are bilaterally symmetrical. Besides these, there are the external genitalia. In the male, they consist of the **'claspers'** and **'penis'**, while in the female is found the **'egg depositor'** or **'ovipositor'**. The external genital openings are generally situated near the posterior end of the abdomen towards the ventral side, usually ventral to the ninth abdominal segment.

The Female Reproductive Organs

The female reproductive system consists of (i) a pair of **'ovaries'** (ii) apair of lateral **'oviducts'**, which unite to form a **'common oviduct'** opening in the vagina (iii) the **'vagina'** and **'genital chamber'** (iv) **'Spermatheca'** or **'receptaculum seminis'** and **'spermathecal gland'** and (v) **'accessory'** or **'collateral glands' (Fig. 34)**.

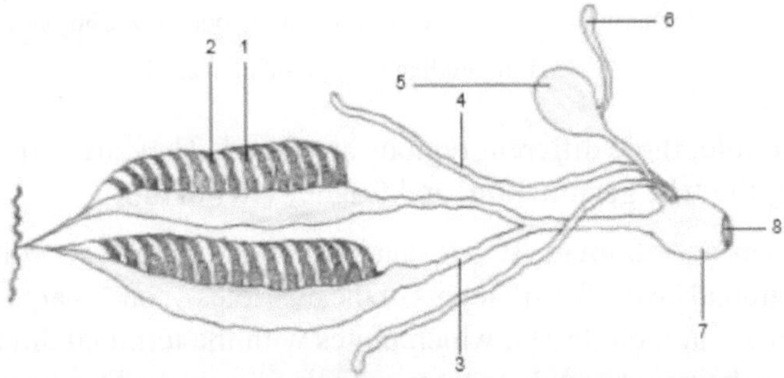

1. Ovary 2. Ovariole 3. Oviduct 4. Accessory gland 5. Spermatheca 6. Spermathecal gland 7. Uterus 8. Gonopore

Fig. 34. Internal reproductive organs of a female (e-g) Grasshopper

i) **The ovaries.** The ovaries are somewhat compact bodies lying in the body cavity of the abdomen, one oneither side of the alimentary canal. Each ovary consists of a variable number of separate **'ovarioles'** or **'egg tubes'**, but usually not exceeding eight in number, which open into the oviduct. However, in some ants there are about 200 ovarioles in each ovary and in some species of white ants the number of ovarioles may be as high as 2400 in each ovary.

A typical **'ovariole'** as in grasshopper is an elongated tube, in which the developing eggs are released one behind the other in a single chain, 1 the oldest **'oocyte'** being situated nearest the union with the oviduct. The wall of an ovariole is a delicate, transparent membrane with its inner coat made up of a layer of epithelial cells and the cells resting upon a basement membrane, which is termed as **'tunica propria'**. Outside this membrane is a peritonial coat of connective tissues, which contain muscle fibres and is covered by a reticulate layer of tracheal end cells **(Fig. 35)**.

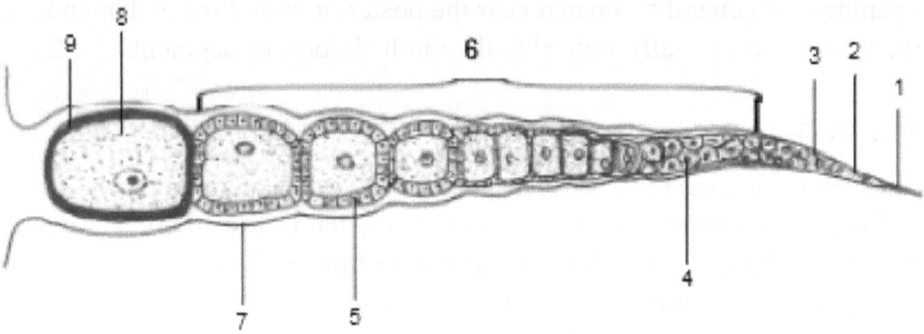

1. Filament 2. Apical cell 3. Germanium 4. Immature oocytes 5. Folicular epithelium 6. Vitellarium 7. Wall of ovariole 8. Mature oocyte 9. Choron

Fig. 35. Longitudinal section of an ovariole

In each ovariole, three different regions are found. They are: (a) the terminal filament (b) the **'the germarium'** and (c) the **'vitellarium'**.

a) **The terminal filament**. It is a slender, thread-like apical prolongation of the peritonial layer. The filaments of the ovarioles of one ovary join together to form a common thread, which unites with the terminal thread from the ovary of the opposite side to form a median ligament. This ligament aids in maintaining the ovaries in position and is attached either to its body wall, the fat body or to the pericardial diaphragm. In some insects, the ovarian ligament is wanting and the filaments end free in the body cavity.

b) **The germarium**. The germarium forms the apex of an ovariole below the terminal filament and consists of a mass of undifferentiated cells from which are differentiated the promordial germ cells and in many insectsthe nurse cells or nutritive cells. The germ cells develop into **'oogonia'** and finally into **'oocytes'** or **'egg cells'**.

c) **The vitellarium**. The vitellarium forms the major portion of the ovariole and contains a large number of oocytes or developing eggs. The epithelial layer of the wall of the vitellarium grows inwards in such a way as to enclose each oocyte in a definite sac or **'follicle'**. The cells of the follicle secrete the **'chorion'** of the egg.

Depending upon the presence or absence of nurse cells and on the location of the nurse cells, the ovarioles are differentiated into three types :

a) **Panoistic type**. In this type no distinct nurse cells are present (e-g) Orthoptera, Dictyoptera, Isoptera and Diptera.

b) **Polytrophic type**. In this type, the eggs and nurse cells alternate in he ovariole (e-g) Neuroptera, Hymenoptera, Lepidoptera and Diptera.

c) **Acrotrophic type**. In this type, the nurse cells are restricted only to the germarium and thus the germarium becomes a nutritive chamber (e-g) Hemiptera.

The polytrophic and Acrotrophic types are often grouped as one and is called **'meroistic type'**.

ii) **The oviducts**. The lateral oviducts or **'genital ducts'** are paired canals leading from the ovaries. These two ducts join to form a **'common oviduct'**.

iii) **The vagina**. The vagina differs in structure from the oviducts in that, it is an invagination of the body wall and is lined with a cuticular layer. The vagina opens into a shallow genital chamber in the ninth sternum. This chamber, which is also known as the **'bursa copulatrix'** is the organ in which the sperms are discharged before entering the spermatheca and is helpful in the copulation process. In certain viviparous insects, the vagina is greatly enlarged to form a chamber or **'uterus'**, which serves as a receptacle for the developing young insects.

iv) **The spermatheca and spermathecal glands**. The spermatheca is a sac for the storage of the seminal fluid received from the male during copulation. As the mating takes place only once in the life of most of the insects, it is essential that provision is to be made for fertilization of all he eggs developed after the union of the sexes. The spermatheca serves this purpose. It varies in form, but mostly is bulbous in shape and opens into the dorsal wall of the vagina by a duct known as **'spermiduct'**.

A spermathecal gland opens into the spermatheca and secretes a fluid, which is said to preserve and lengthen the life of the sperms.

v) **The accessory** or **collateral glands**. One pair of accessory glands is present, which opensinto the distal portion of the vagina. They are fairly large and important organs, which secrete the material for the formation of the **'egg pod'** or **'ootheca'**. In some insects, they supply a cement-like adhesive secretion, which serves to fasten the eggs onto the substratum on which they are laid.

Oogenesis

The cytoplasm of the matured eggs usually forms a reticulum in the centre of which occurs a large nucleus and abundant globules of yolk. The deposition of yolk in the oocyte is called **'vitellogenesis'**. The yolk provides nourishment to the developing embryo. When vitellogenesis is completed, the egg is invested by a delicate homogenous vitellin membrane and an **'egg shell'** or **'chorion'**, which is secreted by the follicle cells of the ovary. The chorion typically consists of two layers, an inner **'endochorion'** consisting of polyphenol and protein and an outer **'exochorion'** consisting of lipoprotein.

Fertilization

After copulation, the spermatic fluid of the male is stored in the spermatheca. When the egg comes down from the oviduct in the vagina, it consists of an opening in its shell for the entrance of the male sperm cell (spermatozoa) into the egg. The opening is termed the **'micropyle'**. One or two spermatozoa coming from the spermatheca enter the egg, but only one succeeds in fertilizing it. After fertilization, the collateral glands secrete a fluid around the egg, which on hardening protects the egg from unfavorable weather conditions and from natural enemies. Some insects such as, the grasshopper lays the eggs in **'egg pods'**. The fertilized egg comes out through the **'gonopore'** and is laid by the ovipositor.

The Male Reproductive Organs

In general, the male reproductive system is similar to that of the female reproductive system and consists of the following parts. (i) A pair of **'testes'** (ii) A pair of **'vasa deferentia'** (iii) **'Seminal vesicle'** (iv) **'Ejaculatory duct'** (v) **'phallus'** or **'penis'** and (vi) **'Accessory glands' (Fig. 36)**.

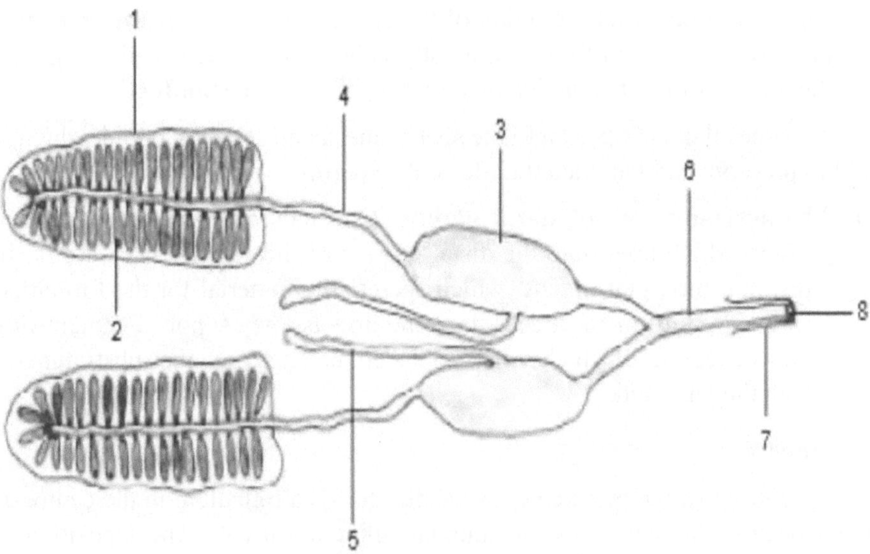

1. Testes 2. Testicular follicle 3. Vesicula seminalis 4. Vasa diferentia 5. Accessory glands 6. Aedeagus or penis 7. Male genital organ 8. genital aperture

Fig. 36. Internal genitalia of a male insect (e-g) Grasshopper

i) **The testes**. Though the testes closely resemble the ovaries in some insects, in the majority of insects, they are much smaller than the female organs of the same species. Their situation in relation to the alimentary canal varies. In some insects, they lie above the alimentary canal, while in others they are placed at the sides of the gut or ventral to it. The testes are maintained in position by the surrounding fat body and tracheae. The suspensory filaments as in the ovaries of females are absent. Each testis is more or less an ovoid body, partly or completely divided into a variable number of follicles or lobes.

The testicular follicle is lined with a layer of epithelium, whose cells rest externally upon a basement basement membrane, outside of which is a peritoneal coat of connective tissues. Each follicle is divided into a series of zones characterized by the presence of sex cells in different stages of development, as indicated : (a) The **'germarium'**, which is the region containing the primordial germ cell **'spermatogonia'** that undergo multiplication (b) The zone of growth, where the spermatogonia increase in size, undergo repeated mitosis and develop into **'spermatocytes'** (c) The zone of division and reduction, where the spermatocytes undergo meiosis and develop into **'spermatids'** and (d) The zone of transformation, where the spermatids become transformed into **'spermatozoa'**. The formation of spermatozoa from germ cells or spermatids is called **'spermatogenesis'**. The spermatozoa from the spermatocyte stage onwards are enclosed in and nourished by the testicular cyst cells and released into the **'vasa deferentia' (Fig. 37)**.

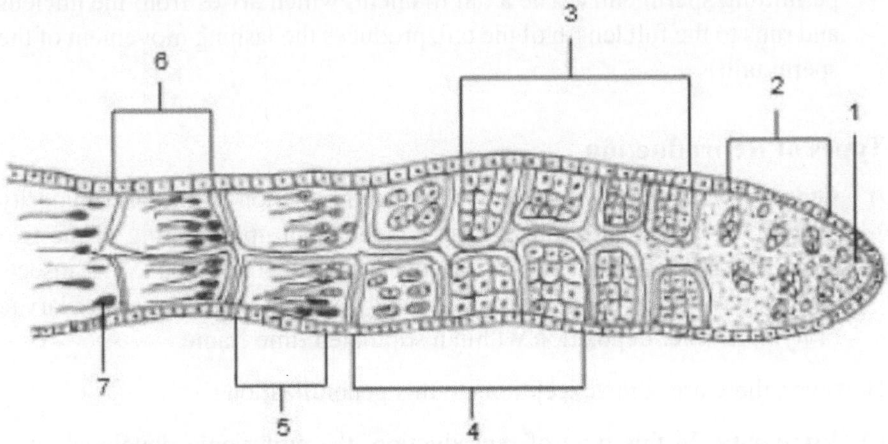

1. Apical cells surrounded by spermatogonia 2. Zone of spermatogonia 3. Zone of spermatocytes 4. Zone of Maturation and reduction division 5. Zone of spermatids 6. Zone of spermatozoa 7. Spermatozoa

Fig. 37. Longitudinal section of testes follicle (e-g) grasshopper

ii) **The vasa deferentia or vas deferens.** The vasa deferentia or the **'genital ducts'** are the paired canals leading from the testes. They vary greatly in length and in majority of insects, each vasa deferentia becomes enlarged along its length to form a sac or **'vesicula seminalis'** or **'seminal vesicle'**, in which the spermatozoa congregate. Posteriorly the vasa deferentia unite to form a short, common canal, which is continuous with a median ectodermal tube or **'ejaculatory duct'**.

iii) **The genital atrium**. The plates at the anal end of the abdomen collectively enclose a cavity known as **'genital atrium'** in which lies the actual copulatory organs, the **'phallus'** or **'penis'**. The genital atrium consists of two parts, a proximal **'phallobase'**, which is the proximal part of the penis and a distal **'aedeagus'**, which is the male copulatory organ, including accessory structures and the penis. The ejaculatory duct opens into it.

iv) **The accessory** or **collateral glands**. One to three pairs of accessory glands are usually present. They are tubular or sac-like in form. Their secretions mix with the spermatozoa and help in preserving them.

v) **Spermatozoa**. The matured **'spermatozoa'** or **'sperms'** re long and filamentous. The head and tail of the sperm are of approximately the same diameter. The greater part of the head region is occupied by the nucleus with only a thin coating of cytoplasm and in front of the nucleus is the **'acrosome'**. The acrosome is probably concerned with attachment of the sperm to the egg and possibly also with lysis of the egg membrane, thus permitting sperm entry. The axial filament, which arises from the nucleus and runs to the full length of the tail, produces the lashing movement of the sperm tail.

Types of Reproduction

i) **Oviparity**. This is the normal type of reproduction in the vast majority of insects. Reproduction in insects is dependent upon mating of the two sexes and the fertilization of the ovum by the spermatozoa. Most insects are **'oviparous'** in the sense they lay eggs, which hatch into either larvae or nymphs after deposition within a stipulated time frame.

However, there are some exceptions to this generalization :

ii) **Viviparity**. In this type of reproduction, the embryonic development is completed within the body of the female parent and hence, instead of laying eggs, the young larvae or nymphs are laid directly.

There are five different types of viviparity :

a) **Ovoviviparity**. Many insect species retain the eggs in the genital tract until they are ready to hatch and so hatching occurs just before or as the eggs are being laid. All the nourishment required for the development of the embryo is present in the eggs and no nourishment is supplied from the parent. It differs from normal oviparity only in the sense that the eggs are retained till the time of hatching (e-g) Some species belonging to Diptera, Coleoptera, Thysanoptera and *Cimex* (Hemiptera).

b) **True viviparity**. In some species, the eggs are retained after fertilization in the vaginal chamber of the female parent and the embryo receives nourishment directly from parent instead of the yolk or in addition to the yolk. In such cases, some anatomical adaptations are provided in the parent or the egg, which facilitate the exchange of nutrients. Viviparous species generally produce fewer number of off springs than related oviparous species and this may be due to the reduction in the number of ovarioles (e-g) Aphids and some flies.

c) **Pseudoplacental viviparity** Some species produce eggs containing very little or no yolk and the eggs are retained in the uterus of the female. The eggs may receive nourishment through embryonic or maternal structures known as **'pseudoplacentae'** (e-g) Some species of aphids.

d) **Adenotrophic viviparity**. In this type, the eggs are ovulated singly and the embryo develops by utilizing nourishment from the yolk. However, after hatching the young larva is retained in the uterus and nourished by secretions from special maternal glands modified into milk glands. The larva moults twice within the uterus and the fully developed larva is then deposited outside for pupation (e-g) Some flies.

e) **Haemocoelus viviparity**. In this type, the development of the embryo takes place in the haemocoel of the female parent. In such cases there are no oviducts and the ovaries lie free among the fat body and are fertilized by the sperms, which enter the haemocoel through the genital canal and when matured they break up and the eggs are dispersed in the haemocoel. The eggs are surrounded only by a trophic membrane through which nourishment is supplied from the parent (e-g) Gall flies.

iii) **Parthenogenesis**. Sometimes eggs develop without being fertilized and this phenomenon is known as **'parthenogenesis'**. Failure of a female to find a suitable male may result in parthenogenesis occasionally, but in a number of insect species parthenogenesis is the normal means of reproduction and is common in a number of insect species.

Parthenogenesis is classified according to the behavior of the chromosomes at the maturation division of the oocyte.

a) **Haplodiploidy**. In this type, normal reduction division occurs in the oocyte and some oocytes are fertilized, while others are not fertilized. The fertilized eggs produce females and the unfertilized eggs produce only males (e-g) Hymenoptera.

b) **Apomietic** or **Ameiotic**. In this type, no reduction division or meiosis occurs, as a result only females are produced (e-g) Aphids.

c) **Apomistic** or **Meiotic**. In this type, normal reduction takes place, but is followed by the fusion of two nuclei, so that diploid condition is restored and hence only females are produced (e-g) Phasmids, Coccids and Psychids.

iv) **Paedogenesis**. In a few cases, In a few cases, Immature insects possess functional ovaries and the eggs produced develop parthenogenetically and this condition is known as **'paedogenesis'**. This is due to hormonal imbalance and most insects reproducing paedogenetically are also parthenogenetic and viviparous (e-g) Some species belonging to Diptera and Coleoptera.

v) **Polyembryony**. In a few cases, the egg gives rise to two or more larvae instead of the normal single larva, which is called **'polyembryony'**. In all such cases, the eggs are relatively free from yolk and nourishment is derived from the host tissue in which they are present (e-g) Some species belonging to Acrididae and some endoparasitic species belonging to Ichneumonidae, which are egg and larval parasitoids of Lepidoptera.

vi) **Hermaphroditism**. In this type of reproduction, in addition to the normal males, which occur very rarely, there are males that are functionally similar to the females, where true females are ansent. The reproductive system is somewhat complex and the gonad, which is formed by the fusion of two originally separate organs, produce both spermatozoa and eggs (e-g) Phorids.

Sense Organs in Insects

Innate behavior referred to as **'instinct'** is defines as 'a fixed response or a series of responses to a particular stimulus or series of stimuli'. With the nervous systems, are closely associated, what is called sensory organs, which receive and transfer impressions received from the effect of changes in the environment. The sensory organs or **'receptors'** are those structures in which the energy of a stimulus is transformed into a nervous impulse, the impulse after transmission to a part of the central nervous system usually results in a change in the behavior of the insect or the maintenance of some existing activity. The sense organs of insects are derived from the cellular and cuticular parts of the integument, which are modified in various ways for the functions they perform. In the simplest form, a sense organ termed **'sensillum'** consists of a sense cell connected with the distal end of a sensory nerve and so situated as to be exposed to stimulus or to the secondary or indirect effects of primary stimuli. The **'sensory neurone'** or **'nerve cells'** of insects are usually regarded as primary sense cells.

The various sense organs in insects are grouped into the following classes viz., (i) Mechano receptors (responsive to pressure or contact) (ii) Auditory organs (iii) Chemo receptors (responsive to chemicals) (iv) Physical receptors viz., Thermo receptors (responsive to temperature) and Hygro receptors (responsive to humidity) such as, temperature and humidity receptors (v) Visual organs or Photo receptors (responsive to light) and (vi) Proprio receptors (responsive to body movement).

i) **Mechano receptors.** Mechano receptors are sense organs, which are excited by processes involving mechanical disturbances of some part of the receptor. The excitement may be due to touch, including contact with solid objects or currents of air or water. Mechano receptors may be distinguished into three main types based on structural modifications :

a) **Trichoid sensilla (Articulated sensory hairs).** A **'trichoid sensillum'** is the simplest form of mechano reception consisting of an articulated sensory hair. The hair is produced by a **'trichogen cell'**, while the socket is produced by a **'tormogen cell'**. The sensillum has a **'bipolar neurone'**, which is in close contact with the base of the hair. Thus, the sensory hair is an innervated hair or **'seta'** having a direct connection with a sensory cell. Such hairs are widely distributed over the insect body, particularly on the antennae, tarsi and cerci, which are the typical tactile organs. A potential is produced when the hair is bent in any direction. The hairs of trichoid sensilla differ in their sensitivity. Long, delicate hairs respond to the slightest force even when air pressure changes, while the shorter, thicker hairs require considerably more force for stimulation.

b) **The campaniform sensilla.** This type of sensillum consists of a dome - shaped, circular structure, often oval in surface view and slightly elevated above or depressed below the general level of the cuticle. The sensory neurone forms a rod- like structure called **'scolopale'**, which is inserted into or touches the cuticular dome. This type of sensilla occurs on cerci, wings, basal parts of appendages etc. They are different from the sensory hairs in that, the external processes (setae) are replaced by the dome-like structure **(Fig. 38)**.

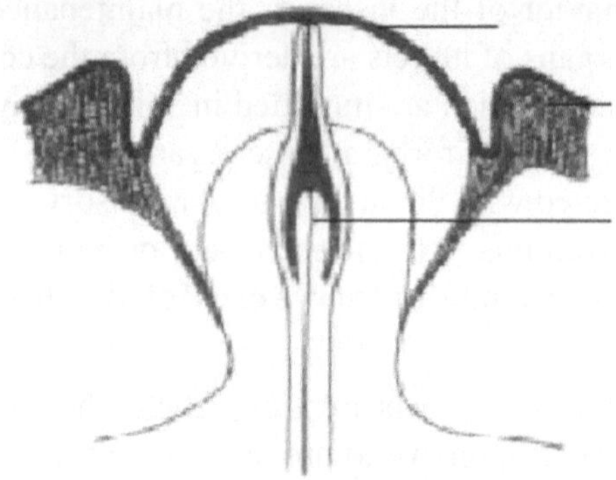

1. Cuticular dome 2. Cuticle 3. Sensory cell process or scolopale

Fig. 38. A campaniform sense organ

c) **The chordotonal organs.** These organs, otherwise called **'scolopophorous organs'** or **'scoloparia'** are made up of a number of specialized sensilla or **'scolophores'**, each containing a conspicuous sensory rod or **'scolopale.** Each chordotonal organ consists of a spindle-shaped bundle of scolophores attached at each end to the integument and frequently ending freely in the body cavity **(Fig. 39)**.

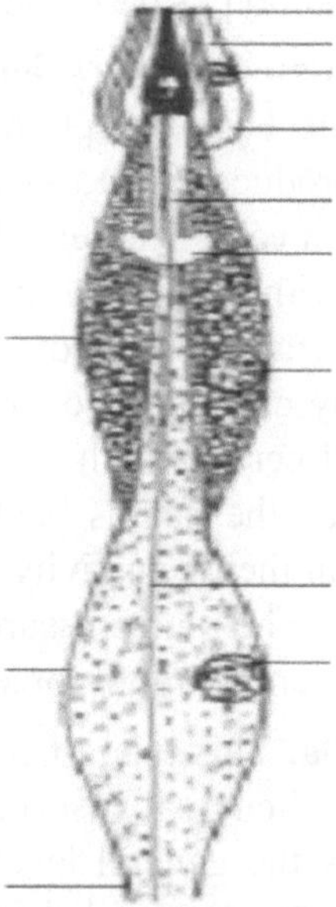

1. Terminal segment 2. Cap cell 3. Cap cell nucleus 4. End knob 5. Scolopale 6. Vacuole 7. Envelop cell 8. Envelop cell nucleus 9. Axial fibre 10. Nerve end cell 11. End cell nucleus 12. Nerve fibre

Fig. 39. A scolophore of the integument type

The scolophore is a complex structure consisting of a bipolar neurone produced basally into a fibre of the chordotonal nerve and drawn out distally into a slender prolongation enclosed by an envelop cell and a cap cell. The scolopale is formed within the envelop cell and its cavity communicates basally with a liquid filled vacuole.

The chordotonal sensilla serve several purposes and respond to stimuli caused by internal pressure changes, mechanical vibrations etc.

A highly specialized type of chordotonal organ located on the second antennal segment of most of the insects is known as the **'Johnston's organ'**, which consists of a variable number of radially arranged sensilla. Their axons enter the antennal nerve. They serve as a type of auditory organ as in mosquitoes **(Fig. 40)**.

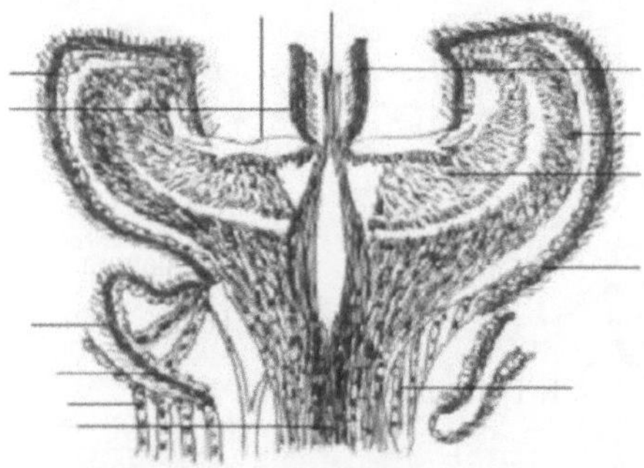

1. Base of clavola 2. Conjunctival plate 3. Process of conjunctival plate 4. Antennal muscle 5. Pedicel 6. Scape 7. Antennary nerve 8. Nerve to clavola 9. Nerve cells 10. First segment 11. Second segment 12. Third segment

Fig. 40. Longitudinal section of the base of the antenna showing Johnston's organ

ii) **Auditory organs.** The auditory organs are replacement receptors, which differ from the mechano receptors only in that they respond to disturbances of much lower intensity due to which the sound is perceived. These receptors are also specialized chordotonal organs and are found in grasshoppers, crickets and cicadas, which also have sound producing organs.

Three types of sound receptors are recognized :

a) **Tympanal organs.** Tympanal organs develop from the fusion of parts of the tracheal dilation and the body wall. The **'scolopidia'**, a chordotonal organ, which is a combination of three cells viz., the cap, scolopale and dendrite, are attached to the tympanum. They are paired structures, which consist of a thin cuticular membrane, the tympanum associated with tracheal air sacs and chordotonal sensilla. In the locust, there is a tympanal organ on each side of the abdominal tergum with an externally visible tympanum, which is surrounded by a circula ring. Inernally, a group of numerous scolophores form a swelling, which is known as **'Muller's organ'** and it is applied to the inner surface of each tympanum and is connected to the metathoracic ganglion by the auditory nerves. Two sclerotized processes and a pyriform vesicle filled with a clear liquid are intimately associated with the Muller's organ and serve to transmit the vibration to the tympanum and to the sensilla. The first abdominal spiracle gives off an air sac to the undersurface of the membrane. Two other air sacs arise from the ventral tracheal trunk on each side of the second abdominal segment, which is in close contact with the first air sac.

In long-horned grasshoppers and crickets, the tympanal organs situated at the base of each fore tibia and in some forms their internal structure is very complex **(Fig. 41, 42, 43, 44)**.

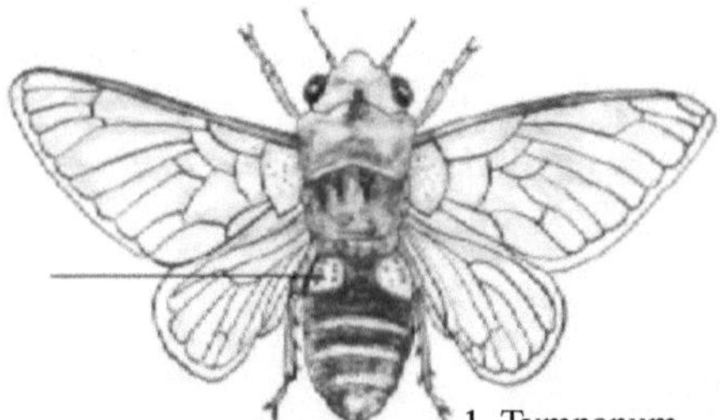

1. Tympanum

Fig. 41. Auditory and sound drums of male cicada

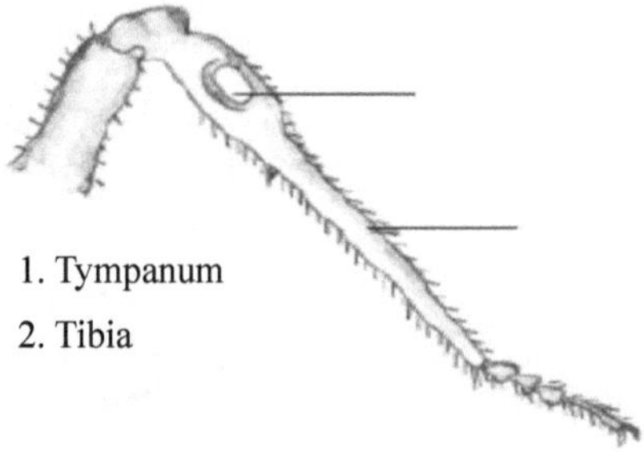

1. Tympanum
2. Tibia

Fig. 42. Tympanum in fore leg of locust

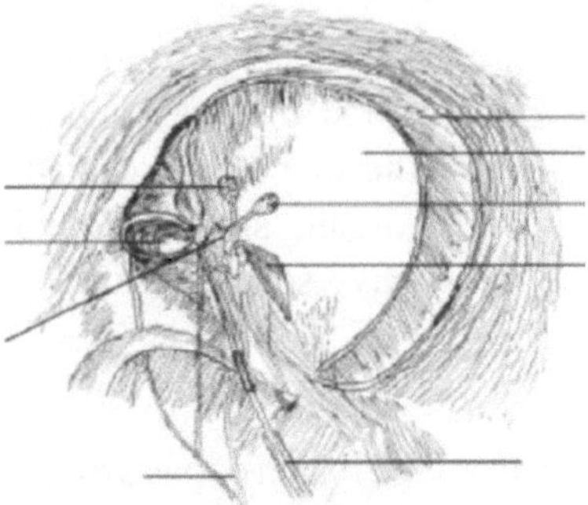

1. Muller's organ 2. Pyriform vesicle 3. 4.Horn-like processes 5.Tensor muscle of tympanum 6. Tympanum 7. Border of tympanum 8. Auditory nerve 9. Spiracle

Fig. 43. Tympanum of a locust viewed from within

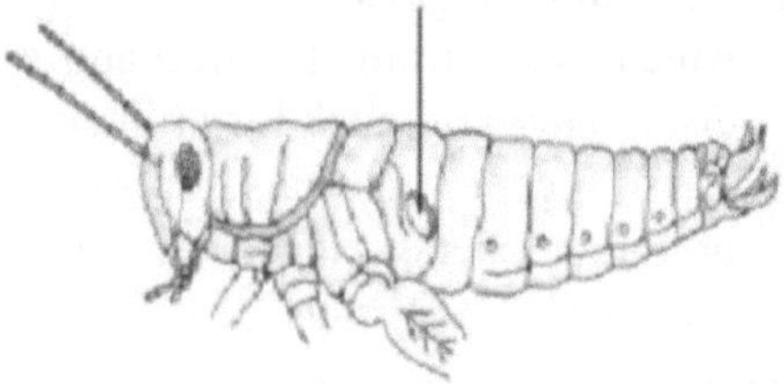

1. Tympanum

Fig. 44. Tympanum on the first abdominal segment of grasshopper

b) **Auditry hairs.** In many lepidopterous larvae, the response to sound are mediated by hair sensilla, which react to low frequencies. On the cerci of some Orthoptera also there are such sensory hairs.

c) **Johnston's organ.** These are specialized organs consisting of a group of chordotonal organs located at the apex of the pedicel of antannae, the sub-genal organ in the tibia, the halteres of Diptera etc. In the males of *Aedes aegyptii*, the organ serves as sound receptors. The sound waves set in motion the hairs of antennal flagellum, which in turn causes the movements of the whole flagellum, thus stimulating the sensilla of Johnston's organ.

iii) **Chemo receptors.** Chemo receptors are found on all parts of the body. Insects can feel the taste of salts, sweets, acids, bitter substances etc. by means of receptors in the mouth as in bees, on the antannae as in some Hymenoptera, on the palpi as in Hemiptera and on the tarsi as in Diptera and Lepidoptera. There may be one or more organs in the same insect. Insects can detect odors, chiefly by sensilla on the antennae and palpi. In sucking insects, especially of Hemiptera, receptors for the perception of specific flavors occur in the pharynx. In some insects, the males have flabellate antennae in which the sense organ is located.

The chemo receptors are characterized by the possession of cuticle, partly thin-walled and innervated at least by one or a group of bipolar neurones, whose associated distal processes contain small refringent bodies. These receptors are derived from sensory hairs and many different types are noticed viz., **'sensilla trichoidea olfactoria'** (thin-walled hairs), **'sensilla basiconia'** (peg-like) **(Fig. 45)**, **'sensilla styloconica'** (cone-like), **'sensilla coeloconica'** (sunken below the integument) and **'sensilla placoidea'** (pore-like).

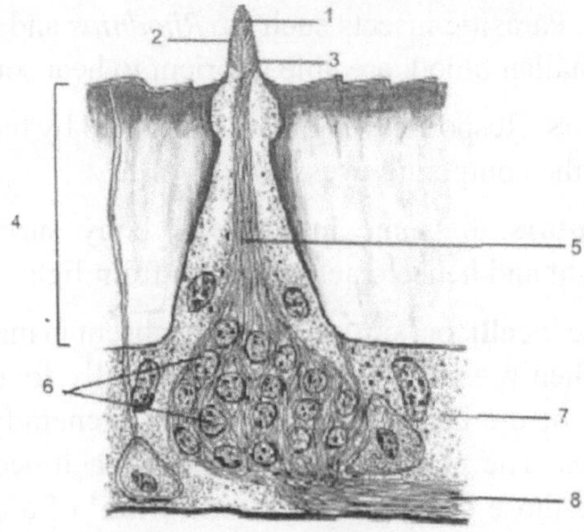

1. Peg 2. Openings in the cuticle 3. Branching dendrites 4. Cuticle 5. Scolopale 6. Sensory neurones 7. Epidermis 8. Nerve

Fig. 45. Chemosensory peg organ in the antenna of grasshopper

Chemo receptors may be differentiated as follows :

a) **Common chemical receptors.** The chemical receptors produce reactions to avoid highly irritant substances such as, chlorine, ammonia etc.

b) **Olfactory receptors**. The sense of smell is stimulated by low concentrations of vapors of substances, which are volatile at ordinary temperatures. Such receptors are found on the palpi and antennae. The queen bee and worker bees are provided with a large number of placoid sensilla on the antennae, which enable them to feel the sense of smell.

c) **Gustatory (taste) receptors**. These receptors produce response to relatively high concentration of stimulants, which in liquid form comes in contact with the receptors. These organs occur on the antennae, surface of the pre-oral food cavity and mouthparts, ovipositortarsus and distal parts of tibia.

iv) **Physical receptors**.

a) **Hygro receptors**. Many insects are capable of monitoring the amount of water vapor in the surrounding air and react to differences in humidity. In *Pediculus*, they are like peculiar tuft-like cuticular sensilla. The red flour beetle *(Tribolium castaneum)* has humidity detectors on the antennae. In *Drosophila* larvae, humidity detectors occur in the anterior sternites. In ants, the humidity receptors help them to know about the coming rains much in advance and they migrate to safer places along with their eggs, larvae and pupae prior to the onset of rains.

b) **Thermo receptors**. Temperature receptors occur on the antennae, maxillary palpi and tarsi. Heat may be transferred to or from an insect by radiation or conduction. Parasitic insects such as, *Rhodnius* and mosquitoes, which feed on mammalian blood, are able to orient to heat source.

v) **Photo receptors.** Response to light are translated by the dermal receptors, the ocelli and the compound eyes.

a) **Dermal receptors**. In some insects, the body surface appears to be sensitive to light and hence react to stimuli from light source.

b) **The ocelli**. The **'ocelli'** or **'simple eyes'** are absent in many species of adult insects, but when present, there are three usually found on the anterior-dorsal surface of the head as minute specks, generally between the two compound eyes. The photo receptor cells on each ocellus are organized somewhat like those of a single **'ommatidium'** of a compound eye and the focus is always fixed, because the immovable lens is continuous with the surrounding skeleton. Ocelli can detect changes in light intensity and may be very sensitive to low intensities. They have a general stimulatory effect on the nervous system and enhance the reception of stimuli by other sensory structures. The simple eyes of larval and pupal forms are termed **'stemmata'** or **'lateral ocelli'**. Though many larvae have no visual organs, if present they are generally arranged along the lateral margins of the head and these lateral ocelli are in many cases different in some features from the dorsal ocelli found in adult insects. The following parts are present in an ocellus:

The cornea. The cornea or lens is a part of the cuticle, which is raised to form the external investment of the ocellus. In this region the cuticle is transparent.

The corneagen layer. This layer is continuous with the hypodermis but differs in its composition. It consists of colorless transparent cells, which are grouped together and form the vitreous body. This layer secretes and supports the the cornea and helps the lens in its function.

The retina. This consists of visual cells, which are the sensory neurons, each being in direct connection with a fibre of the ocellar nerve. These visual cells are associated with a group of cells known as **'retinula'**, which surrounds a longitudinal optic rod called the **'rhabdome'**.

Pigment cells. The ocelli may possess some accessory cells with pigments in some cases and these sometimes form an **'iris'**.

The ocellus is believed to help the insect to distinguish light from darkness.

c) **The compound eyes**. In almost all adult insects, one pair of compound eyes are present conspicuously on the head. In shape, position, size and structural peculiarities, the paired compound eyes exhibit remarkable variations and in some insects, these organs have reached an extraordinary degree of perfection as organs of sight. A compound eye consists of many long, cylindrical units possessing all the elements for light reception. Each unit or **'ommatidium'** cocsists essentially of an optical and sensory part. The optical part gathers light and the sensory part perceives the radiation and transforms it into electrical energy. The ommatadium is covered at its outer end by a translucent cornea derived from the skeletal cuticle, which functions as a lens. The external surface of the cornea or **'facet'** is usually square or hexagonal. Behind the cornea, the ommatadium has a long, cylindrical or tapered element called the **'crystalline cone'**, which functions as a second lens.

The basal part of the ommatadium is formed by the receptor element called the **'retinula'**, around which are arranged elongated **'photo receptors'** or **'retinular cells'**. The retinular cells have their inner surfaces thrown into tubular folds and arranged at right angles to the axis of the ommatidium. The area of the tubules constitutes a **'rhabdomere'** and forms the **'rhabdome'**.

In the rhabdome, there is a clear fluid-filled central area. All the rhabdomeres and the enclosed central space form the rhabdome. The rhabdomeres of an ommatidium function as a single photo receptor unit and transmit a signal that represents a single light point. From each retinular cell extends an **'axon'** and thus a bundle of seven or eight axons leaves each ommatidium. The axons make connections with second order neurone in an **'optic ganglion'** within the brain.

The retinal cells contain black or brown pigment granules comprising of the proximal retinal pigment. Distally the ommatidium is surrounded by a number of special pigment cells forming the distal retinal pigment. Either the proximal or distal pigment or both can migrate centrally or distally depending on the intensity of light.

In bright light, a compound eye produces an **'apposition image'**. Light enters an ommatidium either at an angle or perpendicular to the facet. The proximal and distal retinal pigments are extended and act as a screen to prevent light passing from one ommatidium to another, thereby restricting the light rays to the axial region of the crystalline cone and rhabdome **(Fig. 46A)**.

The crystalline cone is also essentially a lens cylinder. It helps to eliminate oblique rays and direct the more axial light rays to the rhabdome.

The total image formed by the compound eye results from the number of ommatidium found. The image thus formed is somewhat analogous to mthe image produced on a television screen, as the picture is composed of grid formed out of dots of light and the apposition image formed by the compound eye is often called a **'mosaic image'**.

An advantage of the compound eye is its ability to detect movement. A slight shift in the point of light results in a corresponding shift in the ommatidia, which is stimulated. Another advantage of the compound eye is that the total corneal surface is convex, resulting in a wide visual field.

In weak light, the compound eyes function in a rather different manner and form what is called a **'superposition image'**. The pigment is retracted, so no screening effect occurs. Thus, light can pass from one ommatidium to another and one rhabdone responds to light rays entering from several adjacent facets. This condition appears to be an adaptation to see in the weak light, because light is received from many facets to a rhabdome instead of light received from its own facet. However, the term superposition image is not quite apt in the sense that there may be no image formed as in the case of apposition image. In the superposition eye there will be an overlapping of the image formed and hence the image is not distinct. Such eyes are useful to perceive objects and their movements in dim light, but only they can perceive the more general features of the object.

In many insects, the compound eyes are divided into two parts, one apposition type and the other superposition type. This type of eyes, called **'binocular eyes'** is common with the *Mantis* and it is helpful in knowing the distance of the prey **(Fig. 46B)**.

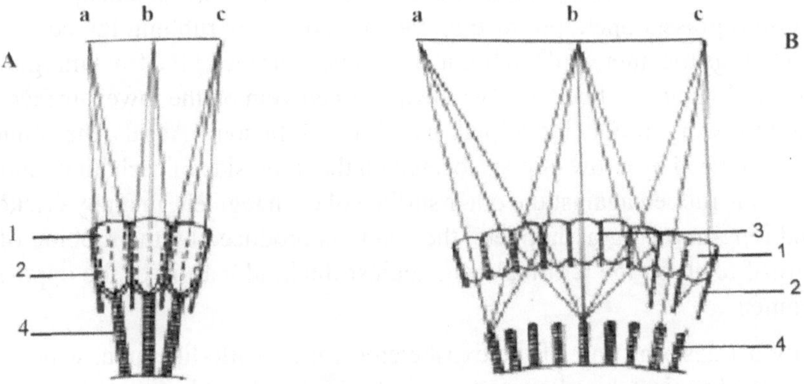

1. Crystalline cone 2. Retinal pigment 3. Retinal pigment retracted
4. Rhabdome 5. 'a', 'b' and 'c' - Light rays

A. Light rays from 'a', 'b' and 'c' received as an apposition image.

B. Light rays from 'a' and 'b' received as a superposition image, as light rays received by a number of ommatidia are concentrated on a single ommatidium and light rays from 'c' received as an apposition image.

Fig. 46A. Compound eye adapted for apposition image
Fig. 46B. Compound eye adapted for superposition image

Sound Producing Organs

Many adult insects and the larvae and pupae of some insects produce sounds and even noise. The cicadas, long-horned grasshoppers, gryllids etc. are well known examples. Insect sound can be very much beyond the range the human ear can perceive. While some sounds are incidental, most of them seem to have a definite purpose such as, to frighten enemies, call the fellow social insects, protect the hive or nest and to attract or charm the opposite sex (e-g) The male mosquito is tuned to the vibration of of wings of the female. Similar is the case with flies and bees. Males of many insects sing when they are mature. Some insects sing in chorus.

Insect sounds are generally produced by rubbing one part of the body against another or against an external object. Sound may be produced by vibrating the wings, by forcing air through spiracles or by the vibration of a special membrane by muscular action.

Different sound-producing organs are met with in insects :

i) **Stridulatory organs**. Sound is produced by stridulation or friction of one part of the body against another part of the body, one of which the **'stridens'** is fixed and is toothed. In some other cases, sound may be produced by friction of the strident against an external object. The movable part rubs like a bow on a violin. Such organs are found in several orders of insects such as, Orthoptera, Coleoptera and Hemiptera. Males of the families Acrididae, Tettigonidae and Gryllidae are best examples. The Acridid grasshoppers produce a cracking sound by rubbing the edges of two wings together while in flight. The upper surface of the front margin of the hind wing is rubbed against the thickened vein on the lower surface of the fore wing, producing a loud, harsh sound. In some Acrids, the sounds are produced by a row of pegs located on the inner side of each hind femur, which is rubbed against the outer surface of each tegmen. In many crickets and long-horned grasshoppers, the sound is produced by the rubbing of a row of teeth on the terminal vein against the hind margin of the opposite tegmen.

In many beetles, one part of the exoskeleton forms a file-like area, which on being rubbed against the adjacent area produces sound and in some beetles this is a sexual character. Several Lepidoptera produce hissing or rustling sound by friction between the fore- and hind wings or between wings and legs. Soldiers of many termites react to the disturbance of their nests by striking their head against the nest walls, thus producing a tapping sound. This is perhaps a danger signal to the colony.

ii) **Spiracular sound organs.** Certain insects force air from the spiracles and in doing so produce a faint noise. Queen bees, blow flies, may beetles etc. produce sound in this way.

iii) **Vibrating membranes**. Cicadas are best known noise makers among insects. Their complicated sound organs consist of a pair of drums or **'tympana'** situated at the base of of the abdomen of the male. The tympanum is set into motion by the muscles attached to it. The males make shrilling noise to attract their mates.

Light Producing Organs

Some insects are self-luminous on account of their possessing special structures. Light producing organs are located beneath the transparent cuticle. The luminous insects popularly called **'glow worms'** are beetles generally belonging to to the families Lampyridae and Cantharidae.

In some species of Lampyridae, the luminous substance is found in all stages of development and is mixed in the plasma. As the insect grows, it gets localized in the photogenic organs. Only the females of some species are luminous, there being apterous and larviform ones. Light may be emitted as a continuous glow or intermittent glow of variable intensity by the luminous insects.

4

Development and Metamorphosis

Insects generally reproduce by laying eggs or **'ova'**. The initiation of growth inside an egg, which in most cases is preceded by fertilization, leads to a long series of changes, finally attaining the adult stage and in due course to sexual maturity and finally to reproduction of a new generation. This chain of events from the egg stage to adult constitutes the **'life history'** or **'life cycle'** of an insect.

Eggs are laid singly or in batches of only a few or up to hundreds or even thousands. Generally eggs are laid under conditions, where food is available for the feeding of the future young ones. The eggs are generally smooth, elliptical or oval, but some have diverse shapes such as, disc-like, hemispherical, conical, globular, spindle-shaped and many other forms. The egg surface may also have a variety of textures. In some aquatic insects, the eggs are covered with a gelatinous secretion, which swells in water forming a jelly-like spawn. In mosquitoes, the lateral sides of the eggs are expanded, which serve as floats. Eggs vary in size from microscopic to about 3.0 mm. in diameter and 5.0 - 6.0 mm .or even more in length.

The development of an insect is a continuous process, but on the basis of external manifestations, it is divided into definite stages. The stage within the egg is termed **'embryonic development'**, that after hatching **'post-embryonic development'** and the changes of form during the post-embryonic development constitute **'metamorphosis'**.

1. Embryonic Development

A typical egg is a cell, enclosed in two coverings, the outer that is tough forming the **'egg shell'** or **'chorion'** and the inner that is delicate forming the **'vitellin membrane'**. The chorion contains a nucleus and cytoplasm. The cytoplasm consists of two portions, the central area or **'deutoplasm'** containing mostly yolk and the peripheral area or the **'cortical area'** or **'periplasm'**, which is relatively devoid of yolk. Since the eggs are formed and enclosed in the chorion within the body of the female before fertilization occurs, it is provided with a **'micropyle'** near the anterior pole, through which the sperm enters the chorion and fertilization takes place.

In the unfertilized egg, the nucleus is situated in the central part of the yolk surrounded by an island of protoplasm. During maturation, the nucleus migrates towards the periphery, where it divides and the polar bodies are formed. After fertilization, the zygote nucleus moves towards the centre and the zygote nucleus divides during the early cleavage. The numerous cleavage nuclei migrate to the periphery, where each gets surrounded by a halo of cytoplasm and a cell wall and gets mixed with the periplasm. Such a unit of nucleus along with cytoplasm is called **'energid'** or **'cleavage energid'**. Thus, a continuous one-cell thick layer of cells called the **'blastoderm'** is formed the egg yolk. On the ventral surface of the egg, the cells soon get crowded to make a thick, ventral plate called **'embryonic primordium'** or **'germ band'**, which forms the future **'embryo'**. The germ band grows by cell multiplication, followed by vertical partitioning of the surface, whereby the body segmentation and appendages are set out.

2. Post Embryonic Development and Metamorphosis

i) **Hatching**. When the embryo is fully developed and ready to hatch out of the egg, the insect has to force its way out through the eggshell. While hatching, the embryo shows rhythmic muscular activity and presses against the eggshell or strikes it repeatedly with its head.

   The process of hatching varies greatly in different groups of insects. In grasshoppers, the shell is ruptured in its anterior part. In bugs, there is a line of weakness near the anterior pole and a cup-like of the eggshell is pushed open like a lid or **'operculum'**. In many moths and butterflies, the larvae eat their way out of the eggshell. The embryonic development may require a few days to several weeks before hatching takes place.

ii) **Moulting**. Since the body wall, which is covered with a hard exoskeleton does not permit increase in size or growth of young insects beyond certain limits, it is shed periodically. Every insect during its growth sheds its skin one or more times and this process is **'moulting'**or **'ecdysis'** and the cast off skin is termed the **'exuvia'**. Before each moult, the insect stops feeding temporarily and a moulting fluid collects between the old and newly formed cuticle. A slit along a predetermined dorsal median line of the head and thorax occurs through which the insect works its way out. The interval or period between the various ecdyses is known as **'stadia'** and the form assumed by the insect during a stadium is called an **'instar'**. The young one after hatching is said to be in the first instar and the adult after emergence is the final instar or **'imago'**. Insects differ greatly in the nuber of moults they undergo, but it is constant for any one species **(Fig. 5)**.

iii) **Metamorphosis**. A very important characteristic of insects is that the newly hatched young one from the egg is anatomically quite different from that of the adult stage. Before reaching the adult stage, the young one undergoes a series of changes of form during its various growth stage, which are collectively called **'metamorphosis'**. During metamorphosis, the adaptive features peculiar to the larva are lost completely.

Different Types of Metamorphosis

a) **Ametabola** or **no metamorphosis**. Some insects on hatching almost resemble their parents in almost all respects, except for their smaller size and underdeveloped state of their reproductive organs. The young ones and the adults live in the same environment, have identical mouthparts and feeding habits. Though such insects undergo slight changes before reaching the adult stage, they are generally considered to have no metamosphosis and the development is termed as **'ametabolous'**. These insects have no wings either in the adult stage or in the young stages and hence are placed under **'Apterygota'** (e-g) Silver fish **(Appendix 4)**.

A vast majority of insects however, pass through quite well marked changes in their life cycle. In these forms, the immature individuals are mainly of two types namely, **'nymphs'** or **larvae'**. A nymph is a young insect, which differs mainly from the adults in that the wings and genitals are present only in an incomplete state of development, whereas, a larva differs fundamentally in form from the adult and it undergoes drastic changes and attains an immobile, pupal state before emerging as an adult. Based on the manner of changes, insects are grouped into two more main categories.

b) **Heterobolous development**. The insects coming under this category undergo a certain amount of changes before attaining adulthood. Further, these insects develop their wings exogenously and are known as **'Exopterygota'**. This is again sub-divided into two groups :

i) **Paurometabolous development** or **gradual metamorphosis**. In this type, the newly hatched **individual** resembles the adult in general body form, but lacks wings and the genitalia are in an underdeveloped state. The young one is called a **'nymph'** and it undergoes a series of feeding instars, which is followed by a moult every time. After each moult the nymph reaches adulthood step by step. However, the nymphs and adults have similar type of mouthparts, food habits and live in identical habitats. After the final moult, the nymph becomes a full-fledged adult with fully developed wings and genitalia (e-g) grasshoppers, bugs, aphids etc. **(Fig. 47, Appendix 4).**

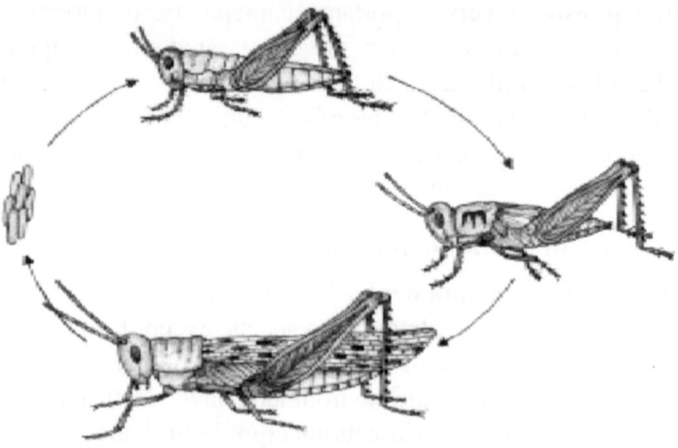

1. Eggs 2. Second stage nymph 3. Fourth stage nymph
4. Full grown adult after the sixth and final stage nymph

Fig. 47. Gradual metamorphosis (e-g) Grasshopper

ii) **Hemimetabolous development** or **Incomplete metamorphosis**. In this type, the changes occurring during the development of the insect is more pronounced than in the case of gradual metamorphosis. The immature stages are called either nymphs or more specifically as **'naiads'**. The naiads are aquatic in nature, while the adults are terrestrial. The naiads breathe by means of tracheal gills and their body structure are variously modified to suit the environment in which they live. They have very slight resemblance to the adults. There is no non-feeding pupal stage. After the final moult adult insect emerges from the naiad. Insects belonging to three orders viz., Odonata (Dragon flies), Ephemeroptera (May flies and Plecoptera (Stone flies) have this type of development **(Fig. 48, Appendix 4).**

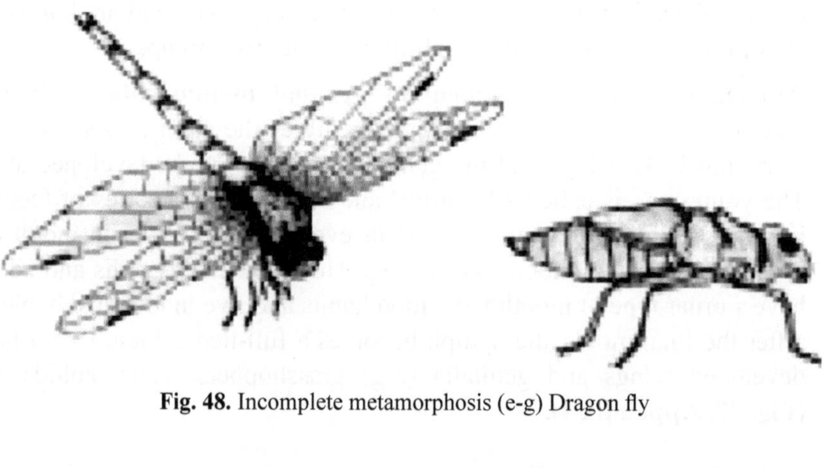

Fig. 48. Incomplete metamorphosis (e-g) Dragon fly

(c) **Holometabolous development** or **Complete metamorphosis**. In this type, the insects undergo drastic changes during their development and hence, it is termed as **'complete metamorphosis'**, which includes four different stages viz., the egg, larva, pupa and adult. The larva differs greatly in form from the adult and passes through a series of growing stages or **'instars'** and moults. In this type , a non-feeding, quiescent stage intervenes between the larval and the adult stages. Further, these insects develop their wings endogenously from the meso- and metathorax of the larvae and the wing formation is completed inside the pupa. The adults energe out from the pupae with full-fledged wings. Insects belonging to the orders Lepidoptera, Coleoptera, Diptera, Hymenoptera etc., undergo this type of metamosphosis **(Fig. 49)**.

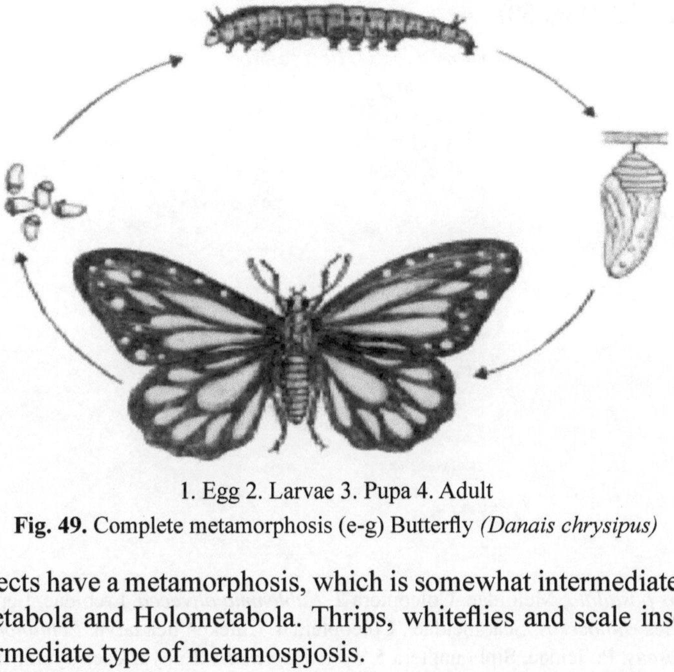

1. Egg 2. Larvae 3. Pupa 4. Adult

Fig. 49. Complete metamorphosis (e-g) Butterfly *(Danais chrysipus)*

Some insects have a metamorphosis, which is somewhat intermediate between Heterometabola and Holometabola. Thrips, whiteflies and scale insects have such intermediate type of metamospjosis.

In thrips, the first two instars are wingless and the young ones are very active and are usually called as **'larvae'**. During the next two instars external wings develop and during these stages, the insects are quite inactive and this stage is termed the **'pupal stage'** and the insect at this stage is called the **'prepupa'**. The final instar is the adult. This type of metamorphosis is called **'complete metamorphosis'**, because wings develop internally in the final instars and the adult stage is preceded by a pupal instar. It is also termed as **'incomplete metamorphosis'** in the sense that the early instars have compound eyes and external wings are present in more than one pre-adult instars.

d) **Hypermetamorphosis**. In some insects, metamorphosis is very complex. In such metamorphosis, there are several types of instars or larvae including a minute, active first instar, a robust or sluggish second instar, followed by a similar but legless instar. The different larval instars are not of the same type. The first instar, active ones are usually campodeiform and the subsequent larval instars are vermiform or scarabeiform. This type of metamorphosis is common in parasitic insects and some others belonging to Meloidae of Coleoptera, Mantispidae of Neuroptera and Strepsiptera.

Types of Larvae

Many tpes of larvae are found in different groups of the holometabolou and hypermetabolous insects. These numerous forms can be grouped into four main categories **(Fig. 50)**.

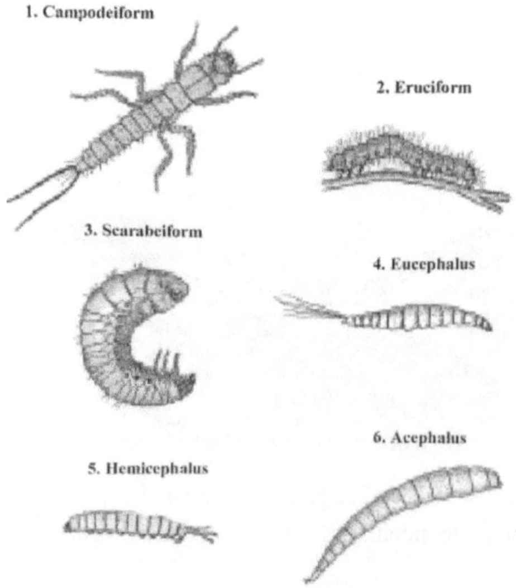

1. *Mylabris postulata*, Meloidae, Coleoptera 2. *Eublemma olivacea*, Erebidae, Lepidoptera 3. *Oryctes rhinoceros*, Scarabeidae,, Coleoptera 4. Chicken flea larva, *Echidnophaga gallinaseus*, Pulicidae, Siphonaptera 5. *Ephydragracilis*, Ephydridae, Cyclorrhapha 6. *Phlebotomus* sp., Psychodidae, Diptera

Fig. 50. Types of larvae of insects

1. **Protopod larvae**. When eggs contain very little yolk, the larvae hatch out, while still in an early embryonic stage. They survive in a highly nutritive medium of the eggs or the bodies of other insects, which serve as their hosts. The larva is just more than a prematurely hatched embryo without any abdominal segmentation, but with rudimentary cephalic and thoracic appendages. This type is found in the primary larvae of some parasitic Hymenoptera.

2. **Polypod larvae**. They are also known as **'Eruciform larvae'** found in Lepidoptera, saw flies and scorpion flies. A well segmented body, small thoracic legs, a number of abdominal prolegs and a tracheal system having a number of spiracles characterize them. The abdominal fleshy legs are present in the third, fourth, fifth, sixth and the last segments. The las pair of prolegs is known as **'claspers'**. The semiloopers have three pairs of abdominal prolegs in the fifth, sixth and tenth segments, while the loopers have only two pairs of prolegs on the sixth and tenth segments. The larvae of Tenthredinidae of Hymenoptera have a maximum of 6 - 9. pairs of prolegs.

3. **Oligopod larvae**. They are characterized by well developed thoracic legs, absence of abdominal prolegs, except a pair of **'cerci'** and usually well developed head capsule and its appendages. Oligopod type of larvae is subdivided into two forms :

i) Campodeiform **larvae**. They are ekongate. More or less fusiform, with depressed body, well sclerotized, prognathous head, long thoracic legs and a pair of cerci. They have no compound eyes or ocelli and are regarded as the most primitive form of larvae (e-g) Larvae of Neuroptera, some Coleoptera, Strepsiptera and Trichoptera.

ii) **Scarabeiform larvae.** They are quite stout, sub-cylindrical, 'C'- shaped with short thoracic legs, soft, fleshy body and without abdominal forelegs (e-g) Some families of Coleoptera.

4. **Apodous larvae**. They are degenerate type of larvae and are common in weevils, wasps, bees and flies. In this type, the body appendages are completely suppressed, legs totally absent and body poorly sclerotized. The apodous larvae are sub divided into three types depending upon the degree of development and sclerotization of the head capsule.

i) **Eucephalus larvae**. These larvae have a well developed, sclerotized head capsule with relatively little reduction of the cephalic appendages (e-g) Most Nematocera, Cerambicidae and Buprestidae.

ii) **Hemicephalus larvae**. These larvae show appreciable reduction of size of the head capsule and its appendages, the head being retracted into the thorax (e-g) Brachycera (Diptera) and Tipulidae.

iii) **Acephalus larvae**. These larvae have no obvious head capsule and appendages. Such larvae are also known as **'maggots'** and are common in flies (e-g) Cyclorraphon, Diptera.

Generally larvae of Lepidoptera are called as **'caterpillars'**, larvae of Coleoptera and Hymenoptera as **'grubs'** and that of Diptera as **'maggots'**.

The transformation of immature insects into reproducing adults is under endocrine control. A hormonal secretion from the brain stimulates a gland in the prothorax, which produces **'ecdysone'**, a hormone that stimulates growth and moulting. During the larval stages another hormone, the **'juvenile hormone'** is secreted by the **'corpora allata'** of the brain. This hormone is responsible for the maintenance of the larval structure and inhibits metamorphosis. The juvenile hormone can exert its effect only after the moulting process has been initiated. It thus must act in conjunction with he prothoracic hormone. When a relatively high level of juvenile hormone is present in the blood, the result is a larva to larval moult. When the level of juvenile hormone is lower, the moult is larva to pupa and in the absence of juvenile hormone, there is a pupa to adult moult.

The larva is the growing and feeding stage, generally covering the largest period in the life cycle of an insect. Usually the young larvae take care of themselves, as the eggs are laid near the food plant or animal host, but in social insects such as, termites, ants and bees, they are looked after by the workers of the colony till they are fully grown.

In the case of some flies, the larvae hatch within the body of the females and the larvae are deposited on meat or in wounds of animals etc., where they fed on host tissues.

The larvae are mostly covered with hairs or setae. Some larvae also have spines, which are branched. In many butterflies and moths, the hairs are of stinging type.

The larvae, which live in seclusion and darkness such as, in the soil or those, which burrow into the stems or branches or mine the leaves are usually white or pale yellow in color, while the free living ones are strongly pigmented and have a variety of colors. Cryptic colors simulating the background or food are common, but in some butterflies the colors are very dense. The latter are generally unpalatable to their enemies and are said to have warning coloration.

The pre-pupa or pharate pupa

It is a quiescent stage between the larval and pupal period. This stage is characterized by non-feeding accompanied by quiescence. It is usually in this period that the cocoon is produced and among dipterous flies the puparium is formed. The larval cuticle gets separated from the hypodermis and the developing pupa with everted wings and appendages lies within the persistent larval cuticle and differs somewhat in appearance from the earlier stage of the larva; its body is depressed and the abdomen relatively shorter. It is however, not a distinct instar. When the old larval skin ruptures, the pupa comes out. In male coccids and thrips, the first instar in which external wing pads appear is sometimes termed pre-pupa. This intermediary stage may last for 2 - 3 days or even less.

The Pupa

The pupa is a resting, dormant instar in all insects with complete metamorphosis. During this stage, it is incapable of feeding and the metabolic activity is kept at minimum level. But it is regarded as a transitional phase during which the larval body and its internal organs are restructured and transformed to meet the adaptational requirements of the imago. One of the most important facts in pupation is in permitting the full development of the wings. The internal development of the wings within the larva is restricted by lack of space. The pupal stage provides sufficient space for the development of wings and the flight muscles. Trichopteran pupae have adaptive modifications , which enable them to swim to the surface of water to allow the emergence of the adults. Less pronounced movements are noticed in the pupae of many insects, which occur in the soil, inside the stem of plants etc.

In the case of females of Strepsiptera and Coccidae, there is no pupal instar, whereas in the males there is a pupal instar. This is because of the fact that the females are wingless, while the males have wings and the pupal instar provides the males the chance for the development of wings.

There is generally a change in the color of the pupa, sometimes before the imago is ready for emergence or **'eclosion'**. During emergence, the insect ruptures the pupal cuticle by means of convulsive movements of its legs and body. A longitudinal slit occurs down the back of the thorax and often there are other small fractures in the region of the legs etc. The insect withdraws its appendages from within that of the pupa or the last nymphal instar and emerges fully formed, except for the wings, which yet resemble a pair of pads. It rests on some kind of support, tree trunk, branches, leaves etc. and in due course the miniature wings expand into their final form by the inflow of blood and pressure exerted by muscular action. In some aquatic insects, the adults is able to fly off almost immediately after eclosion.

The pupal instar is vulnerable to natural enemies, because of their inability to move. So, special methods for their protection are provided. Most pupae are concealed in some way or other from their enemies and from adverse environmental conditions. Many Lepidopterous and Coleopterous larvae construct earthen cells below the ground, in which they pupate. Many others pupate deep inside stems and fruits of plants. A large number of insects construct cocoons either of silken thread alone or of extraneous materials bound together by means of silken thread. Some pupae are covered with body hairs or spines of the larvae for protection.

Types of pupae. The following are the important types of pupae :

i) **Pupa libera, decticous, exarate** or **free pupa**. This type of pupae is found in most beetles. All the appendages are free and thus resemble somewhat the adult condition.

ii) **Pupa adecticous** or **complex**. In this type, the appendages remain in separate sheaths and only partly free or they may adhere to the body and the abdomen is capable of movements (e-g) Mosquito pupae or **'tumblers'**, pupae of most Lepidoptera and Diptera. This group may again be sub divided into two categories :

a) **Pupa obtecta**. Here the appendages are tightly oppressed to the body and are covered by an external skin. Such pupae also known as **'chrysalid'** have a posterior process or **'cremaster'**, which may be furnished with special hooks or anchorage as in some Lepidoptera.

b) **Pupa coarctata**. In this type, the pupa is invisible, being enclosed in the larval skin or **'puparium',** which is functionally comparable to the cocoon or pupal case of other forms. The adult emerges from puparium through a circular rupture near one end by means of special structures. This is found in some Dipterous insects.

The Adult or Imago

After crossing the various stages of metamorphosis, the adult stage is reached, which is the termination of metamorphosis. In the case of hemimetabolous insects, such transformation of external and internal structures occurs gradually. During this process, the nymphal tissues of both the external and internal structures and organs are transformed into the corresponding tissues of the adult forms. In he case of holometabolous insects, the larval tissues are completely destroyed and from the new regenerative cells formed in the place of larval tissues, adult tissues are generated. This type of transformation, in which the larval tissues are completely destroyed is called **'histolysis'**. The new generative cells, called **'histoblasts'** or **'imaginal buds'** that are produced following histolysis generate the adult tissues.

The epidermis is the most important cellular structure that covers the entire body surface of the adult insect and as such it is quite vital that the adult possess the right type of epidermis. So, in the pupal stage drastic changes take place in the formation of the adult epidermis. In the case of holometabolous insects, the epidermis undergoes further changes to form scales and hairs, while in others spikes, spines and spurs are formed. The epidermis also helps in the arrangement and formation of the spiracles and tracheal tubes inside the body. It also helps in the formation and attachment of the various muscles that assist the physiological functioning of the various organs. Epidermis also plays an important role in the segmentation of the body.

The gut of the larval stage remains unchanged and retained as such in the adult stage also. But the entire set of malpighian tubules present in the larval stage is transformed in the pupa without undergoing histolysis.

The glands in the larval stage undergo histolysis and are replaced by a new pair of salivary glands of the adult form in the pupal stage.

The different types of metamorphosis met with in insects is diagrammatically presented in **Appendix-iv**.

5

Classification

Classification of Living Organisms

The classification of living organisms has passed through several changes and has reached the present stage with the gradual assimilation of more and more knowledge about them. The classification is mainly based on the morphology, anatomy, life style, habits, habitats, metamorphosis and such other characteristics. Though the concept of classification of living organisms was perhaps started by **Aristotle** between 384 - 322 BC, it was the Swedish Scientist **Carl Linnaeus** (1788), who presented the classification, which was acceptable to most of the Taxonomists. He propounded the binomial nomenclature of naming each and every organism of animal and plant origin with a genetic and species name. In his classification he was consistent in using Latin names for all the organisms. He was the originator of a **regular system of classification** in which he recognized larger and smaller categories, the larger ones comprising of a varying number of smaller ones and he is rightly called the **'Father of Nomenclature'**. The system was later modified to some extent by **A.D. Imms** in 1861. However, the foundations of the modern system of classification of insects was laid by **Brauer** in 1885. He was the first one to emphasize the fundamental division into primitive wingless insects, the Apterygota and the winged or secondarily wingless insects, the Pterygota. The latest code on classification and nomenclature was approved and published by the **'International Trust for zoological Nomenclature, London'** in 1964.

In the modern system of classification and nomenclature, each living organism is placed under a Phylum, Sub Phylum, Class, Order, Family, Genera and Species according to its allied characters and separate names assigned. The Scientific name thus given to a particular organism denotes its genetic and species name. In each genera several organisms may be present, but in a species only one particular organism is present. Based on the classification of an organism, many of its characteristics can be understood.

The system of classification is of great importance in the systematic study of the organism, to distinguish the differences or similarities between organisms, to identify, name and study newly found organisms, to intensify our knowledge about harmful and beneficial organisms and to adopt suitable measures to control harmful organisms. Among the various living organisms, insects are classified under the class - Insecta.

Living organisms are broadly classified into three **'Domains'** viz., **'Eukarya'**, **'Bacteria'** and **'Archaea'**. Domain 'Eukarya' includes plants, animals, protists and fungi. Domain 'Bacteria' and 'Archaea' include single-celled microorganisms. The organisms under the domains are further classified into six **'Kingdoms'** viz., **'Eubacteria'** (domain Bacteria) - common bacteria, **'Archaebacteria'** (domain Archaea) - uncommon bacteria that live in environments that are devoid of oxygen or are extremely acidic, **'Plantae'** (domain Eukarya) - all plants, **'Animalia'** (domain Eukarya) - all animals including humans, **'Fungi'** (domain Eukarya) - spore producing fungal organisms, **'Protists'** (domain Eukarya) - microorganisms such as, algae, slime moulds etc. The organisms under Kingdoms are further classified into **'Phyla'**. There are multiple phyla under the Kingdoms. The four most important Phyla are **'Arthophyta'** (kingdom Plantae) - flowering plants, **'Arthropoda'** (kingdom Animalia) - invertebrate animals with an exoskeleton, segmented bodies, jointed limbs and a central nervous system, **'Chordata'** (kingdom Animalia) - invertebrates having spinal cord, **'Coniferophyta'** (kingdom Plantae) - plants that stay green year round such as, juniper, fir trees. The organisms under the Phyla are further classified into **'Classes'**.

Classification of Phylum - Arthropoda

Phylum - Arthropoda is divided into seven Classes viz., Onychophora, Crustacea, Diplopoda, Chilopoda, Arachnida, Insecta (Hexapoda) and Triobita. The members of the Class - Triobita have become extinct.

Class - Onychophora *(Peripatus* sp.*)*. Members of this class possess a distinct head and body. The body is many segmented and each segment has one pair of legs. The head bears one pair of antennae. They are terrestrial in habit. Most of the members of this class are extinct today, except for a few species of the Genus - *Peripatus*, which are found occasionally in the remote forest areas of Australia, Africa and South America. *Peripatus* is considered as an Annelid ancestor of the present day Arthropods, as they possess characters of both Annelids viz., smooth skin with a muscular body devoid of chitin, as well as Arthropods viz., segmented appendages.

Class - Crustacea (crabs, shrimps, lobsters, oysters, prawns, cray fish, sow bugs etc.). The class consists of over 30,000 species. Members of this class are characterized by the presence of a cephalothorax and abdomen. The hardened exoskeleton is composed of calcium in addition to chitin. Legs may be numerous, usually five pairs. They have two pairs of antennae, a pair of mandibles and two pairs of maxillae for food manipulation. Respiration is by means of gills or trachea. They are the most diverse among the Arthropods and are either marine, fresh water or terrestrial in habitat.

Class - Diplopoda (Millipedes). Members of this class are terrestrial and have a distinct head and segmented body with 25 - 100 segments. They have a soft, cylindrical body and a pair of antennae. There are two pairs of legs in each body segment. Respiration is by means of trachea. They possess high degree of thigmotropic response and with a slight touch they coil round and lie motionless. Some millipedes release a liquid secretion, which gives off a pungent odor. Millipedes live in moist areas and are generally herbivorous, feeding on living plants or saprophagous, living on dead, decaying plants.

Class - Chilopoda (Centipedes). These terrestrial insects have a distinct head and segmented body with 20 or more segments, which is long, slender, flattened dorso-ventrally and hard in structure. They possess a pair of antennae. There is a single pair of legs in each body segment. Respiration is by means of trachea. The appendages of the first body segment is modified to form large poison fangs by which poison is injected by stinging. They are active predators and are found under logs, stones, debris and such hiding places.

Class - Arachnida (Mites, ticks, scorpions, spiders etc.). The members of this class are all terrestrial in habit. The body is divided into a cephalothorax and abdomen. They usually possess four pairs of legs. However, the tetranychid mites have only two pairs of legs. They do not have antennae. They have biting and chewing or piercing and sucking type of mouth parts. Respiration is by means of book-lungs.

The members of this class are of much economic importance as most of them are non-insect pests of plants and animals, while some such as, a few species of spiders are beneficial and prey on crop pests. The mites are phytophagous. They pierce, suck and feed on the sap from plant parts and cause severe damage to the plants. The ticks are ectoparasites on higher animals and birds. They suck and feed on the blood of their hosts. Further, these organisms serve as vectors of some of the most dreaded diseases.

Class - Insecta or **Hexapoda** (Cockroaches, grasshoppers, butterflies and moths, bees, flies, mosquitoes, lice, beetles and a host of similar small creatures). The name **'Insecta'** is derived from the Latin word **'incised'**, which means the body is incised or divided into three distinct parts viz., head, thorax and abdomen. Class - Insecta is also called Hexapoda. The name **'Hexapoda'** is derived from two Latin words viz., **'Hexa'** = six and **'Poda'** = foot, which means organisms possessing six foot.

The organisms classified under this class are generally known as **'insects'**. Insects are the most abundant, diverse and widespread of all land creatures that can live in dry environment and are the only invertebrates capable of flying.

Excretion is performed by fine malpighian tubules attached to the anterior end of the hindgut. Respiration is by branched, cuticle-lined tracheae that carry oxygen directly to the tissues from paired apertures viz., the spiracles or stigmata, which are located on the lateral sides of the thorax and abdomen. During the process of respiration, they take in oxygen from the air and exhale carbon dioxide.

Another important characteristic of an insect is that its inner body cavity, which is called haemocoel is filled with blood and all internal organs are virtually found floating in the blood. The insect blood lacks haemoglobin and hence it is yellow or greenish in color and not red colored. The primary nervous system lies beneath the alimentary system like a chain and the tubular circulatory system lies above the alimentary system just below the notum.

The respiratory and circulatory systems of insects are so efficient that the supply of oxygen to different organs and their supply of nutrition is rapid and complete. It is for these reasons that insects are so active, develop rapidly and possess rapid power of multiplication.

The classification of insects is presented in Appendix 2 and 3.

Class - Insecta

Sub class - Apterygota or Ametabola

The insects classified under this sub class are the ones, which have no wings and these insects do not undergo any type of metamorphosis and are considered to be most primitive.

Order 1 - **Protura** (Proturans)

('Prot' = first; 'ura' = tail)

The insects belonging to this order are minute, whitish and eyeless with a cone-shaped head. They live in damp, humus soil and beneath bark of trees. These wingless, eyeless insects with only rudimentary antennae have a habit of walking with the forelegs held upward in front of the head. These upraised forelegs apparently serve as sensory organs. They have biting and chewing type of mouth parts. The abdomen is very long and slender. The number of abdominal segments increases after emergence from the eggs. The legs have one-segmented tarsi. Anal cerci are absent and instead, a simple telson is present (e-g) Proturan.

Order 2 - **Thysanura** (Silver fish and Bristle tails)

('Thysano' = fringed; 'ura' = tail)

This order includes wingless insects and they are found throughout the world. These fast running insects are commonly found behind pictures and calendars hanging on the walls and inside books and bindings, where they destroy the paper. Some of them are also found in soil, rotting wood and dead leaf deposits and under stones. Some of them live in the nests of ants and termites. The most common species found in houses is *Lepisma saccharina.* The adults are about 2.0 cm. in length, grayish or whitish in color and scaled forms are metallic and shining. The body is spindle-shaped and compressed dorso-ventrally. The head is small, with prominent compound eyes. Ocelli are also present. The antennae are long and filiform. They have biting and chewing type of mouth parts. The legs are of walking type and the tarsi have 2 - 4 segments. The abdomen is long and has 11 segments. The eleventh abdominal segment has a pair of cerci laterally and its tergum is prolonged into a median caudal filament. There are many instars and some of them moult throughout their lives. The young insects resemble the adult, but do not have scales, cerci and external genitalia.

The order is divided into two families

i) **Family - Lepismatidae**

(e-g) Silver fish *(Lepisma saccharina)* **(Fig. 51)**

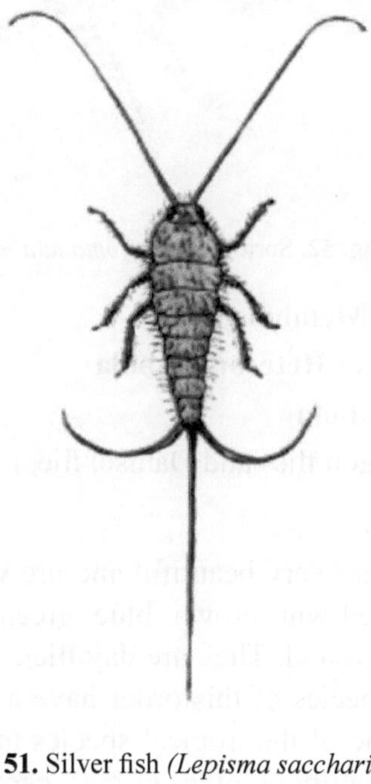

Fig. 51. Silver fish *(Lepisma saccharina)*

ii) **Family - Machilidae**

(e-g) Bristle tails *(Machilis* sp.*)*

Order 3 - Collembola

('Col' = glue; 'embola' = wedge)

The insects of this order are small, less than 0.5 cm. long, wingless and primitive ones and are found under damp dead leaves, bark of trees, moist and dark habitats and are distributed throughout the world. They often feed on pollen grains, fungal spores and mycelium. The body is elongate and slender or short and hump-backed and covered with fine hairs or scales. Usually they are white or gray, but some are purple, yellow or green in color. The antennae are four segmented and rather short. The tarsi are missing and the tibia end in a pair of claws. Compound eyes are absent and simple eyes are present. The abdomen has only six segments. The fourth abdominal segment has a springing organ, which helps the insect to jump suddenly. A few species cause damage to green house seedlings, potted plants and mushrooms (e-g) Spring tails *(Isotoma minos)* **(Fig. 52)**.

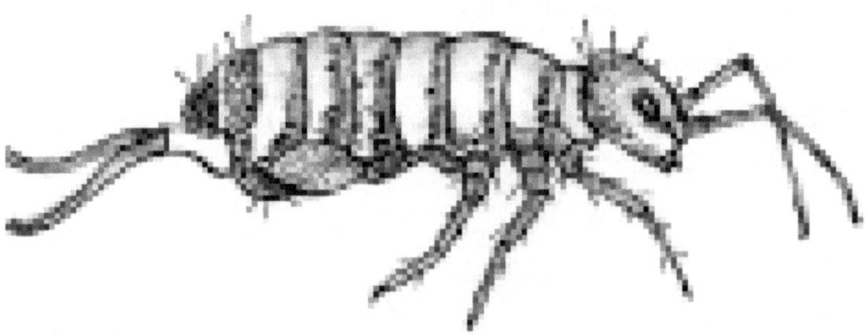

Fig. 52. Spring tail *(Isotoma minos)*

Sub class - Pterygota or **Metabola**

Division - Exopterygota or **Heterometabola**

Sub division - Hemimetabola

Order 1 - Odonata (Dragon flies and Damsel flies)

('Odonto' = teeth)

The adults of this order are very beautiful and are very useful insects. Their bodies are brightly marked with brown, blue, green, red, yellow and shades of these on a black background. They are day fliers and can fly at a speed of 95 km. per hour. Most species of this order have a wing span ranging from 5.0 - 10.0 cm, while some of the tropical species measure 19.0 cm. or even more. The adults are predatory on other insects, capturing the prey during its flight or while it is resting and holding with the forwardly directed legs and transfer it into the mouth.

The spherical or broad head, which is attached to the thorax by a very small and slender neck is very mobile and it has a wide range of vision. The compound eyes are very greatly enlarged. Three ocelli are also present. The antennae are very short, quite inconspicuous, 3 - 7 segmented and filiform. The mouth parts are biting and masticating type and the mandibles are powerful with strong teeth. The body is elongate and the thorax very short in proportion to that of the abdomen. The prothorax is greatly reduced, while the meso- and metathorax are fused together. The legs, which lie close behind the mouth, are long and slender and fitted with long spines. They are unfit for walking, but are modified to capture and hold the prey while eating. They form a basket with their legs and scoop up their prey in flight. The two pairs of wings are narrow, many times longer than broad, almost identical, membranous and net-veined. The wings may be transparent or may have different colored markings. A dark spot is present at the anterior margin of the wings between the 'C' and 'R' veins, called **'stigma'**. The muscular system of the thorax is functionally so modified to enable the adults to fly swiftly and skillfully. The abdomen is elongate, slender, tubular and made up of 10 segments and terminated by short cerci. The copulatory organs of the male are found on the second and third abdominal sterna and the genital aperture of the female opens on the ninth segment.

Metamorphosis is incomplete. The mature males are found hovering along stretches of water and mating often takes place in flight. While mating, the female is grasped firmly by the anal appendages of the male. The adult female drops the eggs in water. The aquatic nymphs of the dragon fly is called **'naiads'**. The labium of the naiad, which is modified as a mask conceals the mouthparts. The legs of the naiad are considerably longer than that of the imago. The naiads are predatory on insects and other organisms found in water. The naiads breath with the help of gills.

The order is divided into three suborders :

Suborder (i) - Anisoptera (Dragon flies)

The wings are not petiolate and are slightly dissimilar in form and venation. They are held horizontally or depressed in repose. The adults are strong fliers. Eggs are usually laid on water or on the surface of aquatic plants. Ovipositor is reduced or vestigial. Nymphs have rectal gills. The abdomen of nymphs is stout and robust (e-g) Dragon flies **(Fig. 53)**.

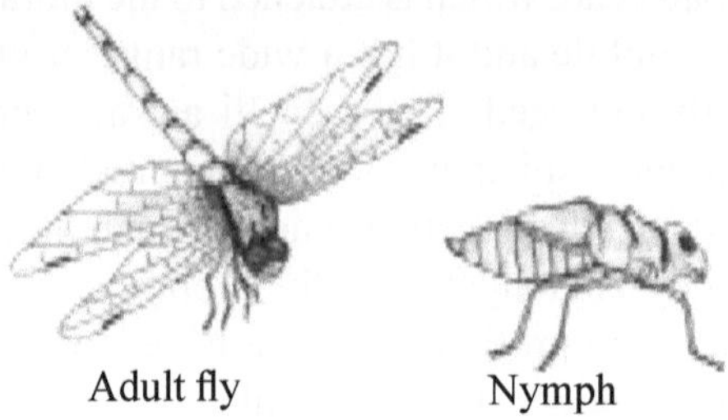

Fig. 53. Dragon fly

Suborder (ii) - Zygoptera (Damsel flies)

The wings of adult insects are more or less petiolate and are similar in form and venation. They are held vertically above the abdomen in repose. The adults are week fliers. Eggs are inserted into succulent stems of aquatic plants. Ovipositor is well developed. Nymphs respire by means of three caudal gills. The abdomen of nymphs is slender and elongate (e-g) Damsel flies **(Fig. 54).**

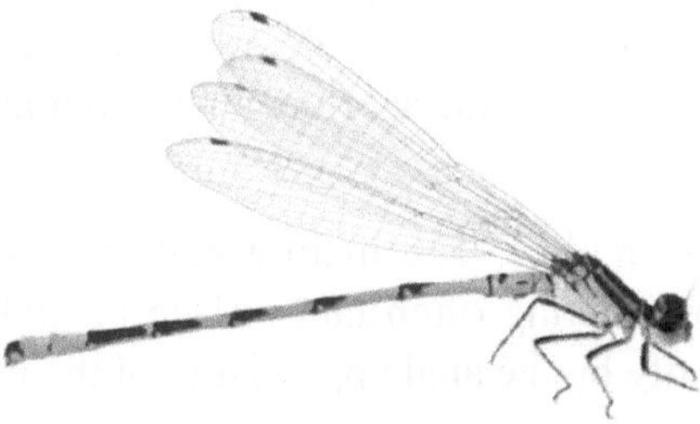

Fig. 54. Damsel fly

Suborder (iii) - Anisozygoptera

The insects belonging to this suborder are almost extinct

Order 2 - Ephemeroptera (May flies, Day flies etc.)

('Ephemero' = living for a day / short lived; 'ptero' = wings)

The insects of this order are the most primitive of the winged insects. The may flies, which belong to this order are shining, elongate and soft-bodied insects, generally seen hovering over the brinks of ponds, lakes, streams, rivers etc. The duration of life of an adult may be a few hours only, but that of the nymphal stages may be as long as three years. May flies form important source of food for fishes. The head of the adult is free with short antennae. Compound eyes are large and sometimes mounted on pillar-like outgrowths. The mouthparts basically of biting type are degenerate, the mandibles being vestigial or absent and the maxillae generally represented by the palpi only. The meso thorax is well developed and the others are insignificant. The forewings are membranous, fragile and generally triangular in shape and are held upright when at rest. The hind wings are small and rounded, but in some species they are completely degenerate. The wings are mostly transparent, but some have white or brown markings. The abdomen has 11 segments, but the tenth and eleventh segments are fused together. The abdomen is terminated by very long cerci with or without a median caudal prolongation. The alimentary canal does not function as a digestive tract, as the may flies do not feed during the adult stage, but is modified for respiratory purpose, the air being taken through the mouth and stored in the stomach, which is in the form of a balloon.

Copulation takes place during flight and it lasts only for a few minutes. The nymphal stages or naiads of the insect are passed in the water. The nymphs of Most of the species breath by tracheal gills. The nymphs are generally campodeiform and feed on aquatic vegetation, chiefly algae. There are many moults during the nymphal stage, which lasts for over three years. The fully-grown nymphs leave the water usually in large numbers by about sunset. Metamorphosis is incomplete and the may flies are unique in having a sub-adult stage From the fully developed nymph, the sub-adult, which is also winged emerges and flies. This stage may last for only a few hours. Later, it again moults and becomes the adult. Swarming may fly adults are eaten by practically all insectivorous birds, bats, dragon flies and leaping fishes (e-g) May flies *(Ephemera vulgata)* **(Fig. 55)**.

Fig. 55. May fly *(Ephemera vulgata)*

Order 3 - Plecoptera (Stone flies)

('Pleco' = folded or plaited; 'ptera' = wings)

The nymphal stage of Plecoptera, which is represented chiefly by *Perlaria* (stone flies) is invariably passed in water. Stone fly adults are soft-bodied, elongate, flattened insects, with long and thread-like antennae. They are poor fliers and are found resting on stones, tree trunks etc. near edges of water collections Some forms, which are generally green in color are found on grasses. Large species are often used as bait for fishing.

The mouthparts are of mandibulate type and week in structure. Tarsi are three segmented. The antennae are long and setaceous. Compound eyes are well developed. The fore- and hind wings are membranous and the latter are generally longer. Wings are many-veined and held flat over the abdomen when at rest. Wing coupling apparatus is lacking. Wing venation is variable. The abdomen is 10 segmented and the eleventh segment is vestigial. There is no ovipositor or male external genitalia. A pair of long segmented cerci is present.

The naiads are generally found under stones in clear water of running streams. They resemble the adults, except for the absence of wings. They have long cerci and antennae. Respiration is cutaneous or through gills. The nymphal stage lasts for 1 – 4 years. Metamorphosis is incomplete.

(e-g) Stone flies *(Perla* species*)* **(Fig. 56)**.

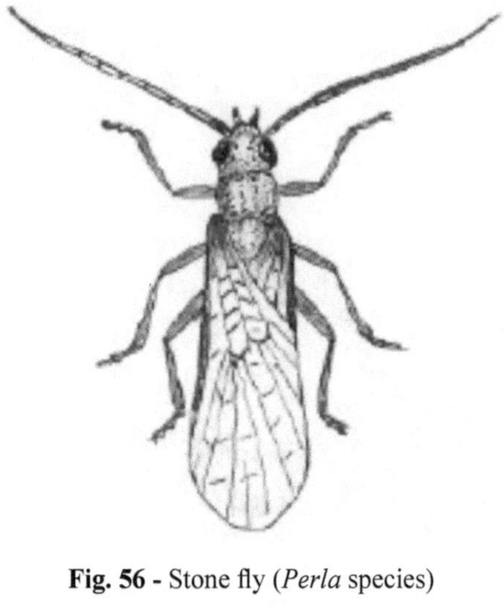

Fig. 56 - Stone fly (*Perla* species)

Sub class - Pterygota or Metabola

Division - Exopterygota or Heterometabola

Sub division - Paurometabola

Order 1 - Dictyoptera (Cockroaches and Mantids)

('Dictyon' = net work; 'ptera' = wings)

The insects of this order are mostly terrestrial forms occurring predominantly in tropical and subtropical regions. They are not good fliers and the wings of many species are reduced or absent, especially in the females. Cockroaches are common in kitchen, bathrooms, bakeries and restaurants and may be carriers of disease organisms. They are omnivorous and mostly nocturnal, hiding in crevices, cupboards, between walls etc. during the daytime and coming out at night to feed on unprotected food and garbage. The praying mantis, which holds up the frontal body with its large front legs in a semblance of prayer, is carnivorous in habit. The front legs are modified in such a way, that they can grasp small insects between the spiny tibia and femora and crush them to death. They are very deceptive and merge with the surroundings.

The head is usually hypognathous and the mouthparts are mandibulate. The hypopharynx is large. The antennae are filiform and many segmented. The compound eyes are usually well developed. The pronotum is large, usually overlapping the neck and head in cockroaches. Walking legs are present in cockroaches and the tarsi are five-segmented. The front pair of wings is sclerotized as the **'tegmina'** and has marginal costal vein. The hind wings are folded in a fan-like manner. The wing pads do not undergo reversal during development. The male genitalia are concealed by the elongated ninth abdominal sternum, which mostly bears a pair of styles. The ovipositor is reduced and concealed under the seventh abdominal sternum. A pair of many segmented cerci is present. The mid gut bears eight tubular enteric caecae and up to about 100 malpighian tubules are found. The salivary glands are large and have conspicuous reservoirs, especially in cockroaches. The insects undergo simple metamorphosis.

The order is divided into two suborders viz., **'Blattaria'** and **'Mantodea'**.

Suborder - Blattaria. The insects belonging to this suborder are mostly nocturnal. The head is rather small and immovable and nearly or almost covered by the large shield-like pronotum from above. Two ocelli are present. The body is oval in shape and flattened dorso-ventrally. Legs are adapted for walking and forelegs not modified. The insects of this suborder are mostly omnivorous. The females do not prey on the males during mating. There is only one family - Blattidae, which is of economic importance.

Family - Blattidae (Cockroaches) The insects of this family measure 1.0 - 5.0 cm in length. Their bodies are oval, flattened and usually brown or black in color. The head is usually partly covered by the pronotum of the thorax. Their eyes are large and the antennae long and filiform. They have well developed biting and chewing type of mouthparts. The abdomen has short apical cerci. The eggs are laid in the form of a capsule or **'ootheca'**, a hard brown structure, which contains a double row of cigar-shaped eggs. The egg capsule in some species is carried in the oviduct almost until hatching. The life history is long and it takes almost a year for the young nymphs to reach adulthood. They generally contaminate food with their excrement, leave an unpleasant odor and are considered as serious house hold pests (e-g) Cockroach - *Periplaneta americana* (Fig. 57); German cockroach - *Blattella germanica*

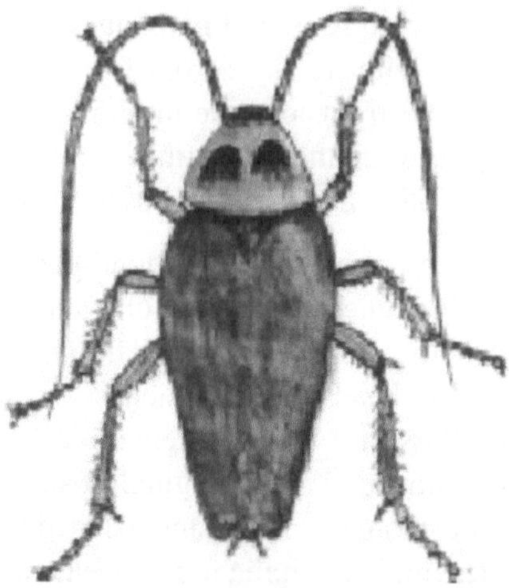

Fig. 57. Cockroach *(Periplaneta americana)*

Suborder - Mantodea. The insects of this suborder are mostly predaceous and feed on a variety of insects. The head is highly movable and is not covered by the pronotum. Three ocelli are usually present. Forelegs are raptorial. The insects are large in size, elongate, have long prothorax, slow moving and have peculiar appearance due to their modified prolegs. The females most often devour the male during mating. There is only one family – Mantidae, which is economically important.

Family - Mantidae (The mantids). These insects are usually large, some 12.5 - 15.0 cm. long. The body is elongate, cylindrical and slightly flattened. The head is large and triangular, with large compound eyes. They are the only insects that can look over their shoulders. The antennae are filiform and the pronotum is elongated. The abdomen has short cerci. The wings are similar to those of grasshoppers. The forelegs are raptorial, the tibia and femora are enlarged, prominently spiny and modified for grasping the prey, while the other two pairs of legs are slender. They are cryptic-colored, many are stick-like and are colored in tones of brown and black, resembling a dry stick. Those living in green vegetation are green - colored. They usually lie in wait for their prey with the forelegs in an upright position in a semblance of prayer. Hence they are generally called as **'Praying mantis'** or **'Soothsayer'**.

The eggs are laid in a characteristic large egg case, which is watertight and fixed to a plant. The case is made up of a frothy gum secreted by the female, which hardens when exposed to air. Within the case there are egg chambers arranged four abreast.The youngest nymphs resemble ants and the life cycle in completed in about one year. In some species, the females attack the males during or after copulation (e-g) Praying mantis - *Mantis religiosa* **(Fig. 58)**.

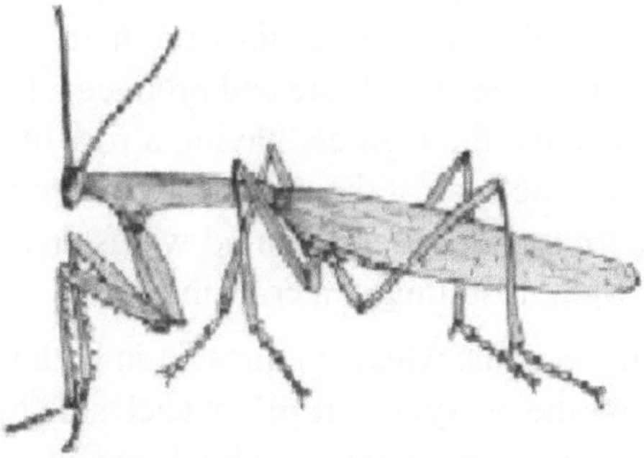

Fig. 58. Praying mantis *(Mantis religiosa)*

Order 2 - Orthoptera (Grasshoppers, locusts, crickets)

('Ortho' = straight; 'ptera' = wings)

The insects of this order flourish best in the tropical and subtropical areas of the world. They are mostly terrestrial. Though generally they jump quickly, only the locusts are well known for their long, intercontinental flights, but not at a stretch. Nymphs and adults are usually phytophagous, while a few are carnivorous. The order includes many notorious pests of crops.

The insects included in this order are diverse in structure, appearance and habits, but all have well developed biting and chewing type of mouthparts. The antennae are generally elongated and filiform. Most of them have a well developed prothorax and two pairs of wings, the anterior pair being thickened and leathery in texture to form the **'tegmina'** with sub - marginal, costal vein. The wings overlap when folded. The hind wings are membranous, larger than the front pair and folded in plaits under the forewings. In some groups, the wings are small and half-grown or completely lacking. The front and middle legs are slender and sometimes the front legs are fitted for digging. The hind legs usually have an enlarged femur adapted for jumping and the tarsi are generally 3 - 4 segmented. The abdomen is usually short and stout. Ovipositor is generally well developed and exposed. Male genitalia are retracted within the ninth sternum. Cerci are usually short. Metamorphosis is simple. The nymphs resemble the adults, but have only wing pads instead of wings.

An important set of structures in the Orthoptera is the presence of stridulatory (sound producing) organs, which seem to bring the sexes together. Generally only the males are capable of producing sound, but the auditory organs are present in both the sexes. The sound producing organs are of two types viz., **'Alary'** and **'Femoro-alary'**. The former is best developed in the families, Gryllidae and Tettigonidae, where stridulation is effected by the friction of tegmina. The femoro-alary organs are best developed in the in the family - Acrididae. Here the stridulation takes place in several ways (a) by means of a ridge bearing a row of many small peg-like projections on the inner side of each hind femur, which is rubbed against the hardened radial vein of the tegmen, thus causing the latter to vibrate and produce a low, hissing sound (b) by rubbing the legs against the tegmen, having a row of peg-like projections on the second vein near the base of the tegmen and a simple longitudinal ridge on the femur (c) by friction between the hind wings and the under surface of the tegmina during flight, resulting in a cracking sound.

The auditory (hearing) organs, which are present in both the sexes, consists of a thin cuticular membrane or **'tympanum'** on each side of the first abdominal segment as in Acrididae or near the proximal end of each fore tibia as in Grillidae and Tettigonidae. Generally, the tympana are covered with a fold of skin under which they repose. The vibrations are transmitted to the sensory centers by means of **'scolopidia'**, which are connected with the nerve endings.

Grasshoppers and locusts lay eggs in cylindrical pods, while those belonging to Gryllidae, Tettigonidae etc. lay eggs separately in loose clusters.

The order is divided into two suborders viz., **Encifera** and **Caelifera**.

Suborder - Ensifera. The insects belonging to this suborder are medium or large-sized and are mostly phytophagous. The antennae are as long or longer than the body and consist of many segments. The tympanal organs when present are situated on the fore tibia. Stridulatory organs when present are located on the tegmina. Ovipositor more or lass elongate.

The more important families under this suborder are :

i) **Family - Tettigonidae** (Long-horned grasshoppers). The antennae are as long or longer than the body and many segmented. The prothorax is large and distinct. The tympanal organs are situated on the front tibiae. The tarsi are generally four segmented. The abdomen is large and fleshy. The females have a well developed ovipositor. Usually the males produce sound by stridulation. The Tettigonids are generally large insects, some more than 5.0 cm. in length , but not very robust. Their color generally blends with that of grass or vegetation on which they are found.

The eggs are flattened in shape and are laid in slits made with the ovipositor at the margins of leaves, stems of grasses, bark of trees and sometimes free on the soil. The Tettigonids are generally nocturnal in habit and while many are herbivorous, some are omnivorous and predaceous, if vegetable food is not available. They are Common during the rainy seasons. Very few Tettigonids are pests. (e-g) Long-horned grasshopper - *Mecopoda elongata* **(Fig. 59)**; *Holochlora indica.*

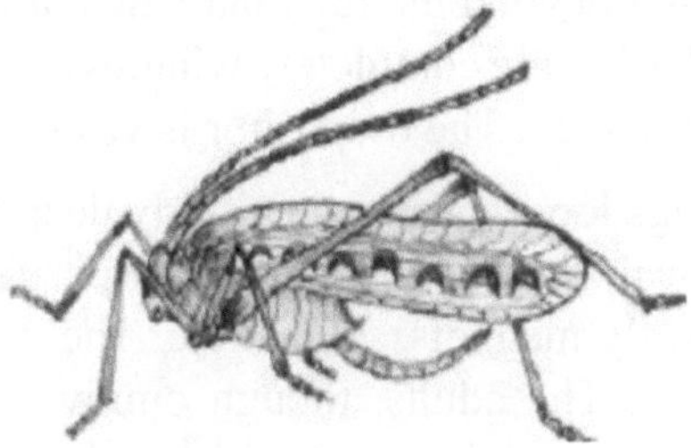

Fig. 59. Long-horned grasshopper *(Mecopoda elongata)*

ii) **Family - Gryllidae** (Crickets). Crickets vary in size from 1.5 - 5.0 cm. in length. They are dull brown or black in color, mostly cryptic. They prefer hot, dry places or live in burrows or under logs, among dry leaves etc. Some live on green plants. The head is large and the prothorax distinct. The hind wings extent back and appear as processes. They have long antenna, longer than the body and an elongated ovipositor, with two cerci on either side. They stridulate by friction of the tegmina and possess tibial auditory organs. Many crickets are entirely wingless. Ovipositor is slender, cylindrical, long and needle-like. The eggs are generally laid singly in the ground, occasionally in masses in underground chambers or rarely in a row in the pith of twigs. Crickets are omnivorous, feeding on living, dead or dry organic matter. Therefore they are both household and field pests. (e-g) Cricket - *Gryllus domesticus* **(Fig. 60)**; *Gryllodes melanocephalus.*

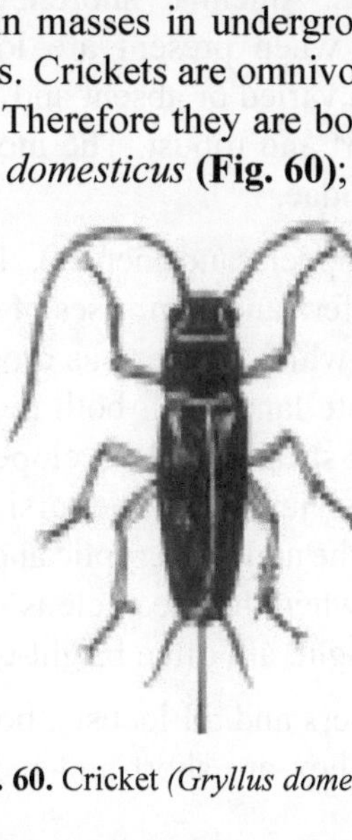

Fig. 60. Cricket *(Gryllus domesticus)*

iii) **Family - Gryllotalpidae** (Mole crickets). The family is allied to Gryllidae and consists of fairly large insects, over 2.5 cm. in length. Their chief character is that the anterior pair of legs are fossorial and are modified for digging and are very strong, with tibiae expanded and digitate. The eyes are reduced. The head and pronotum are hard and the latter is elongate and rounded behind. The wings are tightly wrapped round the soft abdomen. The hind wings are extended backward and appear as slender processes, projecting beyond the tegmina when at rest. Some species are brachypterous and a few are completely wingless. A pair of cerci is present at the end of the abdomen. The ovipositor is vestigial.

The female lays eggs loose clusters in a fairly deep burrow. They are soft, pale in color and oval-shaped. The young nymphs feed on living and dead organisms such as, fly maggots, worms etc. The nymphal stage may last for about two years. The adults, though omnivorous , destroy roots of plants while burrowing and thus are agricultural pest. They are attracted to light. (e-g) Mole cricket - *Gryllotalpa africana* **(Fig. 61)**.

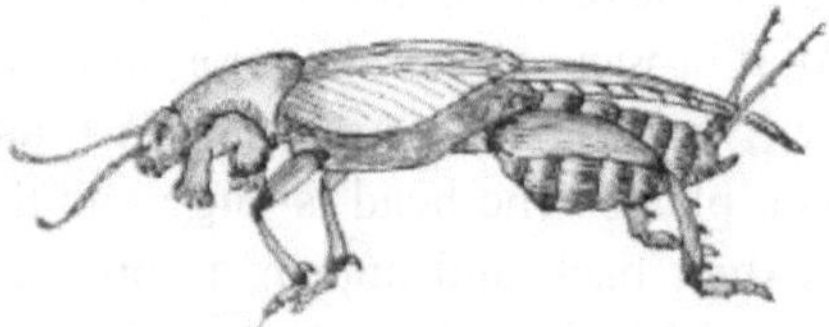

Fig. 61. Mole cricket *(Gryllotalpa africana)*

Suborder - Caelifera. The insects belonging to this suborder are also fairly large-sized and phytophagous. Antennae shorter than the body with fewer segments. Tympanal organs when present are located at the base of the abdomen. Stridulatory organs varied or absent and when present are typically femoro-alary. Ovipositor short and robust. The most important family under this suborder is Family-Acrididae.

Family - Acrididae (Grasshoppers and locusts). This is the most important family of the suborder - Caelifera and comprises of short-horned grasshoppers and locusts, many species of which are serious crop pests. They are found in grasslands, low herbage, waste lands etc., both the plains and hilly terrains. They have short antennae and strong, well developed hind legs, with enlarged femora, which are adapted for jumping. The tarsi are three-segmented. The color in both the nymphs and the adults is cryptic and in some forms, it changes during the different seasons, when the life cycle is very long. The hind wings, which are seen only during flight, are often bright-colored.

In the majority of grasshoppers and all locusts, both pairs of wings are well developed, while in others, they are short and reduced to small lobes. The phenomenon of stridulation is well exhibited by both the sexes in this family and is made possible by different ways. A row of peg-like projections is present on the secondary vein near the base of the tegmen and the femur bears a simple longitudinal ridge. Sound is produced by the rubbing action between these two structures. Some stridulate during flight by friction between the hind wings and the under surface of the tegmina. The auditory organs are on each side of the first abdominal segment. The abdomen is long and extendable in the females, with the help of which eggs are laid deep in the soil. Eggs are deposited in the soil in a cluster having sticky material, which gradually hardens and makes the cluster compact. The oviposition hole in the soil generally contains a frothy peg. Each female lays several clusters of eggs during its life span. These insects are voracious feeders of vegetation, both in the nymphal and adult stages. Locusts normally have two phases at various periods viz., solitary phase and gregarious phase. In the latter phase, they congregate together in plague form and destroy all vegetation totally along their path of flight. (e-g) Migratory locusts - *Locusta migratoria*; Desert locusts - *Schistocerca gregaria*; Grasshopper - *Hieroglyphus banian*. This grasshopper is a pest of rice, sugarcane etc. **(Fig. 62)**; *Hieroglyphus nigrorepletus*. It is a pest of sorghum and other millets.

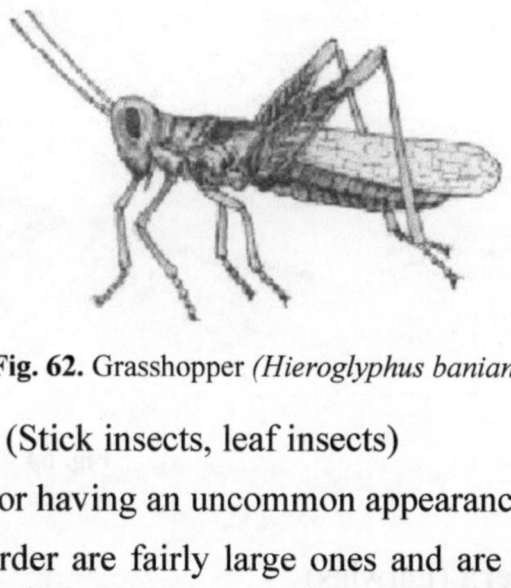

Fig. 62. Grasshopper *(Hieroglyphus banian)*

Order 3 - Phasmida (Stick insects, leaf insects)

('Phasm' = Phantom or having an uncommon appearance)

The insects of this order are fairly large ones and are most common in the tropics. They have small prothorax, very large mesothorax and small tegmina, which may be totally absent in some forms, and have single jointed cerci. They present a great variety of form and color, some found among leaves being green and leaf-like with lamellate expansions, while others are stick-like, resembling twigs or dry branches or a blade of grass. In coloration also, they are very cryptic and both shape and color give them a close resemblance to their habitats. Stick insects can emit protective chemicals from special glands located on the thorax as an effective defense mechanism against birds or other predators.

The head is small and prognathous, with small compound eyes, but the ocelli are absent sometimes. The antennae are very long, filiform or moniliform and many segmented, sometimes up to 100 segments are present. Mouthparts are of biting and chewing type and a hypopharynx is present. The forewings are tegmina and mostly reduced in size. Wing pads do not undergo reversal during development. In the leaf insects, the venation of the forewings resembles the veins of a leaf. The legs are long and very slender. The tarsi are generally five-segmented. There are 11 distinct segments in the abdomen, the first being fused with the metathorax. Specialized auditory and stridulatory organs absent. Ovipositor small and is concealed by the eighth abdominal sternum. Copulation lasts for several hours. The eggs are laid singly, usually falling like seeds on the ground. Metamorphosis is incomplete and the nymphs resemble the adults, except for their lack of wings. They are herbivorous, but none of the species is an established pest.

The order has two important families :

i) **Family - Phyllidae** - Leaf insects (e-g) *Pulchriphyllium (Phyllium) scythe* **(Fig. 63)**.

ii) **Family - Phasmidae -** Sick insects (e-g) *Carausius* species **(Fig. 64)**.

Fig. 63 - Leaf insect *(Phyllium scythe)*

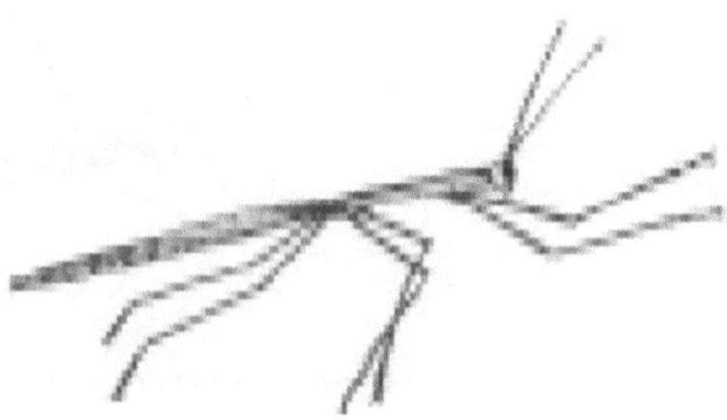

Fig. 64 - Stick insect *(Carausius* species*)*

Order 4 - Dermaptera (Earwigs)

('Derma' = skin; 'ptera' = wings)

Members of this order are medium-sized insects, mostly nocturnal, hiding during the day under barks, stones or soil or among vegetation and are mostly brownish in color. They are elongate insects with biting mouthparts. The head is broad and horizontal. Compound eyes are large, ocelli not developed and in some forms eyes are degenerate or even absent. The antennae are many segmented and about half the length of the body. The forewings are generally short and truncated, modified into leathery tegmina, devoid of veins and meeting along the median line, but sometimes absent. The hind wings are membranous and semi circular, with the veins disposed radially. They seldom fly. The tarsi are three-segmented. The abdomen is eleven-segmented. Ovipositor is reduced or absent. The most characteristic feature is the presence of highly sclerotized, forceps-like appendages at the tip of the abdomen in the majority of forms, which are the modified cerci and are chiefly used as organs of offence and defense, for opening and closing of the wings, to pierce the prey.

while feeding and also in the process of copulation. When alarmed or disturbed, the posterior part of the abdomen is often upraised and the forceps widely opened in a threatening gesture. The forceps are variable in size even within the same species and are generally shorter in the females and in some species they are as long as the body.

Another interesting feature in the biology of earwigs is the parental care, which is shown for the eggs and the young ones. The spherical, white eggs are laid in clusters in the soil and the female sits on them. The young ones also remain for some time crowded under the body of the mother. The earwigs feed on decaying vegetable matter or insects or other small organisms. The insects overwinter in the adult stage. They are attracted by light.

There are a few families under this order. However only one or two are of economic importance.

Family - Labidurudae (e-g) Earwigs - *Euborellia stali* **(Fig. 65)**. It has been recorded as a pest of groundnut pods in recent years.

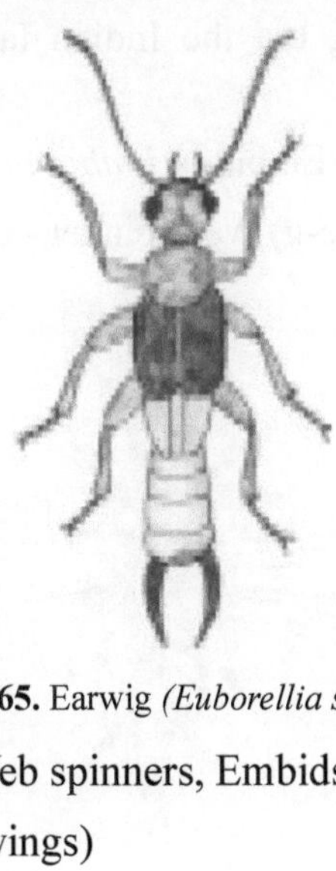

Fig. 65. Earwig *(Euborellia stali)*

Order 5 - Embioptera (Web spinners, Embids)

('Embio' = lively; 'ptera' = wings)

This is a small order comprising largely of tropical and subtropical, gregarious insects living inside silken tunnels and galleries in debris, in cracks in the soil, under stones or barks and among epiphytic plants, lichens etc. They have a soft, thin cuticle and are weak fliers. All are somber-colored, being either brown or yellowish-brown, with smoky wings. They generally avoid day light. The males are winged and are attracted by light, while the females are apterous. The characteristic feature in the biology of these insects is their habit of constructing silken tunnels in which they live. The front legs have enlarged first segment of the tarsi containing silk glands and with the help of these spinning tarsi these insects construct the silken tunnels. The tunnels provide protection from natural enemies and prevent excessive loss of moisture from the body. The eggs, which are large and cylindrical are laid inside the galleries and are covered with chewed food particles and are attended by the female. The web spinners often feign death when disturbed, but they can move very fast usually running backward.

The head of the insects is small, prognathous and with filiform antennae. The compound eyes are reniform and ocelli are absent. They have biting and chewing type of mouthparts. The forelegs are stout and are used for construction of tunnels. The two pairs of wings are long and narrow, with few cross veins and are almost identical. The females are apterous. The abdomen is ten-segmented and ends in a pair of cerci. In the males, the cerci are asymmetrical, with one shorter than the other. The insects feed on dead vegetative matter only and are of no economic significance.

There are about 6 families, but the Indian fauna are mostly found in the following two families :

i) **Family-Embiidae** (e-g) Embids - *Embia minor.*

ii) **Family-Oligotomidae** (e-g) Web spinner - *Oligotoma saundersi* **(Fig. 66).**

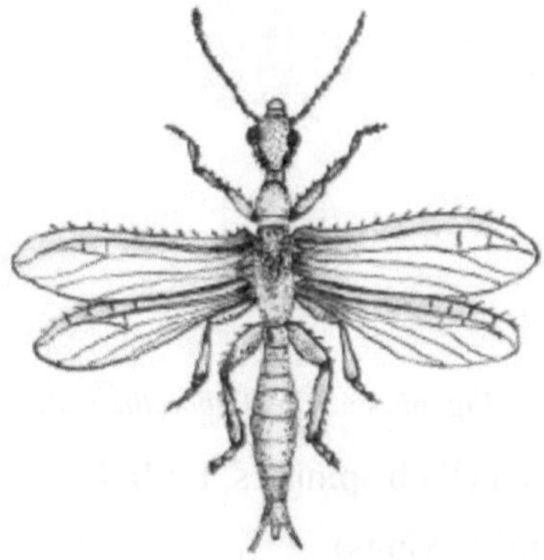

Fig. 66. Web spinner *(Oligotoma saundersi)*

Order 6 - Isoptera (Termites)

('Iso' = equal; 'ptera' = wings)

The termites commonly called white ants because of their whitish color are not related to ants, which belong to the Order - Hymenoptera. But they resemble ants in that, both have somewhat similar life, living in colonies in which there are a variety of forms. Termites are mostly tropical or subtropical. They live in large communities, mostly in underground nests. A termite colony has four types of individuals or castes, which can be grouped into reproductives and sterile forms. The reproductive forms include fully winged forms, whose function is the formation of colonies. They are the **'king'** and the **'queen'**, which have developed winged forms and have subsequently lost their wings before settling down to form a new colony. In the winged forms, the wings are superposed flat over the back when at rest. The sterile castes include soldiers and workers. They are wingless males and females and they perform all other functions except reproduction. The soldiers protect the colony from enemies and intruders, while the workers do the entire work in the nest such as, construction of nest, cleaning the nest, procuring food, caring the eggs, feeding the young ones and queen etc.

The termites are soft-bodied insects, except for the head, which is hard and well sclerotized. In the soldiers, the head may be as large as the rest of the body. In the other castes, the head is small, rounded and well developed. Compound eyes are present only in the winged forms, whereas they are absent in the soldiers and workers. The antennae are short, straight, moniliform and many-segmented. The mouthparts are mandibulate type and the mandibles are large in size in the soldiers. The thoracic segments are distinct. The fore- and hind wings are similar in size, form and texture and venation. After the wings are shed, the remnant or stump of the wing remaining behind is called **'scale'**. The legs are adapted for running. The abdomen has 10 distinct segments. A pair of cerci is present in the primitive forms. The cerci, when present are very short. Termites have a characteristic frontal gland on the head. The sexual organs in the soldiers and workers are degenerated.

Castes of termites. The reproductive castes include (a) primary reproductives and (b) supplementary reproductives. The primary reproductives are winged adults with two pairs of large, membranous wings of almost equal size. The body is dark and the compound eyes are well developed. They lead a short terrestrial life and their main function is to locate suitable places so as to develop new colonies. Their genital organs are well developed.

The supplementary reproductives have no aerial life, the body is straw-colored and the compound eyes are generally reduced in size. They have wing pads and the sexual organs are smaller. Supplementary reproductives generally appear in a colony after the death of the primary reproductives. The queens developed from the supplementary reproductives do not attain such a large size as those, which develop from the primary reproductives, the latter may reach a length of about 10.0 cm., the increase affecting only the abdomen.

The queen is the mother of the entire colony and lives in a specially prepared royal chamber situated almost at the centre of the nest at a depth of 0.30 - 1.80 m. below the ground level. She is very long, creamy white in color, wingless and measures 5.0 - 7.5 cm. in length and her whitish abdomen is marked with transverse, dark brown bands. She always remains in the chamber and her only function is to lay eggs. It lays 40,000 - 80,000 eggs in a day. She is constantly being taken care of and fed by the workers. There is usually a single queen in each colony.

The workers are soft-bodied, straw-colored with prominent head and well developed antennae. Compound eyes are usually absent. The workers are responsible for the damage caused to crops, wooden structures etc. The soldiers can be easily detected by the large size of their head and mandibles. But there are forms of soldiers in which the jaws are not so well developed and vestigial, and they secrete a viscid fluid through the frontal pore situated in the anterior region of the head, which emits a repulsive odor. Soldiers are of both the sexes.

Formation of a new colony. Swarming generally takes place after the first showers of monsoon rains, when the weather is hot and moist and the primary reproductives leave the old nests to form new colonies. In many species, the winged reproductives are attracted to light. Within the old nests the individuals of the reproductive castes appear sometime before they actually swarm. They wait for the weather conditions to be suitable before actually leaving the old nest. They soon attain sexual maturity. On completing the flight, they cast off their wings by means of basal fractures and meet a similar individual of the opposite sex. Then the male and the female seek a suitable place in wood or in the ground, excavate a small chamber in which mating takes place. Copulation takes place frequently several times. Soon after copulation, egg laying starts. The number of eggs laid increases with the age of the queen, reaching several thousands per day and a colony comprising of over a million individuals develops. The earlier broods of nymphs are generally sterile castes of workers and soldiers and they attend to the king and queen. The life of termites is very long and ranges from 15 - 50 years. The death of a queen is followed by the development of supplementary reproductives. The life of a termite colony may be 100 years or more. The capacity of egg laying by the supplementaries is not so large as that of the primaries and more than one supplementary queens may be found in a colony. Rarely both kinds of queens may be found in a colony.

Feeding habits. Two main types of feeding habits are found, as the food requirement of the various castes differ. (a) Feeding on sound or decaying wood and other plant parts such as, stems, roots etc., as well as humus, grass, fungi or any material containing cellulose. The workers feed on such materials mainly. The food of termites consists mostly of cellulose materials, which are very hard for digestion. The cellulose in the food of termites is digested by the presence of varieties of flagellated protozoan present in the termite's digestive tract. This symbiotic association is known as **'mutualism'**. Some termites harbor bacteria rather than protozoa for this purpose. (b) Feeding on a diet prepared by the members of the colony. This includes **'stomodaeal feeding'**, in which proportions of salivary secretions and regurgitated intestinal contents is received from the workers, as also the **'proctodaeal feeding'**, in which the contents of the rectal pouch is obtained from the anus of the workers in response to tactile stimulation by the soliciting termite castes. Soldiers and all young nymphal stages live entirely on such food. The reproductives of all ages also have the same feeding habit.

Many species of Termitidae and Homotermitidae feed outside the nest. The workers of *Hodotermes*, which have well developed compound eyes, go out and collect grass and other woody material. The nests of some of the Termitidae have regular fungus gardens, which are composed of spongy, dark, reddish-brown, coral-like combs constructed by the workers from vegetative matter on which the fungi grow. Such fungi are considered to be the source of vitamins and organic nitrogen for the termites.

Forms of nests. The nests of wood-eating species are comparatively simple. Some primitive members of the group live in moist, decaying trunks and logs of trees, in which they form a series of galleries without any external opening. These termites have no worker caste. Species of many genera such as, *Kalotermes* bore into dry wood such as, furniture, wooden structures etc. The species of genera *Coptotermes* live in the ground and infest wood through the soil and damage woodwork of buildings in contact with the ground. They also emerge above the ground to get access to other building materials and for this purpose construct covered galleries made of their faecal matter as well as earth and wood and reach upper stories of buildings or tall trees.

Many species of Termitidae build mounds or **'termitaria'**, which are built of earth excavated while making underground chambers. They vary in size and some may be as high as 3.0 meters.

Some termites live in the ground without constructing mounds. They damage roots of crops, grass and other vegetation. Wheat, sugarcane, sunflower, groundnut etc. are some of the crops subjected to attack by termites. The common Indian species *Odontotermes obesus* may form conical mounds or live in nests at the surface or may live underground at depths up to about 3.0 meters.

There are five families under this order :

i) **Family-Kalotermitidae** - Wood dwellers **(e-**g) *Kalotermes indicus*; *Cryptotermes domesticus* - The colonies of these termites have no workers.

ii) **Family-Hodotermitidae** - Subterranean or wood inhabiting. (e-g) *Hodotermes macrocephalus* - In the colonies of some of the genera workers are present.

iii) **Family-Rhinotermitidae** - Subterranean. (e-g) *Coptotermes ceylonicus*; *Stylotermes Fletcheri* - Workers are present in the colonies.

iv) **Family-Termitidae** - Ground inhabiting, (e-g) *Odontotermes obesus* **(Fig. 67)**; *Microtermes obesi; Microtermes anandi; Microtermes serrulatus* - Ground inhabiting with a wide range of food habits and colony structure. Worker caste well developed.

v) **Family-Indoermitidae** - Ground Inhabiting termites. (e-g) *Indotermes maymensis*.

1 2 3

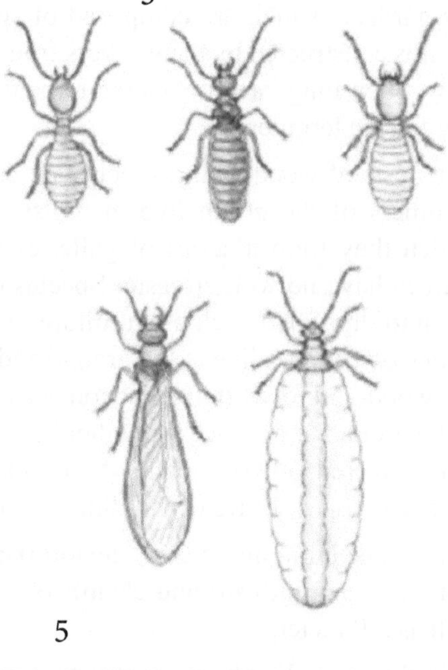

4 5

1. Worker 2. King 3. Soldier 4. Winged adult 5. Queen

Fig. 67. Ground inhabiting termite *(Odontotermes obesus)*

Order 7 - Psocoptera (Psocids, book lice and bark lice)

('Psoco' = small; 'ptera' = wings)

Psocids or book lice are small, soft-bodied insects with membranous wings , which are reduced or lacking in many species or fully formed. Brachypterous and apterous forms may occur within the same species. Some species are found in books, feeding on book-binding materials, in old papers or in dry animal and vegetable matter. They also feed on starchy materials and infest insect collections and museum specimens. They may also found outdoors, living on small fungi, lichens and vegetable matter. The winged forms are not active fliers, but may float in the air like aphids, sometimes in swarms.

Compound eyes and ocelli are well developed in the winged forms, but reduced or absent in apterous forms. The antennae are long, filiform, generally 13 segmented, but may be more. The prothorax is much reduced, particularly in the winged forms and the wings are membranous, with clear venation. The wings are held roof-like over the abdomen when at rest. Cerci are absent. Tarsi are 2 or 3 segmented. There are two pairs of labial glands, the dorsal pair secretes silk with which the insects make webs to cover their colonies, while the ventral pair is the salivary gland.

External genitalia are inconspicuous. The female lays eggs singly or in clusters of 20 - 100 in numbers, either exposed or concealed with webs. Some forms are viviparous and some species are parhtenogenetic. Metamorphosis is simple.

There are about a dozen families in the order of which the family - Psocidae is important.

Family - Psocidae *(e-g)* Book lice - *Liposcellis transvaalensis* **(Fig. 68) -** The insects damage book bindings, dry insects, museum specimens etc.

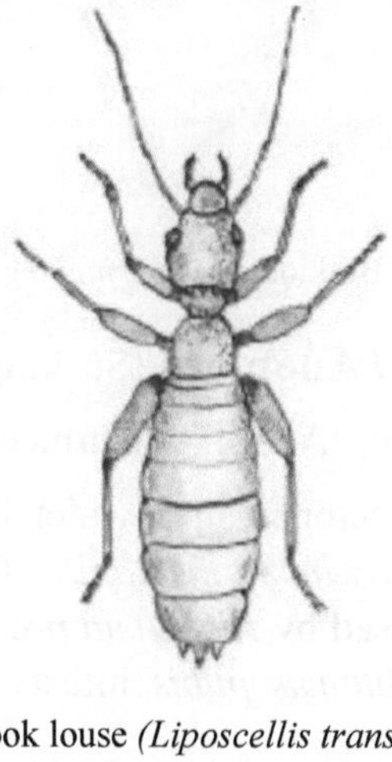

Fig. 68. Book louse *(Liposcellis transvaalensis)*

Order 8 - Mallophaga (Bird lice, elephant lice)

('Mallo' = wool; 'phaga' = to eat)

The insects of this order are small, flat-bodied and are wingless. All of them are ectoparasites, mostly on birds and mammals. Many of these forms feed on feathers, hairs and other horny materials and some of them feed on blood in addition to such materials, while a few others live on blood alone. The hosts of these parasites do not die because of the infestation, but may become unhealthy as a result of irritation and itches caused by them.

The head capsule is large and horizontal. Eyes are reduced and ocelli are absent. The antennae are short and 3 - 5 segmented. The mouthparts are of modified biting type. The prothorax is distinct and the meso- and metathorax are imperfectly separated. Cerci are absent. Tarsi 1 or 2 segmented, that terminate in a single or paired claw. Metamorphosis is slight.

The order is divided into about half a dozen Families, of which only a few are quite important.

i) **Family - menoponidae** (e-g) Bird lice - *Osborniella guiraensis* **(Fig. 69)**

ii) **Family - Philopteridae** (e-g) Bird lice - *Lipeurus tropicalis*

iii) **Family - Haematomyzidae** (e-g) Elephant lice - *Haematomyzus elephantis*

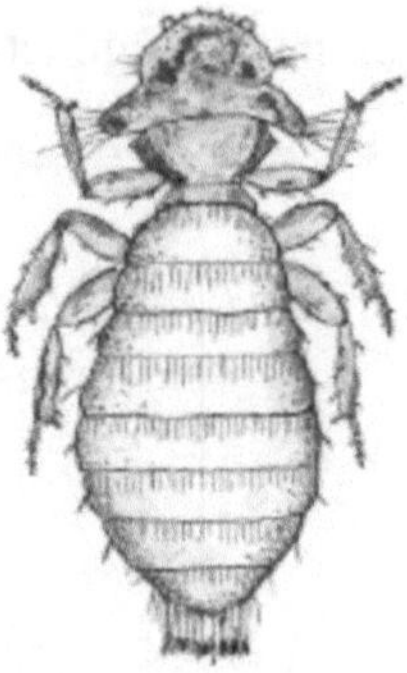

Fig. 69. Bird louse *(Osborniella guiraensis)*

Order 9 - Siphunculata / Anoplura (Sucking lice)

('Siphunculus' = a little tube; 'Anopl' = unarmed; 'ura' = tail)

The insects of this order are ectoparasites, which live on the blood of mammals. The human louse - *pediculus humanus capites* transmits several diseases, of which epidemic typhus caused by *Rickettsia prowaseki* is the most important. Another human louse - *Phthirus pubis* infests hairs in the pubic and anal regions.

The insects are apterous and the body is flattened dorso - ventrally. The head is conical and pointed and narrower than the thorax, with short, 1 - 5 segmented antennae. Eyes reduced or absent. Ocelli are also absent. The thoracic segments are fused. The mouthparts are modified for piercing and sucking purposes. The legs are modified for clinging to the hair of the host. The single - segmented tarsus has a powerful claw. The female louse lays about 300 eggs in about 10 days, which are cemented to the hairs. There are three pupal instars.

The order has about half a dozen families, of which Family - Pediculidae is the most important one

i) **Family - Pediculidae** (e-g) Human head louse - *Pediculus humanus capites* **(Fig. 70)**; Human body louse - *Pediculus humanus corporis.*

Fig. 70. Human head louse *(Pediculus humanus capites)*

ii) **Family - Phthiridae** (e-g) Human pubic louse or Crab louse - *Phthirus pubis*

iii) **Family - Haematopinidae** (e-g) Buffalo louse - *Haematopinus tuberculatus* **(Fig. 71)**

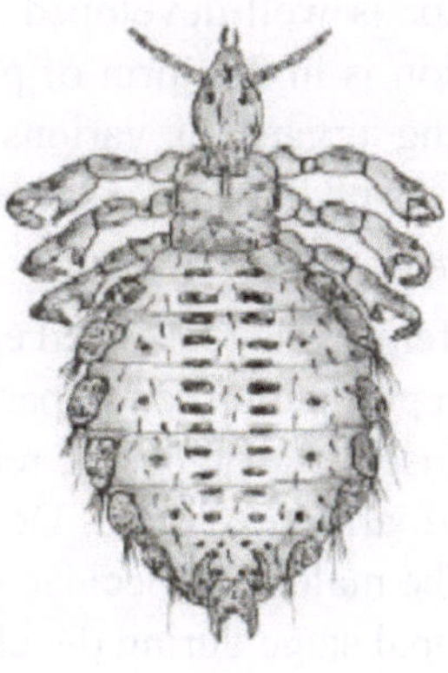

Fig. 71. Buffalo louse *(Haematopinus tuberculatus)*

Order 10 - Hemiptera (Bugs, hoppers etc.)

('Hemi' = half; 'ptera' = wings)

The insects of this order possess mouthparts, which are adapted for piercing and sucking. The vast majority of members feed on the sap of plans and include many economically important crop pests such as, aphids, white flies, leaf and plant hoppers, scale insects etc. Bed bugs, which feed on human blood is also included in this order. Some other bugs of this order feed on the blood of animals and birds. Many of the plant bugs have a remarkably rapid rate of multiplication. The order is divided into two suborders viz., Heteroptera and Homoptera, based on the consistency of the wings. In Homoptera, of the two pairs of wings, the forewings are often of harder consistency and uniformly so in the entire wings. But in Heteroptera, the apical portion of the forewings are membranous, while the basal portion is of a harder consistency and so is termed **'hemelytra'**.

The head is usually opisthognathous in the Homoptera. Compound eyes are well developed and two or three ocelli are present. The mandibles and maxillae are modified into two pairs of bristle-like stylets, which rest in a grooved sheath formed by the labium to form the beak. The two maxillae fit together in the beak to form two channels viz., a food channel and a salivary channel **(Fig. 9)**. The palpi are atrophied. In Heteroptera, the pronotum is always large. The two pairs of wings show marked variations in size and in some species, one of the sexes may be apterous. In some forms such as, the bed bugs, both the sexes are apterous.

A large number of Heteroptera have characteristic odoriferous or repugnatorial glands. The glands usually open on the sides of the thorax. The abdomen consists of 11 segments, some of which, the anterior or the posterior are fused with one another. Cerci are absent. In most of the Homoptera and some of the Heteroptera, ovipositor is well developed. Many Homoptera possess wax glands and their secretion is in the form of powder or threads. In many Heteroptera, sound producing organs of various types are found. Among Homoptera, the Cicadidae are well known for these organs, which are situated on the ventral aspect of the base of the abdomen.

In most of the Hemipterans, females lay fertilized eggs of various shapes, sizes and colors, but in the aphids reproduction can be parthenogenetic and viviparity is also found. In the scale insects, all types of reproduction, from oviparity through ovoviviparity to viviparity are found. Development in Hemiptera is by simple metamorphosis. The males of Coccidae and both the sexes of white flies (Aleyrodidae) have a pupal stage during development.

Distinguishing features of Heteroptera and Homoptera

Heteroptera	**Homoptera**
The forewings are harder and heavily sclerotized at the base, while apical portion remains membranous and are called 'hemelytra'	The forewings are of slightly harder consistency and uniformly so in the entire wings
Wings are held flat over the back at rest	Wings are held like a tent over the back at rest
Beak arises from the anterior end of the head	Beak arises from the posterior end of the head And extends backward between the legs
Pronotum usually very much enlarged and quite Prominent	Pronotum usually not quite prominent
Odoriferous glands present	Wax glands usually present
Tarsi usually 3 segmented	Tarsi 1 - 3 segmented
Nymphs either terrestrial or aquatic	Nymphs or larvae terrestrial
Phytophagous, predacious or blood sucking	Only phytophagous

Both the suborders - Heteroptera and Homoptera have several families and only the most important families are discussed below :

Suborder - Heteroptera (True bugs).

i) **Family - Reduviidae** (Assassin bugs, Kissing bugs or cone-nose bugs, Reduvids). Unlike The vast majority of Hemipterans, a large number of species of this family are predacious, sucking blood of other insects and animals and even sucking blood of man. *Triatoma* and *Rhodnius* species, besides sucking blood, are the main carriers of *Trypanosoma cruzi*, which causes human **'Trypanosomiasis'**.

The insects belonging to this family are medium to large-sized, usually black or brownish in color and mostly predacious on smaller insects. They pierce the prey with the beak, suck the body content and kill them. Hence they are called **'Assassin bugs'**.

The head is narrow and elongate, constricted behind the eyes forming a neck. The beak is short, 3 segmented, usually curved and the tip fits into a groove in the prosternum. The abdomen is broad at the middle portion and exposes the margins ofm the segments beyond the wings. The tarsi are 3 segmented (e-g) *Reduvius cincticrus, R. personatus, Pirates sanctus, Acanthaspis flavipes, Platymeris laevicollis.*

ii) **Family - Gerridae** (Pond skaters, water striders). The members of this family are semi-aquatic and run about or skate with a fast jerky movement on the surface of water. Most of the species are dark brow or black and the body is long and narrow. Wings are mostly reduced or absent. The forelegs are short and used for capturing food, while the other two pairs are long and used for locomotion. They feed onany living or dead insects that fall into the water, grasping them with their front legs. The tarsi are clothed with fine, hydrofuge hairs, that trap a cushion of air, which prevents the insect from sinking and thus enable it to run and skate on the surface of water. (e-g) Pond skater - *Gerris spinolae* **(Fig. 72)**; Water strider - *Gerris remigis* **(Fig. 73)**; Sea skaters - *Halobates* species.

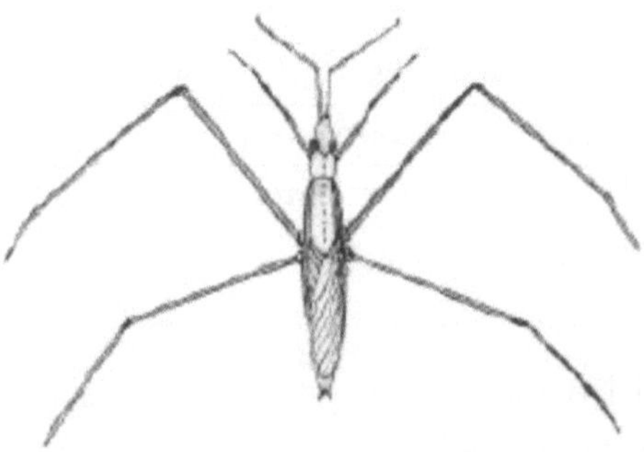

Fig. 72. Pond skater *(Gerris spinolae)*

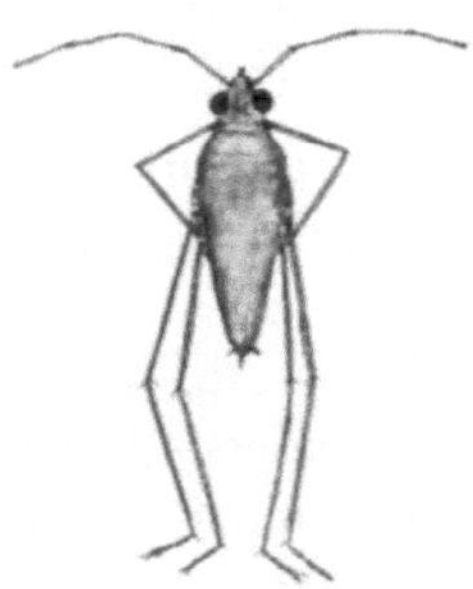

Fig. 73. Water strider *(Gerris remigis)*

iii) **Family - Miridae** (Mosquito bugs, mirid bugs. leaf bugs, or plant bugs)- This is the largest family in the order. Most of the Mirids aresmall and delicate insects. They are usually brown or greenish in color. They are characterized by their hemelytra having a cuneus. Antennae and beak are 4 segmented. Ocelli are absent. Tarsi 3 segmented. The females have a distinct ovipositor. A few are predacious, but the majority is phytophagous. They feed mostly on plant sap and some of them are very serious crop pests. While feeding, the insects inject a toxic substance along with the saliva, which causes **'Phytotoxemia'** leading to drying and death of leaves and branches. (e-g) *Helopeltis theivora*; *H. antonii*.These tea mosquito bugs are very destructive to tea plants **(Fig. 74)**; *Calocoris angustatus* - Earhead bug of cholam. It is a serious pest of sorghum crop; *Disphinctus politus* - It is a serious pest of betel vine plants; *Disphinctus humeralis* - It is a pest of cinchona plants.

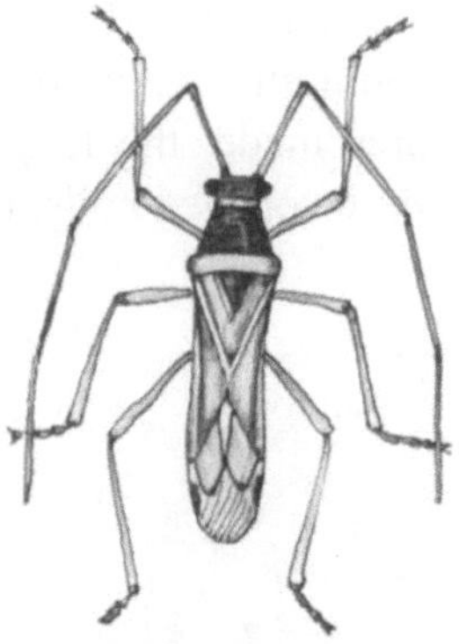

Fig. 74. Cholam earhead bug *(Calocoris angustatus)*

iv) **Family - Cimicidae** (Bed bugs) - This is a small family of ectoparasites feeding on the blood of man, animals and birds (Haematophobic). They are dull, reddish - brown in color, with a tough skin, broadly oval in shape, about 6.0 mm. in length, flattened dorso - ventrally, with very short, diminutive hemelytra, the suctorial proboscis lying in a ventral groove. Ocelli are absent. Legs short and tarsi 3 segmented. They are particularly prevalent in dirty houses, lodging houses, buses, trains and in places frequented by man, especially in large towns and cities. They are nocturnal in habit and hide in cracks, crevices and joints in the walls, floors, furniture etc. The female may lay 100 - 250 eggs. Adults may live for several months and can survive long periods without food. They have a repulsive odor. (e-g) Bed bug - *Cimex hemipterus* **(Fig. 75)**; *Cimex lectularis*

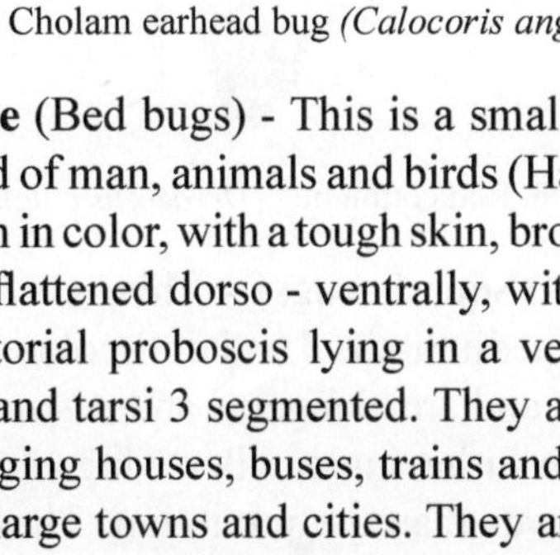

Fig. 75. Bed bug *(Cimex hemipterus)*

v) **Family - Pyrrhocoridae** (Stainers or red bugs) - The bugs are medium - sized, usually bright -colored insects. They have quite conspicuous eyes, but no ocelli are present. Rosrum is slender and 4 segmented. A few are serious pests of cultivated crops. (e-g) The red cotton bug - *Dysdercus* *cingulatus* **(Fig. 76)** - The red cotton bug is a serious sucking pest of cotton. The bugs pierce the immature cotton bolls, suck and feed on the sap. Through the punctures made, the fungus *Nematospora* enters the bolls and stains the lint. In the opened bolls, the lint is also stained by the excrement of the insects.

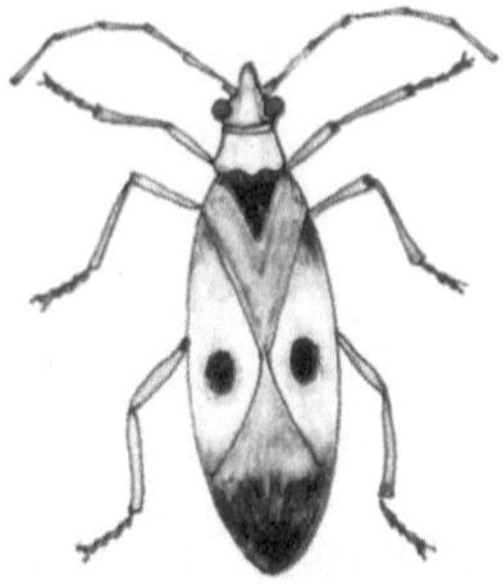

Fig. 76. Red cotton bug *(Dysdercus cingulatus)*

vi) **Family - Coreidae** (Squash bugs) - This is a moderately large family comprising mostly medium-sized and dull-colored insects. They are more or less oval in shape and are elongated and resemble the pentatomid bugs, but differ from them in having smaller wings and a large abdomen. The head is much narrower than the pronotum. They have well - developed scent glands that emit a bad, nauseating odor. Numerous veins are present in the hemelytra. Some species have their hind tibia dilated and appear leaf-like. While many of them are crop pests, a few are predacious. (e-g) Rice earhead bug - *Leptocorisa varicornis*, a serious pest of rice crop **(Fig. 77);** Sap sucking pest from leaves and tender shoots of red gram, green gram, black gram and allied pulse crops - *Riptortus pedestris*.

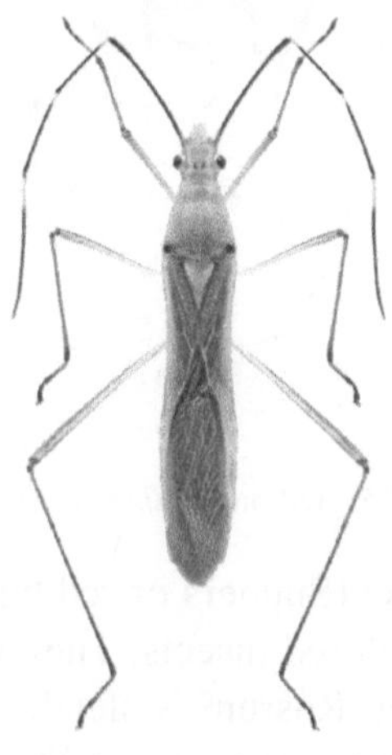

Fig. 77. Rice earhead bug *(Leptocorisa varicornis)*

vii) **Family-Pentatomidae** (Shield bugs or stink bugs) - Most of the species of this family are large-sized, with the scutellum very prominent and forming a shield that covers about half of the abdomen and hence are called by the popular name, **'shield bugs'**. Antennae are long and usually 5 segmeted. Tarsi are 2 or 3 segmented. Hemelytra well developed In the middle aged and older nymphs, there are four pairs of dorsal abdominal odoriferous glands, which emit an abnoxious smelling fluid. These glands are atrophied in the adults and in the adults the usual metathoracic scent glands are found. The vast majority of species feed on plant sap, while a few are predatious on insects (e-g) The green stink bug - *Nezara viridula* and *Dolycoris indicus* are sucking pests of cotton; *Bagrada cruciferarum* is a sucking pest of cruciferous crops; *Canthecona furcellata* **(Fig. 78)** is predacious on many insects; *Andrallus spinidens* **(Fig. 79)** is predacious on American cotton boll worm.

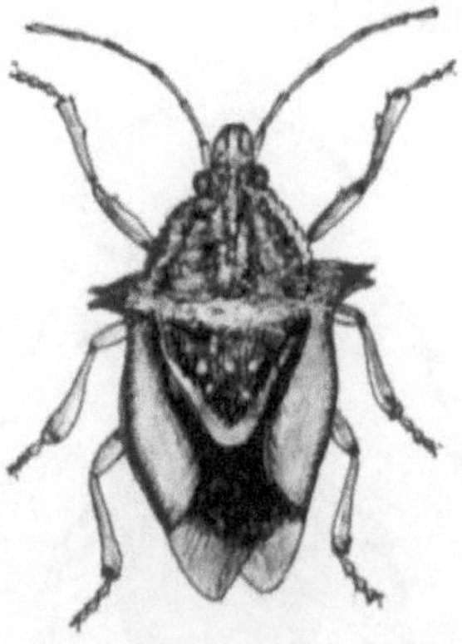

Fig. 78. Predatory bug *(Canthecona furcellata)*

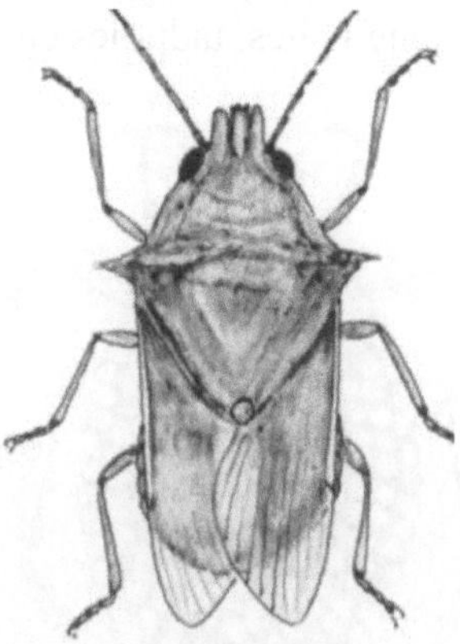

Fig. 79. Predatory bug *(Andrallus spinidens)*

viii) **Family - Tingidae** (Lace bugs, Lacewing bugs) - These bugs are easily recognized by the sculptured, lace - like pattern on the upper surface of the body, head and the lateral expansions of the thorax. The wings usually present a pattern of elevated ridges and sunken membranous areas. The lace - like appearance is found only in the adults. The nymphs are usually spiny. The pronotum usually extends posteriorly and conceals the scutellum. Ocelli absent. A few are important pests of cultivated crops. They suck and feed on the sap from the under surface of leaves of plants. (e-g) Banana lace wing bug - *Stephanitis typicus* **(Fig. 80)** - It is a serious sucking pest of banana.

Fig. 80. Banana lacewing bug *(Stephanitis typicus)*

ix) **Family - Belostomatidae** (Giant water bugs) - This family consists of some species, which are the largest among insects. The giant North American bug - *Belostoma grandis* is quite large and 10 - 12 cm. long. They are all flattened insects, with the posterior legs specially adapted for swimming. In some species such as, *Sphaeroderma molestum* **(Fig. 81)**, the female forcibly attaches her eggs on the dorsum of the male, which carries the eggs till the time of hatching (e-g) Giant water bug - *Belostoma indicum* **(Fig. 82)**; *Lethocerus grandis* and *L. indicus* - These bugs feed on young fishes, tadpoles etc. They are attracted to light.

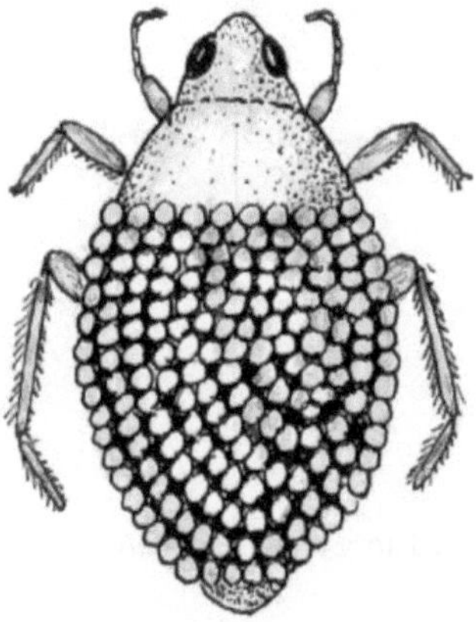

Fig. 81. Giant water bug *(Sphaeroderma molestum)*

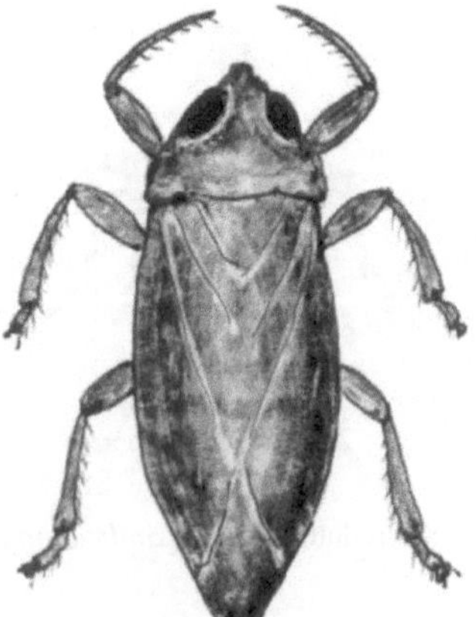

Fig. 82. Giant water bug *(Belostoma indicum)*

x) **Family - Nepidae** (Water scorpion) - Members of this family are predacious aquatic bugs, with raptorial front legs,bent like pincers of a scorpion and the posterior pairs of legs adapted for walking. They are characterized by the presence of a long, caudal breathing tube resembling a sting formed by the cerci. They have well developed wings , but they rarely fly. They are predacious on small insects and other aquatic organisms. (e-g) Water scorpion - *Ranatra elongata* **(Fig. 83)**; Water scorpion - *Laccotrephes maculatus* **(Fig. 84)**; *Nepa cinerea* - Water scorpion.

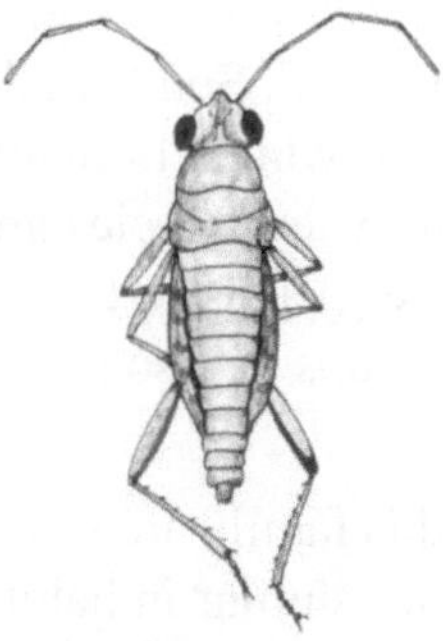

Fig. 83. Water scorpion *(Ranatra elongata)*

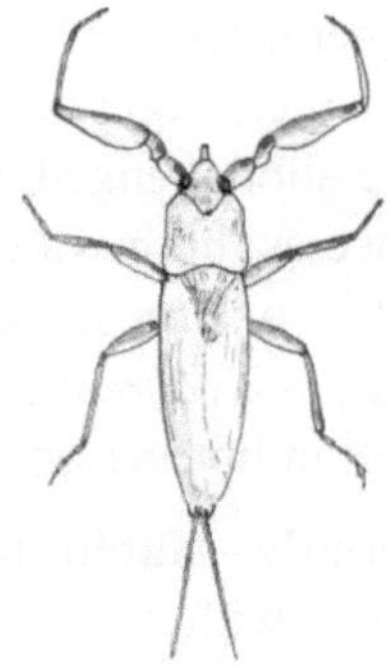

Fig. 84. Water scorpion *(Laccotrephes maculatus)*

xi) **Family - Notonectidae** (Back swimmers) - These insects differ from other aquatic Hemiptera in that they swim on their back, which is shaped like the bottom of a boat. Therir powerful hind legs are outstretched and are flattened oar-like, fringed with hairs and adapted for padding. They can dive into the water and fly occasionally. They are predatory on aquatic insects and other small organisms. (e-g) *Notonecta glauca*; *Anisops sardia.*

xii) **Family - Cirixidae** (Water boatman) - These insects live submerged in water, clinging with their hind legs to various objects and water plants beneath the water surface. They are oval, somewhat flattened, usually dark gray in color and have elongated, oar - like hind legs. These fresh water insects can swim very fast. They lack gills, but come to the surface of water to get supply of air for breathing. They feed on algae and minute aquatic organisms. (e-g) *Micronecta striata* **(Fig. 85)**; *Sigara hieroglyphica*; *Corixa punctata.*

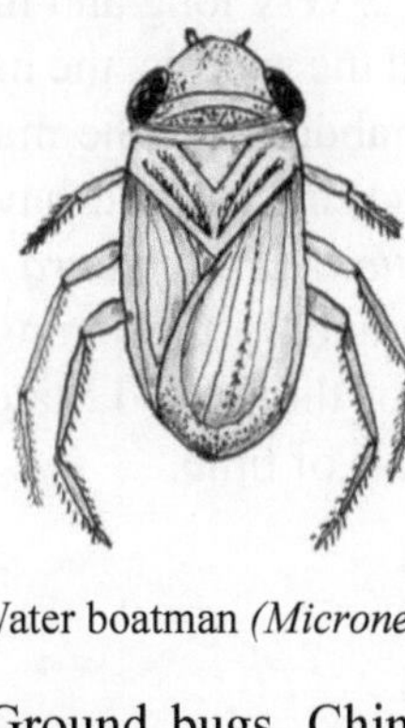

Fig. 85. Water boatman *(Micronecta striata)*

xiii) **Family - Lygaeidae** (Ground bugs, Chinch bugs or cotton stainers) - These bugs are small, elongated, soft-bodied, dark or brightly colored insects and most of them are plant sap-feeders, and occasionally feed on other insects. Many species are marked with conspicuous spots or bands of other bright colors. Ocelli are present. Tarsi 3 segmented. Some kinds are short-winged and flightless, while others have long wings. In many species, both winged and wingless forms are present. A few species are pests of crops, especially cereal crops and cotton. (e-g) Dusky cotton bug - *Oxycarenus hyalinipennis.* It is a pest of cereal crops. It also stains cotton lint; Sesame seed bug - *Elasmolomus sordidus.*

xiv) **Family - Nabididae** (Damsel bugs) - The insects of this family are small bugs, with body somewhat anteriorly narrower and are similar in habits to assassin bugs. Antennae are 4 or 5 segmented. Like the assassin bugs, they are predacious and prey on plant bugs, plant lice, caterpillars and other insects, using their front legs for catching the prey and their sharp, beak-like proboscis for piercing them (e-g) *Alloeorrhynchus* species.

xv) **Family - Veliidae** (Water crickets, water striders or ripple bugs) - It is a family of predatory bugs, similar to the pond skaters, but shorter and broader with thickened hind legs. They are brown or black in color, often with silvery markings. They are gregarious and mostly found near ripples. Some species are wingless. (e-g) *Microvelia capitata.*

Suborder - Homoptera (Aphids, Leaf hoppers etc.)

i) **Family - Cicadidae** (Cicadas) - The cicadas are well known for the shrill noise the males make. The noise is sometimes almost deafening. The females are silent. The cicadas have transparent wings and 3 ocelli are present. The front femora are thickened. The eggs are laid in small holes made by the females on the twigs of trees, but on hatching, the young nymphs fall down and pass a subterranean life, feeding on plant roots etc. The life history is very long and the nymphal stage may last for several years. In almost all the species, the males possess sound producing organs at the base of the abdomen. The males live only for a few days. The insects are fairly large in size and have long, piercing mouthparts. (e-g) *Platypleura octoguttata*; *Platypleura machinoni* **(Fig. 86)** - These cicadas are commonly found in the plains; *Magicicada septendecim* - These periodical cicadas of the United States of America appear in large numbers after long intervals of time.

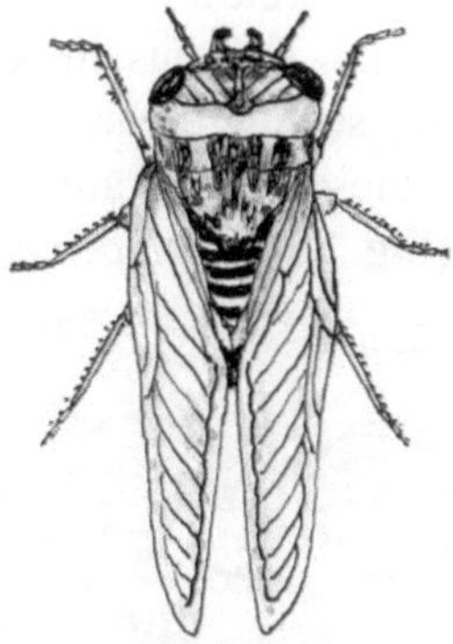

Fig. 86. Cicada of the plains *(Platypleura machinoni)*

ii) **Family - Cercopidae** (Spittle bugs, Cuckoo spit or frog hoppers) - The plant bugs of this family possess broad head, frog-like appearance, with hexagonal pronotum and are capable of jumping. The name cuckoo-spit is given, because the nymphs of many genera secrete a foamy, spit-like fluid mostly out of their anus and live hidden inside the spit attached to leaves of grasses or ow vegetation, which protect them from the hot sun and predators. If they are separated from the spit, they shrivel and die. Nymphs of some genera live underground. The adults feed on wild and cultivated plants. (e-g) Common cuckoo spit insects - *Poophilus costalis* and *Clovia puncta* **(Fig. 87)**; *Phymatostetha deschampsi* and *Machaerota planitae* found on ber, cotton etc.

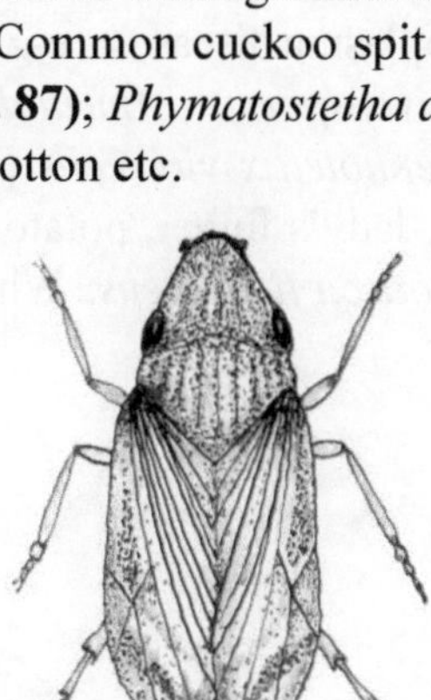

Fig. 87. Spittle bug *(Clovia puncta)*

iii) **Family - Membracidae** (Cow bugs or tree hoppers) - These mostly small, protectively colored insects can be easily recognized by their pronotum, which is extended backwards in the form of a hood over the abdomen giving the insect a bizarre appearance. The head is modified with an almost vertical crown and horizontal face. Ants commonly attend the nymphs and adults of this family, as they secrete a sweet fluid from their anal tube, which is much relished by the ants. The ants stroke the membracids with their antennae, whereupon they secrete the sweet fluid. A few species exhibit maternal care. None of these insects are of any economic importance. (e-g) *Leptobelus dama* **(Fig. 88)**; *Leptocentrus taurus*; *Tricentrus bicolor.*

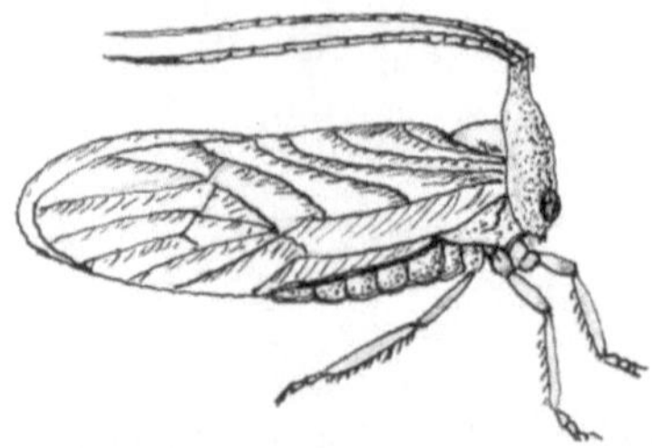

Fig. 88. Tree hopper *(Leptobelus dama)*

iv) **Family - Jassidae** or **Cicadellidae** (Jassids, leaf hoppers) - Next to aphids, the insects of this family are most abundant of all hoppers. These attractively colored, plant sucking insects are met with on all kinds of vegetation, cultivated and wild. They are slender, wedge - shaped insects, with tapering posterior end, very active and hop off at the least disturbance. They walk diagonally. They are attracted to light. Many of them are serious crop pests and are of considerable economic importance. Some of them transmit virus and mycoplasma diseases. (e-g) Mango hoppers- *Idioscopus clypealis*, *Idioscopus niveosparsus*, *Amritodes atkinsoni* **(Fig. 89)**; Rice green leaf hoppers - *Nephotettix virescens* **(Fig. 90)** and *N. nigropictus*; Sucking pest of cotton, lady's finger, potato etc. - *Empoasca devastans*; Green fly of tea - *Empoasca flavescens*; White jassid of rice - *Tettigella spectra.*

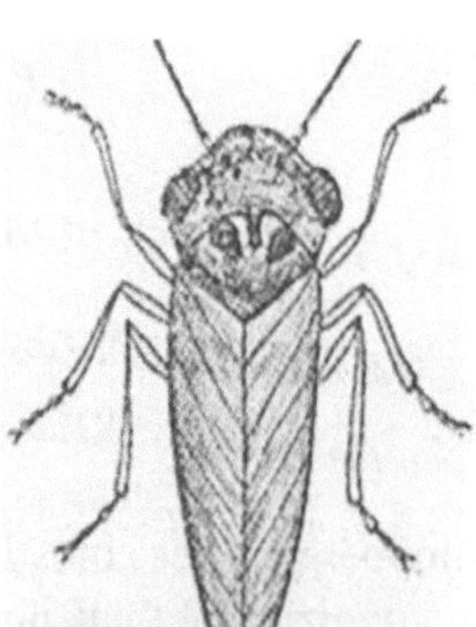

Fig. 89. Mango hopper *(Amritodes atkinsoni)*

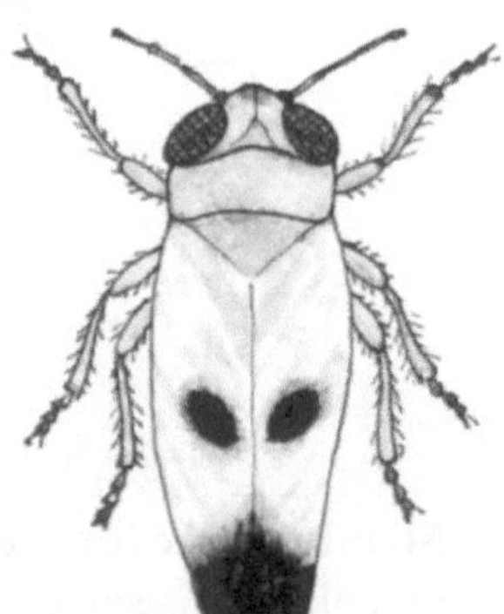

Fig. 90. Rice green leaf hopper *(Nephotettix virescens)*

v) **Family - Fulgoridae** (Lantern flies) - The family includes brilliantly colored insects, often of large size. In many genera, the front of the head is drawn out to form a hollow, proboscis-like prolongation. Some species secrete a fluorescent, white, waxy substance. The insects are of no economic importance (e-g) Lantern fly - *Pyrops chennelli.*

vi) **Family- Delphacidae** or **Araeopidae (**Plant hoppers) - It is the largest of the fulgorid families and includes a number of serious crop pests. They are mostly small, dull - colored, wedge-shaped insects, with median ocellus. The apex of the hind tibia possesses large, mobile, flattened spur. These sucking pests are responsibl for causing hopper burn. The nymphs and adults are vectors of a few virus diseases. (e-g) Rice brown plant hopper - *Nilaparvata lugens* **(Fig. 91)**; Sugarcane leaf hopper of Queensland and Hawaiian Islands; Bamboo hopper - *Purohit cervina*; White-backed rice plant hopper - *Sogatella furcifera.*

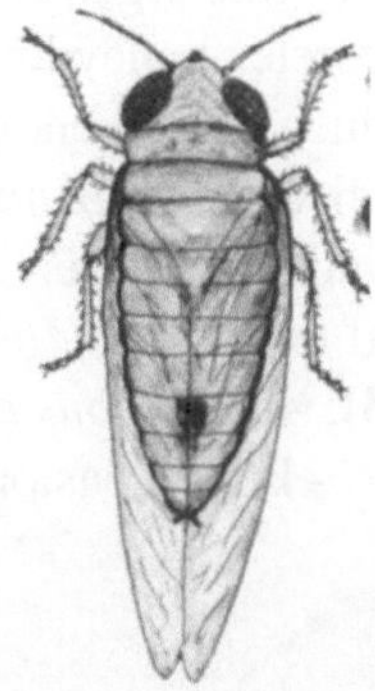

Fig. 91. Rice brown plant hopper *(Nilaparvata lugens)*

vii) **Family - Lophopidae** (Leaf hoppers) - The leaf hoppers of this family have elongated beak or rostrum and long hind basitarsus. Some of them ar serious crop pests (e-g) Sucking pest of sugarcane, maize, jowar, wheat, wild grasses etc - *Pyrilla perpusilla* **(Fig. 92)**.

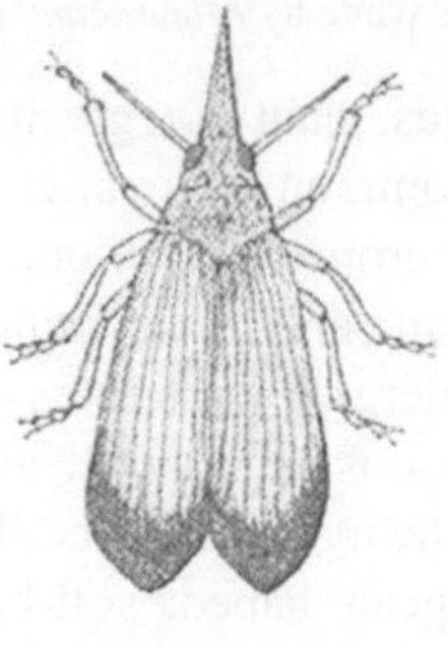

Fig. 92. Sugarcane hopper *(Pyrilla perpusilla)*

viii)**Family - Psillidae** or **Chermidae** (Jumping plant lice) - Members of this family are small, sucking insects resembling cicadas and active in habit. They generally leap about and cannot fly long distances. The hind legs, like those in Cicadellidae are long and stout t and help in jumping. Cross veins are absent in both the pairs of wings. The nymphs are flattish, with wing buds. (e-g) Citrus psilla - *Diaphorina citri* - It is a serious pest of citrus and curry leaf plants.

ix) **Family - Aleyrodidae** (White flies) - The members of this family are very small, 2 - 3 mm in length and ook whitish on account of the white, mealy, powder or waxy materials, which cover their wings and body. The nymphs are scale-like and sessile and there is an immobile, inactive stage called pseudopupal stage in the life cycle. The most characteristic organ of these insects is the vasiform orifice, which opens on the dorsal surface of the last abdominal segment. The opening is covered with an operculum under which there is a tongue - like organ termed ligula. The anus opens at the base of the ligula and secretes honeydew, which collects in the ligula. Wing venation is much reduced. Parthenogenesis is common. The eggs are provided with a long pedicel. (e-g) *Bemisia tabasi* - It is a serious pest of cotton, tobacco, tomato and many other cultivated and wild plants. It is also a vector of tobacco leaf curl virus; *Aleurocanthus spiniferus* - It is a pest of citrus plants **(Fig. 93)**; *Aleurolobus barodensis* - It is a serious pest of sugarcane; *Aleurodes citri* - Infests castor.

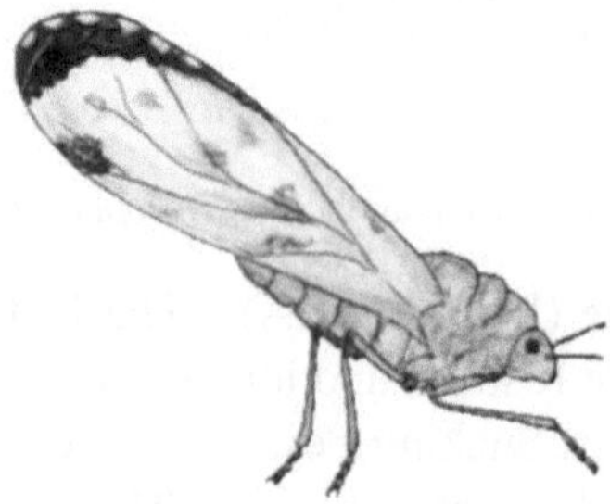

Fig. 93. Citrus white fly *(Aleurocanthus spiniferus)*

x) **Family - Aphidiae** (Aphids, plant lice, green flies) - The insects belonging to this family are quite common on a variety of plants and are important pests of many cultivated crops and vectors of many plant viruses. They mostly occur in colonies and are found young shoots, foliage, flowers and pods, while a few live underground on roots or live both on the shoots and roots during different seasons as in wooly aphids. Some species produce galls on the plant parts. The nymphs and adults suck and feed on the plant sap. Aphids are small, pear-shaped, soft-bodied, delicate insects. The rostrum is long and many jointed. Compound eyes are present. The most characteristic feature of these insects is the presence of a pair of honey tubes at the anal end, which produce a waxy substance that covers the aphids. The waxy secretion is exuded when the aphids are attacked by a predator and act as alarm pheromones. Honeydew is secreted through the anus. Presence of a pair of cornicles at the posterior end of the abdomen is typical to aphids. Polymorphism is common and alate, aptrous and brachypterous forms are found. In the winged forms, the wings are held vertically above the body while at rest. Parthenogenesis, viviparity and oviparity are all common. The sexual females reproduce by laying eggs and parthenogenetic females give birth to young ones. Several generations may appear in a year. The sexually matured females lay eggs during autumn and the egg stage lasts through the winter. With the advent of spring, the eggs hatch and give rise to apterous, parthenogenetic, viviparous females. These females produce a new generation, which may have a few winged females. By the end of summer or in the autumn, their progeny gives rise to sexual females and males. The winged females may migrate to other places and infests new hosts. In non-migratory aphids, the entire life cycle is spent on the same host plant or on plants of the same species. The sexually matured oviparous females from the subsequent generations lay eggs in the ensuing autumn and the cycle continues. (e-g) *Myzus persicae* - It is a polyphagous and serious pest and infests several economically important crops such as, potato, tobacco, cruciferous crops, peach etc. **(Fig. 94)**; *Aphis gossypii* - It is a common and serious pest of cotton; *Aphis crassivora* - It infests lab lab and other pulse crops; *Eriosoma lanigera* - The apple wooly aphid, which is a serious pest of apple **(Fig. 95)**; *Lipaphis erysime* - It infests cabbage and other crucifers; *Pentalonia nigronervosa* - The banana aphid that infests banana; *Macrosiphum rosae* - It infests rose plants; *Tetraneura nigriabdominalis* - The root aphid that infests roots of finger millet.

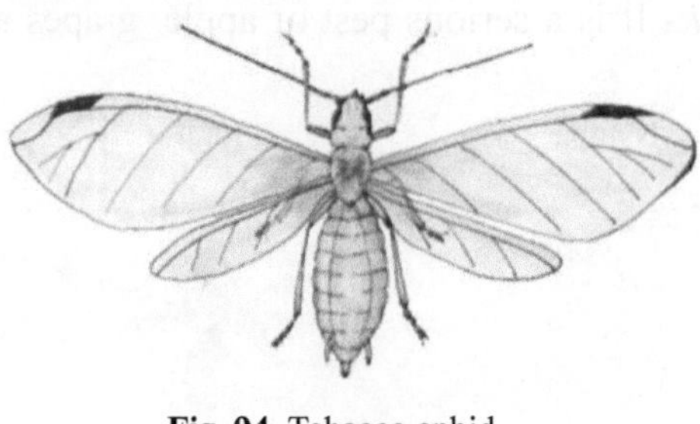

Fig. 94. Tobacco aphid *(Myzus persicae)*

Fig. 95. Apple wooly aphid *(Eriosoma lanigera)*

xi) **Family - Pseudococcidae** (Mealy bugs) - The family consists of many species injurious to various crops, especially horticultural crops. They suck and feed on the sap. The males are usually alate and the forewings are larger, while the hind wings are reduced to haltere-like structures. The females are wingless. Females are more common and the common name **'mealy bugs'** refer only to the females. The first stage nymphs, known as crawlers are provided with functional legs and are mainly responsible for dispersal. Subsequently the nymphs and adult females becomestationary, being attached to the host with the mouthparts. The body of female is elongate - oval, flattened and segmented and is covered with a waxy, mealy or filamentous coating. They are oviparous or viviparous and they secrete honey dew. (e-g) *Ripersia oryzae* - It is a serious pest of rice crop **(Fig. 96)**; *Pseucoccus citri* - Infests citrus plants and causes extensive damage; *Ferrisia virgata* -It is a pest of safflower' jute etc.

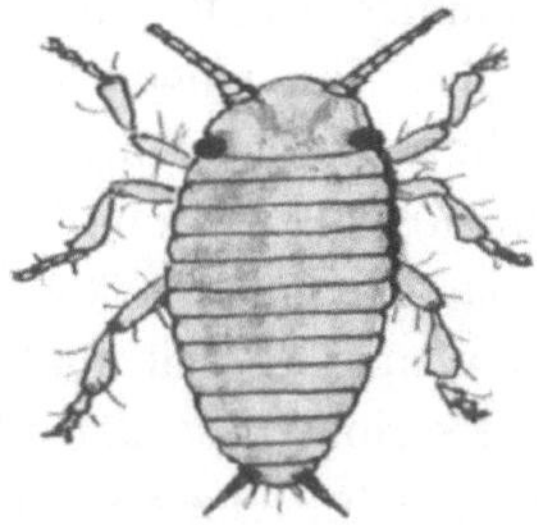

Fig. 96. Rice mealy bug *(Ripersia oryzae)*

xii) **Family - Margarodidae** (Giant coccids) - The females are large and rounded, with distinct body segments and simple eyes. The body is often covered with waxy secretion. The females live underground for many years, enclosed in layers of shining waxy secretions. Adult males are winged and they possess compound eyes. (e-g) *Drosicha mangiferae* - Mango mealy bug. It infests mango and citrus plants **(Fig. 97)**; The cottony cushion scale, *Icerya purchasi* - It is a serious pest of apple, grapes and citrus plants **(Fig. 98)**.

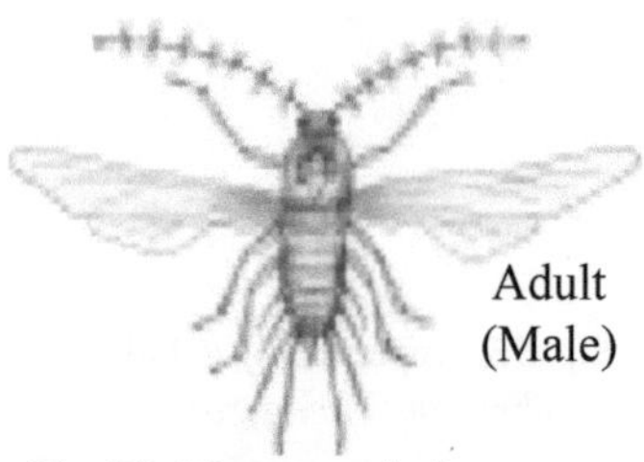

Fig. 97. Mango mealy bug *(Drosicha mangiferae)*

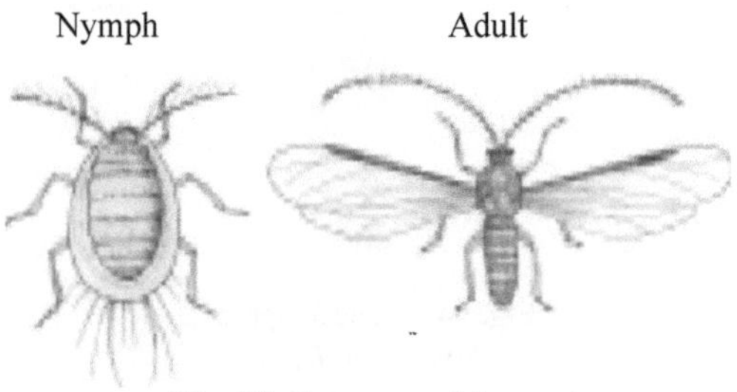

Fig. 98. Cottony cushion scale *(Icerya purchasi)*

xiii) **Family - Lacciferidae** (Lac insects) - This Male family includes the most important lac Female insect in India. The females are highly degenerated with an irregularly globular body enclosed in a thick resinous cell. Wings, antennae, and legsare much reduced or absent. The dermal gland secretions provide stick lac. (e-g) *Laccifer lacca* - It is the Indian lac insect of great commercial importance. Its secretion of lac provides the stick lac from which 'shellac' is produced **(Fig. 99)**.

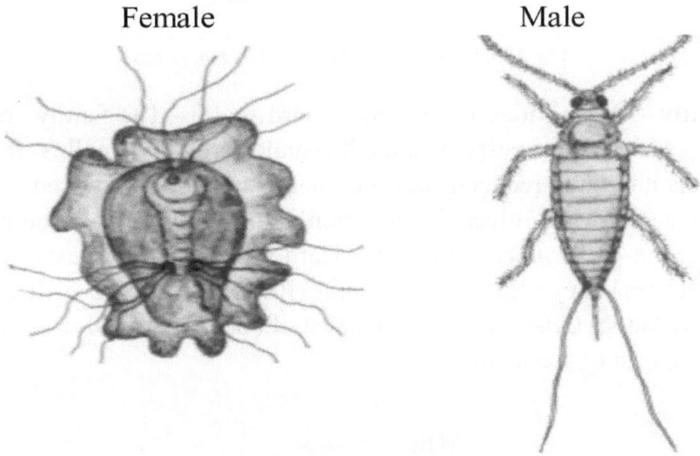

Fig. 99. Indian lac insect *(Laccifer lacca)*

xiv)**Family - Dactylopiidae** (Cochineal insects) - The insects of this family resemble the mealy bugs in appearance and habits. The females are minute, red in color, elongate-oval and convex with small legs and antennae. The body is distinctly segmented and is covered with a waxy, mealy, plate-like coating. Legs and antennae are short. (e-g) *Dactylopius opuntiae* - This species infests cactus plants. It has been used widely for the control of cactus weeds. The red dyestuff, cochineal extracted from it is used in the manufacture of cosmetics; *D. Tomentosus* also produces the dyestuff cochineal.

xv) **Family - Coccidae** or **lecaniidae** (Soft scale insects, wax scales and tortoise scales) - This family consists of many species, which are injurious to various crops. The females are flattened, elongate - oval, somewhat tortoise – shaped and are covered with a coating of wax. The males are alate or winged, while the females are wingless and remain stationary and are covered by a hard, scaly covering A few species yield wax in large quantities (e-g) *Coccus viridis* - It is generally called the green scale of coffee and is a serious pest of coffee plants **(Fig. 100)**; *Coccus floridensis* - It is a pest of guava plants.

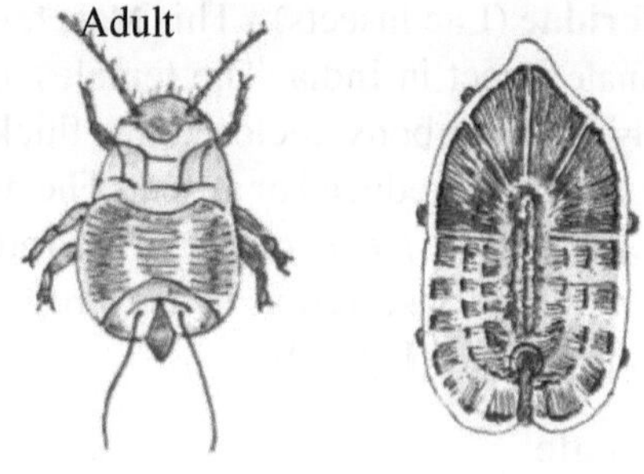

Fig. 100. Green scale of coffee *(Coccus viridis)*

xvi)**Family - Diaspididae** (Armored or hard scale) - The family consists of some serious crop pests. The adult females are very small, soft - bodied insects and have reduced legs and antennae and are covered by a hard, waxy scale, which gives the common name **'hard scale'**. The males are winged. e-g) *Aonidiella aurantii* - Hard scale of rose. Infests rose plants and causes heavy damage to the plants **(Fig. 101)** ; *A. citrina* - It is a pest of citrus plants; *A. destructor* - It is a pest of coconut; *Melanaspis glomerata* - It is a pest of sugarcane.

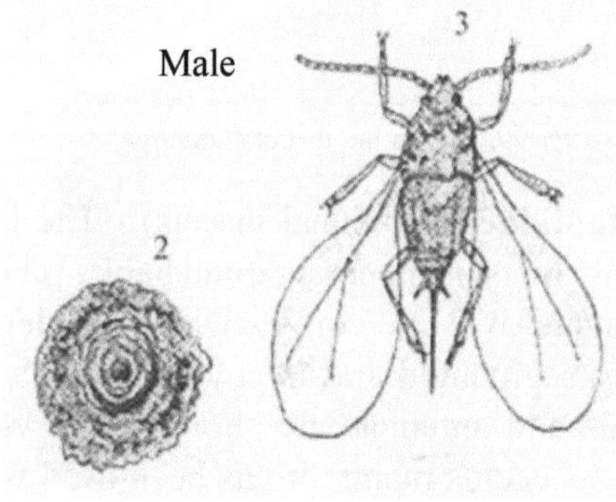

Fig. 101. Hard scale of rose (*Aonidiella auranii)*

Order 11 - Thysanoptera (Thrips)

('Thysano' = fringed; 'ptera = wings)

The insects of this order are very minute, at the most 0.8 mm. in length, mostly yellow, yellowish-brown or black in color. They are found in many kinds of vegetation, sucking and feeding on the sap from the foliage and flowers. A few forms are occasionally predacious and suck the body fluid from small insects such as, aphids, mites etc. Many of them exhibit of curving the apex of the abdomen upward, especially preparatory to flight.

The head is somewhat quadrangular in shape, with a with a pair of small, prominent compound eyes, three ocelli in the winged forms and their antennae consisting of 6 - 10 segments. Ocelli absent in the apterous forms. The mouthparts, the components of which are asymmetrical, are adapted for rasping and sucking. The right mandible is present, while the left mandible is vestigial. Prothorax well developed. The wings are narrow and strap-shaped, with long marginal setae like a hair comb and with very few veins and without cross veins. The wings of each side are interlocked by means of several hooked spines near the base of the hind wings, which hook on to a membranous fold on the fore wings. The size of the wings may vary in the same species or within the two sexes of the same species. Apterous and brachypterous forms are also common. The external genitalia of both the sexes are well developed in many forms. The abdomen is elongate, tapering posteriorly and consists of 11 segments. Cerci are absent. The newly hatched nymphs resemble the adults, but without wings . In some forms, there are prepupal and pupal stages, which have conspicuous wing pads, but these do not feed. Parthenogenesis is common. Some species are vectors of virus diseases.

There are about a dozen families in this order, but only a few are of much economic importance :

i) **Family - Thripidae** (Common thrips) - Common insects associated with many important crops. With their rasping and sucking mouthparts, they damage the plant parts, suck and feed on the sap. (e-g) *Thrips tabaci* - Infests cotton, onion etc.; *Stenchaetothrips biformis* - It is serious pest of of rice, especially the young plants **(Fig. 102)**; *Scirtothrips dorsalis* Infests chillies **(Fig. 103)**.

ii) **Family - Phioeothripidae** (Flower thrips) - They are large and stout-bodied thrips, dark brown or black in color, with light-colored or mottled wings. (e-g) *Haplothrips ceylonicua* - Infests the floral parts of inflorescens of coconut trees.

iii) **Family - Panchaetothripidae** (e-g) *Panchaetothrips indicus* - Infests turmeric plants.

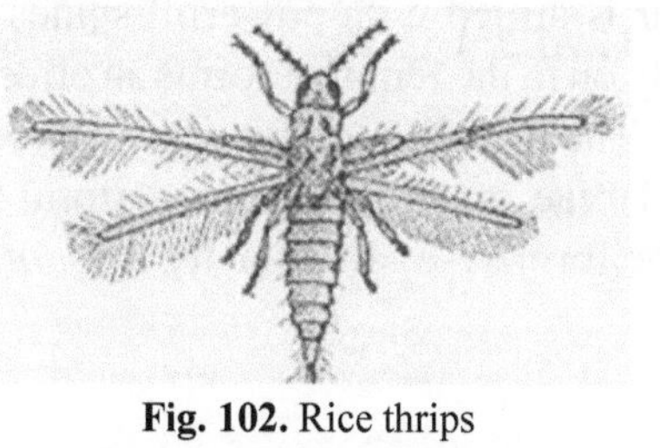

Fig. 102. Rice thrips (*Stenchaetothrips biformis*)

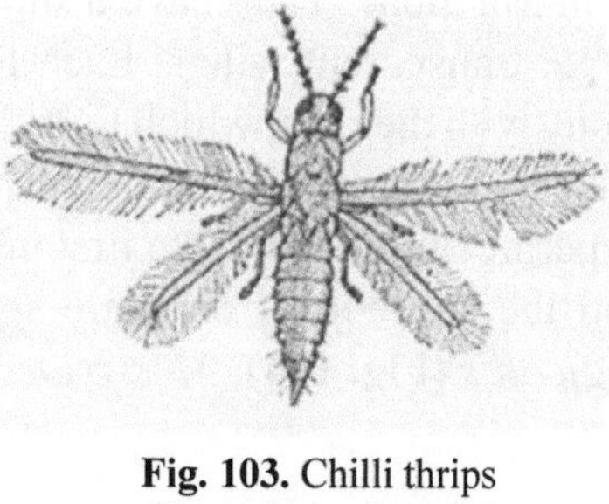

Fig. 103. Chilli thrips (*Scirtothrips dorsalis*)

Sub class - Pterygota or **Metabola**

Division - Endopterygota or **Holometabola**

Order - 1. Neuroptera (Alder flies, lacewings, antlions)

('Neuro' = nerve / wing veins; 'ptera' = wings)

The order includes small to large-sized, soft-bodied insects, which are rarely found in abundance. They mostly occur as individuals and are generally not strong fliers. They mostly have long, segmented and club-shaped antennae and posses mandibulo-suctorial type of mouthparts, with strongly developed mandibles, but feed mostly on liquid matter such as, honeydew or soft-bodied insects. Both the pairs of wings are very similar, membranous and are held roof-like covering the abdomen when at rest. The wing venation is primitive with many accessory veins. The wing-coupling apparatus is of the jugo-frennate type. The tarsi are five-segmented and the abdomen is without cerci.

The insects undergo complete metamorphosis. The larvae are generally campodeiform and often have long, sickle-like jaws. In many cases, the larvae are aquatic and have abdominal gills. They are mostly predacious, both as larvae nd adults and feed on aphids, ants etc. Feeding is done by sucking the body fluids of the victim through a fine channel formed between the mandibles and maxillae. The larvae generally pupate inside a silken cocoon. Pupae are exarate and decticous. The silk is produced by the malpighian tubules and is spun from the anus instead of from the modified salivary glands as in most of the insects.

A few species are of economic importance :

i) **Family - Hemerobiidae** (Brown lacewing flies) – Predatory insects, smaller and more hairy than the closely related green lacewing flies. They are mostly found on the bark of trees such as, pines and other conifers. (e-g) *Hemerobius stigma*; *Eumicromus australis*

ii) **Family - Mantispidae** (Mantid flies or Preying lacewing flies) - The members of this family have elongate prothorax and large raptorial prolegs as in Mantidae. They have a superficial resemblance to mantids, but have large, transparent wings. Each femur is armed with powerful spines and along with the tibia, which fold closely on to the femur, it forms an effective prehensile organ for seizing the prey. The members of this family undergo hypermetamorphosis, the first instar larvae are active. campodeiform type and the subsequent ones are scarabeiform (e-g) Mantid fly - *Mantispa rugicollis* **(Fig. 104)**; *M. styriaca.*

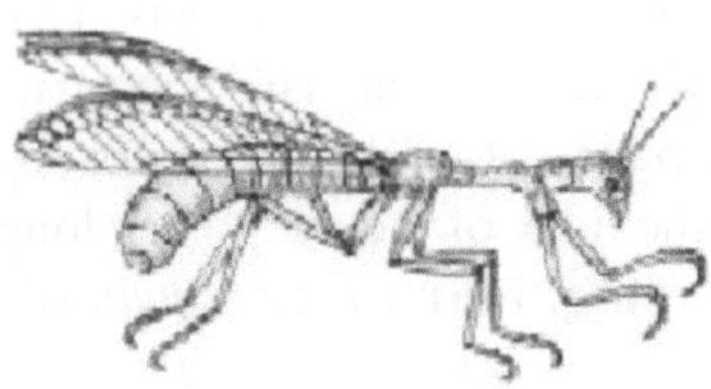

Fig. 104. *Mantid fly* (Mantispa rugicollis)

iii) **Family - chrysopidae** (Green lacewing flies, aphid lions) - These insects are commonly found preying on aphids and some of them infest plants. Many of them have bright green bodies and their prominent reddish golden eyes exhibit a metallic lustre and hence are called **'golden eyes'.** They are also known as stink flies as they emit a foul odor from a pair of prothoracic lands when they are disturbed. The adults have filiform antennae. The eggs are laid in bunches on the foliage. A small amount of viscid fluid is deposited on the leaf surface, which is drawn into a fine stalk by the abdomen and then the egg is deposited at the tip of the pedicel that hardens rapidly even before the egg is deposited. In some species, a number of eggs have a common stalk. The larvae are shorter and broader and they vary in color from whitish to greenish with red, chocolate or black markings. The body is covered by setae that Arise from dorso - ventral tubercles. They posses weak mouthparts and are predacious mostly on aphids and are called **'aphid lions'**. Besides aphids, they prey on jassids, psillids, white flies, coccids, thrips, mites etc. The larvae often carry the remains of their victims and plant debris on their backs. (e-g) Green lacewing fly *(Chrysopa cymbela)* - Feeds on cotton white flies **(Fig. 105)**; *C. scelestes*; *Chriysoperlacamea.*

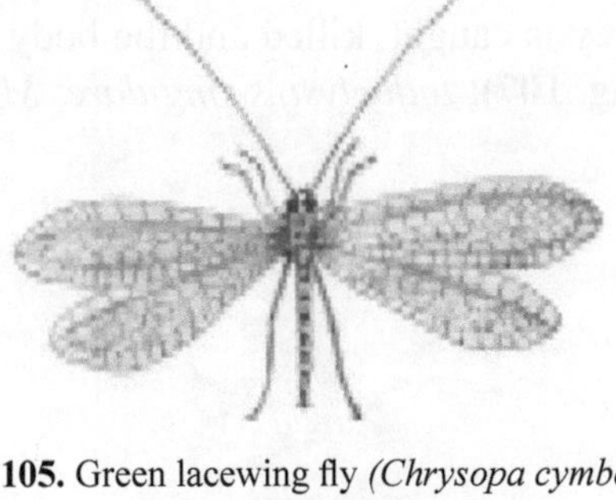

Fig. 105. Green lacewing fly *(Chrysopa cymbela)*

Family – Nemopteridae (Thread lacewing flies) - The insects belonging to this family are quite beautiful with enormously elongated ribbon-like hind wings. They are all predatory insects. The eggs are laid in refuse on floors. The larva has a large, quadrate head and long, curved, dendate mandibles.

They prey upon psocids and other minute insects. They cover themselves with dust and refuse and are hard to detect. The wing pads in the pupa are coiled like watch springs. The pupa is found inside a cocoon composed of sand and debris. In some forms, the neck of the adult is as long as the rest of the body. (e-g) *Nemoptera sinuata* **(Fig. 106)**; *Croce filipennis*; *Halter halteratus*

Fig. 106. Thread lacewing fly *(Nemoptera sinuata)*

Family - Myrmeleontidae (Ant lion flies) - The adults resemble dragon flies. They usually hide during the day and come our at dusk. The adults have short, knobbed antennae. Their wings are long and narrow with many accessory and cross veins, The eggs are usually deposited in the sand. The larvae known as **'ant lions'** are stout, flattened and ovoid with large head and bristles on the body, which are directed forwards. They have mandibulo - suctorial type of mouthparts and are voracious feeders. They excavate conical pits in th e ground for the purpose of securing their prey . They lie buried at the bottom of the pits with their mandibles slightly protruding. When an ant or any other small insect happens to come near the border of the pit, the sand is shifted and the insect falls into the bottom and is covered by more sand by the larva and eventually the prey is caught, killed and the body fluids sucked up. (e-g) *Palpares infirmus* **(Fig. 107)**; *Indoclystus singulare*; *Myrmelonformicarius.*

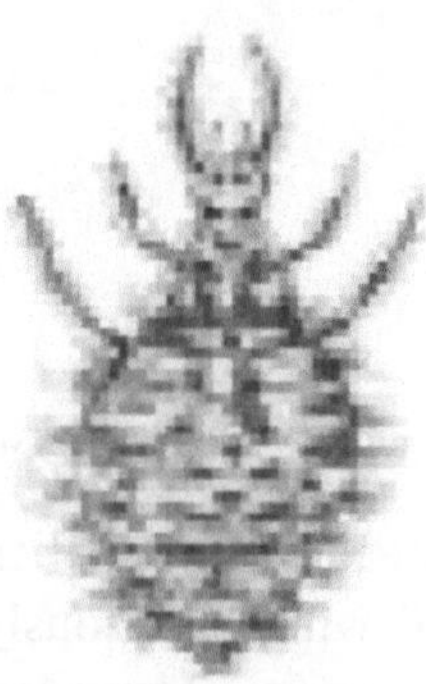

Fig. 107. Ant lion *(Palpares infirmus)* - Larva

Order 2 - Mecoptera (Scorpion flies)

('Meco' = long; 'ptera' = wings)

The members of this order have a long, beak-like prolongation of the head capsule. They are small to medium-sized insects, slender and have elongated filiform antennae. Males of many species carry the terminal segments of the abdomen curled upwards like a scorpion. They are mostly terrestrial insects, found commonly in rank herbage in shaded situations. They are usually carnivorous and have biting mouthparts. The compound eyes are well developed. The legs are long and slender and adapted for walking. The two pairs of wings are almost similar in form and size, membranous and often with large markings in the form of spots or bands. The wings are held longitudinally and horizontally while in repose. Wing venation is primitive. In some forms, the wings are reduced. The abdomen is elongate with short, single-segmented cerci. The males have very prominent genitalia. Metamorphosis is complete and the larvae are eruciform, with biting mouthparts. Pupae are exarate and decticous.

There are about half a dozen families under this order, of which only the families - Bittacidae and Panorpidae are important :

Family - Panorpidae (Scorpion fly)

(e-g) Scorpion fly *(Leptopanorpa furcata* **(Fig. 108)**

Family - Bittacidae

(e-g) *Bittacus latipennis*

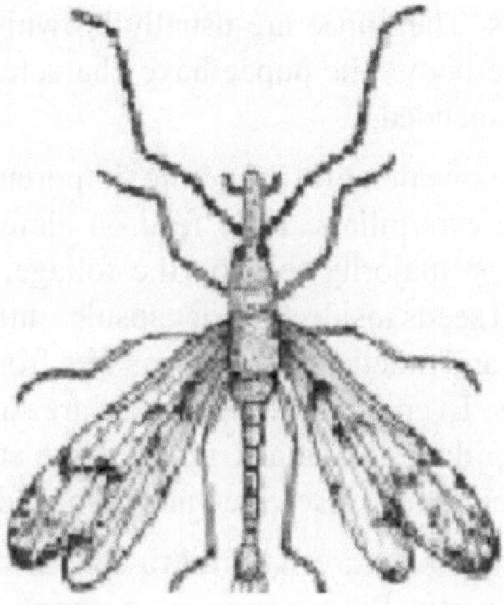

Fig. 108. Scorpion fly *(Leptopanorpa furcata)*

Order 3 - Lepidoptera (Butterflies, Skippers, Moths)

('Lepido' = scale; 'ptera' = wings)

Among the insects, butterflies, skippers and moths belonging to this order are the most common and easily recognizable ones. They are quite familiar on account of their beautiful color patterns, especially among butterflies. They are abundant in vegetable and flower gardens and are cosmopolitan. It is the second largest order of class Insecta and at least on lakh species have been recorded all over the world. Some of them are quite small, whereas some others are very large and have a wingspan of up to 30 cm., but they are very light in weight.

Metamorphosis in this order is complete. The larvae usually referred to as caterpillars, differ entirely in structure and food habits from the adults. Like the adult forms, the caterpillars also exhibit a great variety of form and coloration. They exhibit many adaptations to the general coloration of their environment. Their mouthparts are typical biting and chewing type and they feed mostly on green vegetation. Most of the adults have a suctorial proboscis formed by the fusion of the maxillae, while some such as, he fruit-sucking moths have piercing and sucking type of mouthparts. There is a definite pupal or resting stage, after the caterpillars are full grown, during which a complete transformation takes place and ultimately the adult butterfly or moth emerges.

The members of this order are easily recognized at all the different growth stages and there is a greater uniformity of structure and appearance. Eggs are of three main types namely, upright dome-shaped as in the case of butterfly eggs, rounded as in the case of moth eggs and flat, scale-like as in the case of some small moth eggs. The vast majority of larvae are more or less of one type, with usually three distinct pairs of thoracic legs and two to five pairs of abdominal or prolegs. The pupae are usually brownish with the appendages firmly opposed to the body. The pupae have characteristic hooks at the apex by which they are suspended.

Lepidoptera are of considerable economic importance on account of the feeding habits of the caterpillars. They feed on almost all parts of plants of various crops; the vast majority feed on the foliage, roots and fruits, while others eat flowers and seeds inside pods or capsules, mine into leaves, bore into roots, shoots or stem and practically no crop is free from damage of one part or other by these insects. Even grains after harvest are subjected to attack during storage. These caterpillars also attack most of the stored products. Timber, carpets, clothes, paper etc. are also eaten and damaged by some caterpillars.

While most of the members are harmful, a few are beneficial to mankind, either directly or indirectly. Silkworm is one among such caterpillars. It gives valuable silk. Many butterflies and moths indirectly help man in effecting pollination in a number of crops. Many lively and colorful butterflies form the basis for art and design.

The adult insect has a small head, the major part of which is occupied by the large, globular compound eyes, consisting of a large number of opmmatidium. Two ocelli are present, one on each side of the head close to the compound eyes. They are not clearly visible as scales generally conceal them. The antennae vary in length, structure and in the number of segments. The mouthparts of butterflies and many moths are of siphoning type, while in some species the mouthparts are vestigial and they do not feed in the adult stage. In a few species, the adults have piercing and sucking type of mouthparts. When not in use, the proboscis in the siphoning type is coiled like a watch spring and stowed away beneath the head. There is considerable variation in the length of the proboscis among the species. The prothorax and metathorax are small, while the mesothorax is larger. Three pairs of legs are present and the tarsi are 5 segmented. Generally two pairs of large, triangular wings are present. They are membranous and are fully covered by scales, which give the characteristic color patterns to the wings. The wing-coupling apparatus are highly developed. The abdomen consists of 10 segments. The first segment is mostly reduced. The last 2 or 3 segments are modified into external genital organs.

Originally the order was into two groups viz., Rhopalocera (Butterflies) and Heterocera (Moths), based on their distinct difference in their morphological and othe characters.

The butterflies are diurnal in habit with less denser covering of scales, bright colored, having a slender body, club-shaped or capitate antennae and siphoning type of mouthparts. They feed on nectar, honey or plant juices. When at rest, both the pairs of wings are held vertically upwards.

The moths on the other hand are nocturnal, with profuse scales, mostly dull-colored, stout and with pectinate or setaceous antennae. Their mouthparts are atrophied and they do not feed generally. A few species have piercing and sucking type of mouthparts. When at rest, their wings are held horizontally or folded over their back like a roof.

Some Entomologists divided Lepidoptera into two groups viz., Macro and Micro Lepidoptera based on their size. Subsequently several systems of classifications were propounded on the basis of the characters of the adults, larvae, pupae etc. **A.D. Imms** followed a system of classification, which was the most acceptable one. He divided the order into three suborders viz., Zeugloptera, Monotrysia and Ditrysia. Based on the general features and other characters, the order is divided into three groups viz., Butterflies, Skippers and Moths.

Distinguishing features of Butterflies, Skippers and Moths

Butterflies	**Skippers**	**Moths**
Active by day (Diurnal)	Active by night Nocturnal	Active by night Nocturnal
Wings held vertically upright when at rest	Wings held vertically upright when at rest	Wings held horizontally or folded over the back like a roof
Wings large in relation to the body size	Wings comparatively small	Wings comparatively small
Body slender	Body intermediate in size	Body stout
Antennae capitate	Antennae club-shaped with a terminal hook	Antennae variable
Flight erratic and bouncing	Flight erratic, but darting	Flight mostly straight
Color bright with beautiful patterns	Color dull or dark, with distinct patterns	Color pale or dark with dull patterns
Frenulum absent	Frenulum absent	Frenulum absent
Dorsal ocelli present	Ocelli absent	Ocelli absent
Larvae smooth and not hairy	Larvae not hairy	Larvae mostly hairy
Pupae obtect (Chrysalid)	Pupae obtect (inside spun cocoon)	Pupae obtect (Pupation in concealed form)

There are more than 1,20,000 recorded families under the order Lepidoptera, however, a vast majority of them are harmless and are not of much economic importance. The more important families of economic importance are described below :

i) **Family - Papilionidae** (Swallow tails) - This family includes many large, conspicuously colored, beautiful butterflies found most commonly in the tropical countries. The antennae are slender, capitate and abruptly knobbed. The wings are very variable in shape and in many species the hind wings are prolonged into tail-like processes . The ground color is generally black, marked with conspicuous shades of yellow, orange, red, green or blue. The wings have amplexiform wing-coupling mechanism. The larvae are smooth and provided with fleshy dorsal tubercles. The larva possesses an eversible, forked scent gland (osmeterium) on the upper part of the prothorax and when the larva is disturbed, an offensive odor emanates from the gland. The pupae have a silken girdle and the head region bears two lateral cephalic projection (e-g) *Papilio demoleus* - The citrus or lemon butterfly **(Fig. 109)**.

Fig. 109. The lemon butterfly *(Papilio demoleus)*

ii) **Family - Nymphalidae** (Brush-footed butterflies) - This is a large family of butterflies. The butterflies are usually large - sized, beautiful and brightly-colored. The forelegs are much reduced in size and the tibia are short and clothed with long hairs and hence are called brush-footed butterflies. Larvae are smooth and cylindrical. The pupae are suspended from the tail end (e-g) *Danaise chrysippus* - It is a beautiful butterfly found throughout the year and have warning coloration **(Fig. 110)**; *Ergolis merione* - Castor butterfly or spiny caterpillar, a pest of castor crop.

Fig. 110. *Danaise* butterfly *(Danaise chrysippus)*

iii) **Family - Pieridae** (Cabbage butterfly) - These are butterflies found very commonly everywhere. They are white, yellow or orange with characteristic black markings. They have six functional legs with a pair of forked claws on each foot. The larvae are usually green or yellowish, long and have a covering of setae. The pupae have silken girdle, with which they are held on to the support to which only their caudal extremity is attached and are thus suspended in a upright position. The butterflies are noted for their long migratory flight and hey are mostly found in fields where cruciferous crops are grown.(e-g) *Pieris brassicae* - The cabbage butterfly, the larvae of which feed on the leaves of cabbage and other cruciferous crops **(Fig. 111)**; *Eurema hecabe* - Infests daincha crop; *Pieris rapae* - Infests rape and mustard.

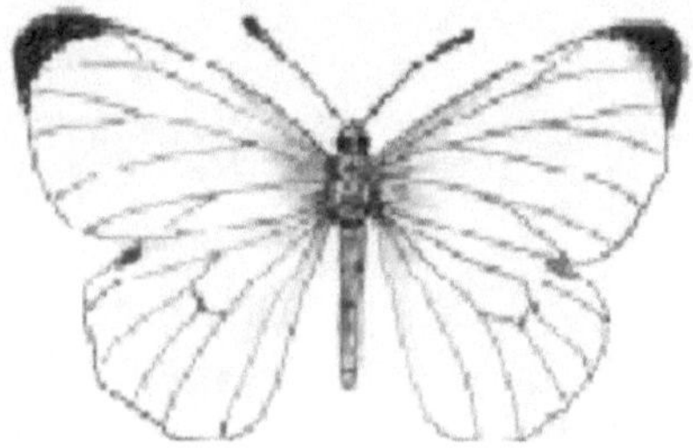

Fig. 111. The cabbage butterfly *(Pieris brassicae)*

iv) **Family - Lycaenidae** (Blues, Coppers) - These are mostly small, delicate, brightly Colored butterflies with slender body. The upper surface of the wings is prominently colored metallic blue, coppery, dark brown or orange, but the under surface somber-colored with dark-centered eye spots or delicate markings. The hind wings have often a tail - like prolongation. The eyes have a rim of white scales. The larvae are slug-like, somewhat flattened with retractile head. No osmeteria are present. Cephalic or anal horns are present. Body bears numerous secondary setae. (e-g) *Virachola isocrates* - The Anar butterfly or pomegranate fruit borer **(Fig. 112)**; *Lampides boeticus* - The blue butterfly, the caterpillars of which bore and damage the pods of red gram, bean, lab lab etc. **(Fig. 113)**.

Fig. 112. Anar butterfly *(Viracola isocrates)*

Fig. 113. The blue butterfly *(Lampides boeticus)*

v) **Family - Hesperiidae** (Skippers) - These are generally small, stout-bodied, butterly-like insects and are generally called skippers because of their fast, darting and erratic flight, unlike butterflies. They appear moth-like with wide, hairy bodies and wide heads. The antennae are widely separated at their bases and dilated apically to form a gradual club with the tips usually recurved or hooked. When at rst, most skippers hold front and hind-wings at a different angle. The larvae are moderately stout, smooth and their body taper towards both the extremities. The larva pupates inside a cocoon made up of leaves fastened together with silken thread (e-g) *Paranara mathias* - Rice skipper.

vi) **Family - gelechiidae** (Palp moths) - This is a large family and is of much economic importance and includes some serious crop pests. The moths are all small-sized with prominent palps and with fringed hind wings. The larvae vary in habits considerably. Some are leaf miners, some form galls, while some ohers tie the terminal buds and leaves of plants into silk-bound nests, within which they live securely and feed on the plant parts (e-g) *Aproaerema modicella* - Groundnut leaf miner **(Fig. 114)**; *Gnorimoschema operculella* - Potato tuber moth; *G. heliopa* - Stem borer of tobacco and other solanaceous crops; *Pectinophora gossypiella* - Cotton pink boll worm; *Sitotroga cerealella* - Rice grain moth **(Fig. 115)**; *Holcocera pulverea* - Predacious on lac insect.

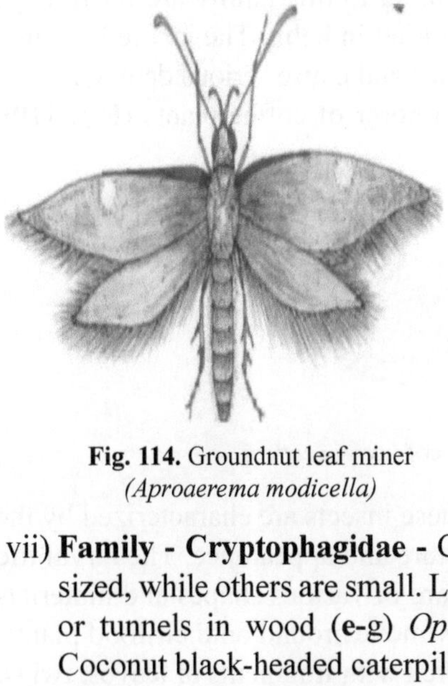

Fig. 114. Groundnut leaf miner *(Aproaerema modicella)*

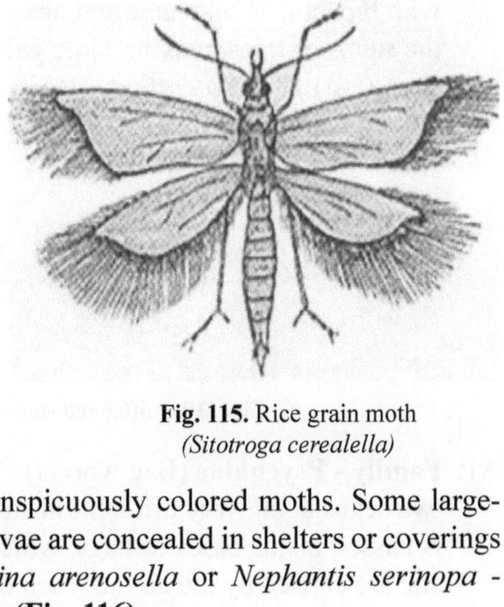

Fig. 115. Rice grain moth *(Sitotroga cerealella)*

vii) **Family - Cryptophagidae** - Conspicuously colored moths. Some large-sized, while others are small. Larvae are concealed in shelters or coverings or tunnels in wood (e-g) *Opicina arenosella* or *Nephantis serinopa* - Coconut black-headed caterpillar **(Fig. 116)**.

viii) **Family - Gracillariidae** - The insects of this family are mostly small and delicate moths with long, narrow and fringed wings and can easily be recognized by their habit of resting with the fore part of the body upraised by the widely separated fore legs. The larvae are apodous and are mostly leaf miners. The larva pupates in a delicate cocoon inside the mine (e-g) *Phyllocnistis citrella* - Citrus leaf miner **(Fig. 117)**; *Gracillaria theivora* - Leaf miner pest of tea crop.

ix) **Family - Plutellidae** - This is a family consisting of small moths. The larvae feed from within a slight web in the leaves or occasionally mine the leaves or stem. (e-g) *Plutella maculipennis (xylostella)* Diamond-back moth, which attacks cabbage and other cruciferous crops **(Fig. 118)**.

Fig. 116. Black-headed caterpillar *(Opisina arenosella)*

Fig. 117. Citrus leaf miner *(Phylocnistis citrella)*

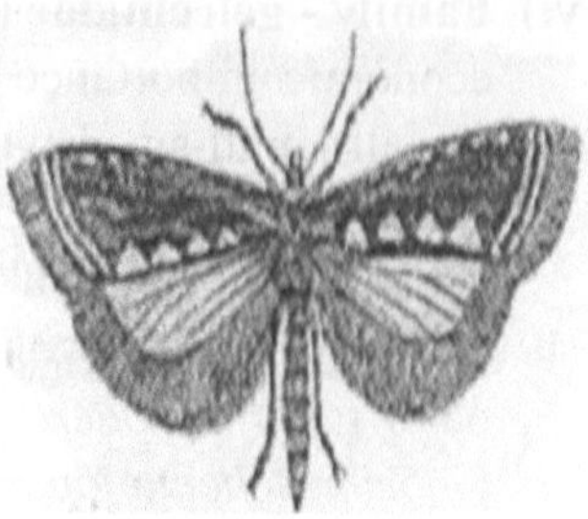

Fig. 118. Diamond-back moth *(Plutella maculipennis)*

x) **Family - Cossidae** - The moths belonging to this family are fairly large with bipectinate antennae and are nocturnal in habit. The larvae bore into the stems of trees, making large galleries and cause serious damage to the trees (e-g) *Zeuzera coffeae* - Red stem borer of coffee plants **(Fig. 119)**; *Cossus cadambae* - Teak stem borer.

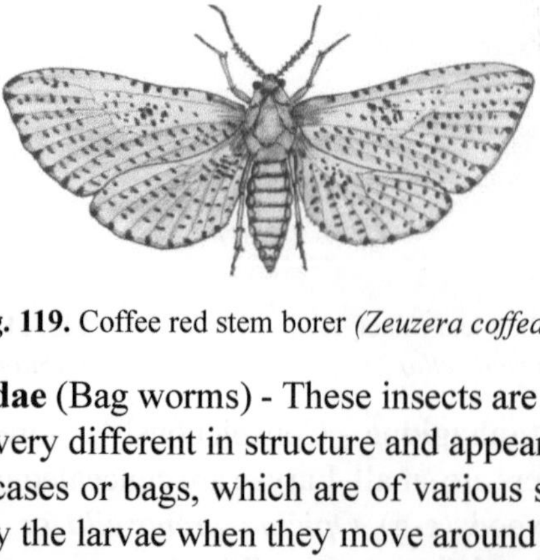

Fig. 119. Coffee red stem borer *(Zeuzera coffeae)*

xi) **Family - Psychidae** (Bag worms) - These insects are characterized by the two sexes being very different in structure and appearance. The larval life is passed inside cases or bags, which are of various shapes and materials and are carried by the larvae when they move around on their food plants. The bags are made of silk and are covered with fragments of leaves, twigs, grass etc. Pupation takes place inside the larval cases. The males are swift fliers with wings thinly clothed with hairs and imperfect scales and devoid of markings. The frenulum is very large. The females are always apterous with the antennae, mouthparts and legs degenerate or absent and are worm-like. The female lays eggs inside the bag and the eggs overwinter in the bags and hatch during the spring and each larva constructs its own bag. Full-grown larva attaches the bag to a twig and pupates inside it (e-g) *Monatha albipes* - Bag worm of coconut.

xii) **Family - Metarbelidae / Teragriidae** (Bark and stem borers) - the moths are stout, light brown in color, shining and fringed with fine, dense hairs. The dark brown caterpillars construct long, zigzag galleries in the tree barks with silken thread, frass and bark tissues, stay hidden inside the galleries and feed on the bark and stem tissues (e-g) *Indarbela tetraonis*- Stem and bark borer of curry leaf trees, mango trees, citrus plants etc.; *Indarbela quadrinotata* - Stem and bark borer of cashew, guava, jack, cocoa trees etc. **(Fig. 120)**; *Indarbela theivora* - Bark borer of tea plants.

Fig. 120. Stem and bark borer of cashew *(Indarbela quadrinotata)*

xiii)**Family - Limacodidae / Cochlididae** (Slug caterpillars) - The moths of this family have a somewhat rounded fore wings and a shorter and rounded hind wings. The body is thick set or short and the male moths have pectinate antennae. Most of these moths are dull brown in color, while some are Larva Adult moth greenish in color. The larva is slug- like with the head , legs and sucker legs retractile, the ventral surface forming a peculiar smooth, sucker-like attachment along which the larva slides like a slug and hence are called slug caterpillar. The larvae of some species bear spiny tubercles and the spines are often very irritant (e-g) *Parasa lepida*- Castor slug caterpillar; *Contheyla rotunda* - Coconut slug caterpillar **(Fig. 121)**.

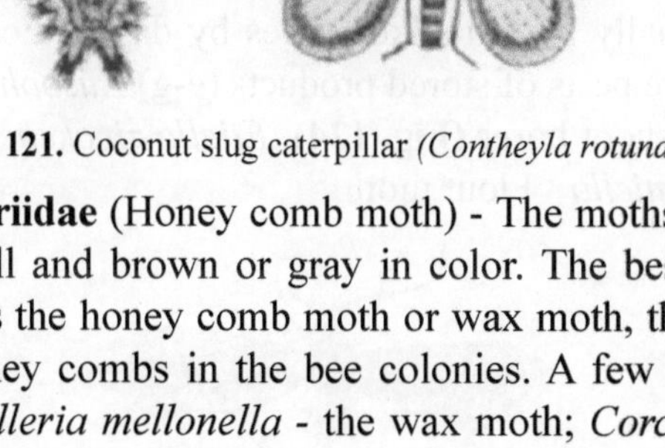

Fig. 121. Coconut slug caterpillar *(Contheyla rotunda)*

xiv)**Family - Galleriidae** (Honey comb moth) - The moths belonging to this family are small and brown or gray in color. The best known member of this family is the honey comb moth or wax moth, the larvae of which destroy the honey combs in the bee colonies. A few are pets of stored grains (e-g) *Galleria mellonella* - the wax moth; *Corcyra cephalonica* - Rice meal moth, the larvae of which infest paddy as well as grains of many other crops. They also cause damage to shredded and *powdered* grains **(Fig. 122)**.

Fig. 122. Rice meal moth *(Corcyra cepalonica)*

xv) **Family - Crambidae** (Grass moths, snout moths) - The larvae feed grasses, reads and some cultivated graminaceous plants such as, sugarcane, maize, jowar etc. Many are borer pessts. The moths are small with narrow and elongated fore wings. They are whitish, pale brown or grayish in color and are crepuscular or nocturnal in habit. The maxillary palp and labial palp are long and straight and together they form a snout-like protuberance in front of the head. In the day time they rest on the stems or leaves of the host plants with the wings folded over the back, (e-g) *Chilo infuscatellus* - Sugarcane early shoot borer **(Fig. 123)**; *Chilo sacchariphagus indicus* - Sugarcane internode borer; *Chilo zonellus* - Jowar borer.

Fig. 123. Sugarcane early shoot borer *(Chilo infuscatellus)*

xvi)**Family - Phycitidae** - The moths of this family are small and delicate insects with elongate fore wings and the hind wings having well defined pattern o f hairs on the dorsal side and a simple frenulum in both the sexes. The larvae usually live in silken tubes by day and come out to feed by night. A few are pests of stored products (e-g) *Euzophera perticella* - Egg plant / brinjal shoot borer **(Fig. 124)**; *Etiella zinckenella* - Pea pod borer; *Anagastra kuhniella* - Flour moth.

Fig. 124. Brinjal shoot borer *(Euzophera perticella)*

xvii)**Family - Pyralidae** - The moths of this family are small and delicate insects with triangular or elongate fore wings and broad hind wings. The larvae generally feed on dry or decaying vegetable matter. Some are fruit borers (e-g) *Leucinodes orbonalis*- Brinjal shoot and fruit borer **(Fig. 125)**.

Fig. 125. Brinjal shoot and fruit borer *(Leucinodes orbonalis)*

xviii)**Family - Pyraustidae** - This is a large family consisting of many crop pests of economic importance. The moths are small and delicate. The larvae are borers feeding on the inner tissues of stalks an crowns of cereals and other plants. The larvae feed from inside a delicate web among spun - up leaves or in the stems (e-g) *Tryporyza incertulas* - Rice stem borer; *Gnaphalocrocis medinalis* - Rice leaf folder; *Antigastra catalaunalis* - Gingelly leaf and shoot webber; *Nymphula depunctalis* - Rice case worm; *Dichocrosis punctiferalis* - Castor shoot and capsule borer; *Maruca testulalis* - Spotted pod borer of pulses **(Fig. 126)**.

Fig. 126. Red gram spotted pod borer *(Maruca testulalis)*

xix)**Family - Schoenobiidae** - It is a family of small, nocturnal moths. The caterpillars are mostly borer pests (e-g) *Scirpophaga nivella* - Sugarcane top borer **(Fig. 127)**.

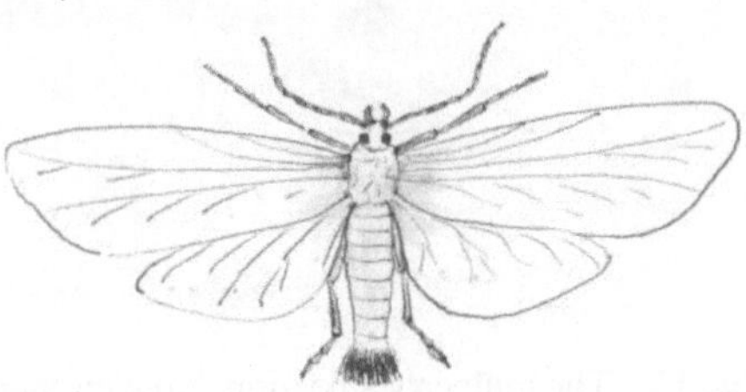

Fig. 127. Sugarcane top borer *(Scirpophaga nivella)*

xx) **Family - Pherophoridae** (Plume moths) - The moths of this family are very lightly built, chiefly characterized by having the wings deeply fissured and fringed and appear as feathers. The fore wings are longitudinally cleft two to four divisions and the hind wings into three divisions. The fore wings are much elongated. The legs are long and slender and have prominent tibial spurs. Maxillary palpi are absent (e-g) *Exelastis atomosa* - Red gram plume moth; *Sphenarches caffer* - Bottle gourd plume moth **(Fig. 128)**.

Fig. 128. Bottle gourd plume moth *(Sphenarches caffer)*

xxi)**Family - Saturniidae** (Giant silkworm moths or Emperor moths) - The family includes large-sized moths, some of which have wing expanse of about 25 cm. The caterpillars are stout and smooth. Several species yield silk (e-g) *Antheraea paphia* - The Tassar silkworm; *Antheraea assamia* - The Muga silkworm; *Philosamia ricini* - The Eri silkworm; *Actias selene* - The moon moth.

xxii)**Family - Bombycidae / Eupteroptidae** (Silkworm moths) - The moths of this family are medium - sized, heavy and very hairy. They have pectinate antennae, hairy, spineless legs and lack proboscis. The larvae are fleshy and are voracious feeders. Silkworms are beneficial insects and are reared commercially for their silk. In the genus *Eupterote*, the larvae are tufted with long hairs (e-g) *Bombyx mori* - The mulberry silkworm **(Fig. 129)**; *Eupterote mollifera* - Moringa hairy caterpillar.

Fig. 129. The mulberry silkworm *(Bombyx mori)*

xxiii)**Family - Geometridae** (Loopers or measuring worms) - The moths of this family are slenderly built, delicate and have relatively broad wings with fine wavy lines, but are rather weak fliers. When at rest, the wings are often laid horizontally. They are nocturnal in habit. The two sexes are different in color and in a few species, the females are wingless or have only rudimentary wings. The larvae, also known as **'geometers'**or **'earth measurers'** are elongate, very slender and have only two pairs of abdominal legs on the sixth and tenth segments and they walk by drawing the posterior part of the body closeto the thoracic legs forming a loop and then moving the anterior end of the body, thus progressing in a characteristic looping fashion. Many loopers, when disturbed, stand erect on the prolegs and remain motionless resembling small twigs (e-g) *Hyposidra successaria* - Pest of daincha, a green manure crop; *Biston suppressaria* - Tea looper, a pest of tea plants **(Fig. 130)**.

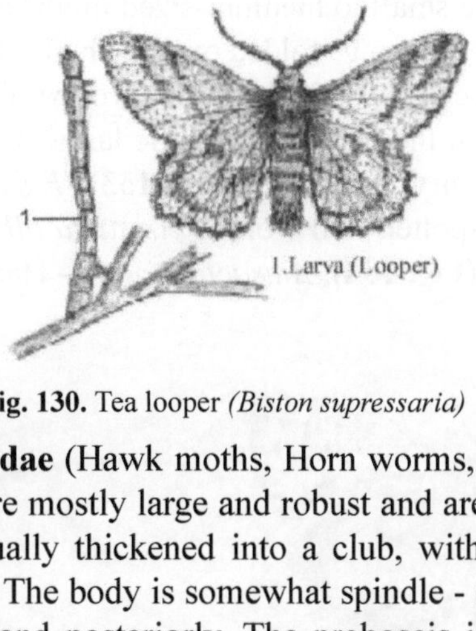

Fig. 130. Tea looper *(Biston supressaria)*

xxiv)**Family - Sphingidae** (Hawk moths, Horn worms, Sphinx caterpillars)- The hawk moths are mostly large and robust and are powerful fliers. The antennae are gradually thickened into a club, with the terminal region pointed or hooked. The body is somewhat spindle - shaped, tapering/ and pointed anteriorly and posteriorly. The proboscis is strongly developed and very long. The fore wings are much larger than the hind ones, narrow and pointed. The moths fly with a very rapid wing beat. Some are day-fliers, but most of them are active at dusk or twilight. The moths feed on nectar from flowers and appear like humming bird moths while feeding and are sometimes called humming bird moths.The larvae of most species have a conspicuous horn or spine-like process on the dorsal surface of the eighth abdominal segment. Some of he larvae assume a sphinx – like position and hence they arereferred to as sphinx caterpillars. The larvae mostly pupate in the ground and in some cases form pitcher - like pupae, with the proboscis appearing as a handle of a pitcher (e-g) *Acherontia styx* - The horned caterpillar or sphinx caterpillar of gingelly **(Fig. 131)** ; *Herse convolvuli* - The hawk moth **(Fig. 132)**.

Fig. 131. Horned caterpillar *(Acherontia styx)*

Fig. 132. Hawk moth *(Herse convolvuli)*

xxv)**Family - Arctidae** (Tiger moths, Wooly bears, Hairy caterpillars) - It is one of the largest group of Lepidoptera. The moths are rather stout, often with moderately broad wings, which are frequently spotted or banded in bright colors. The small to medium-sized moths are mostly nocturnal and when at rest hold the wings folded over the body like a roof. The larvae are often hairy and are called hairy caterpillars or wooly bears. The cocoons are made largely from the body hairs of the larvae (e-g) *Utitheisa pulchella*- The sunnhemp hairy caterpillar **(Fig. 133)**; *Earias insulana* and *Earias vitella* - Cotton spotted boll worms; *Amsacta albistriga* - Groundnut red hairy caterpillar **(Fig. 134)**; *Amsacta moorei* - Hairy caterpillar of millets, castor etc.

Fig. 133. Sunnhemp hairy caterpillar *(Utitheisa pulchella)*

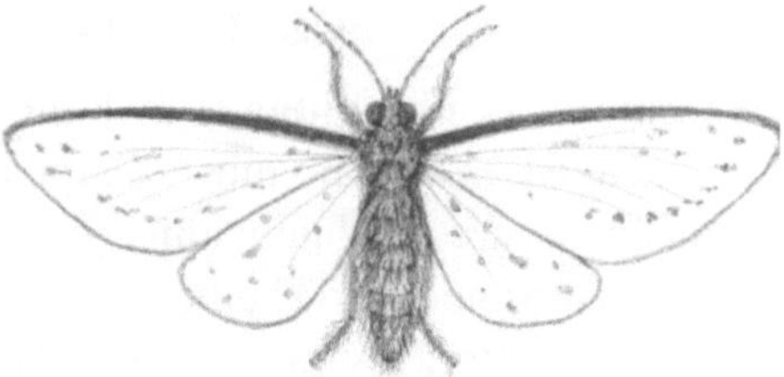

Fig. 134. Groundnut red hairy caterpillar *(Amsacta albistriga)*

xxvi)**Family - Noctuidae** (Nocturnal moths) - It is the largest group of Lepidoptera and has several species. It includes the cutworms, army worms and many other serious crop pests. The moths are mostly nocturnal in habit and several of them are attracted by artificial light. The noctuids vary greatly in size and color, but most of them are of medium size, measuring 2.5 - 4.0 cm. across the wings, cryptic and somber-colored. The moths are mostly stout , with the fore wings somewhat narrow and the hind wings broader. The antennae are generally filiform. A frenulum is always present. They have well developed proboscis with which they suck nectar. The larvae are usually smooth and dull-colored and most of them posses five pairs of prolegs. Some are semiloopers with only three pairs of prolegs. Many are polyphagous. The caterpillars are also nocturnal in habit and do the damage generally during the dark hours (e-g) *Helicoverpa (Heliothis) armigera* - The American cotton boll worm **(Fig. 135)**; *Agrotis ipsilon* - The gram cutworm **(Fig. 136)**; *Plusia signata* - The gingelly and snake gourd semilooper **(Fig. 137)**; *Eublemma olivacea* - Bringal leaf roller; *Spodoptera (Prodenia) litura* - The tobacco caterpillar **(Fig. 138)**; *Cirphis compta* - Rice army worm; *Sesamia inferens* - Ragi pink borer; *Achaea janata* - Castor semilooper **(Fig. 139)**; *Othreis materna* - The fruit sucking moth **(Fig. 140)**; *Trichoplusia ni* - The cabbage green semilooper.

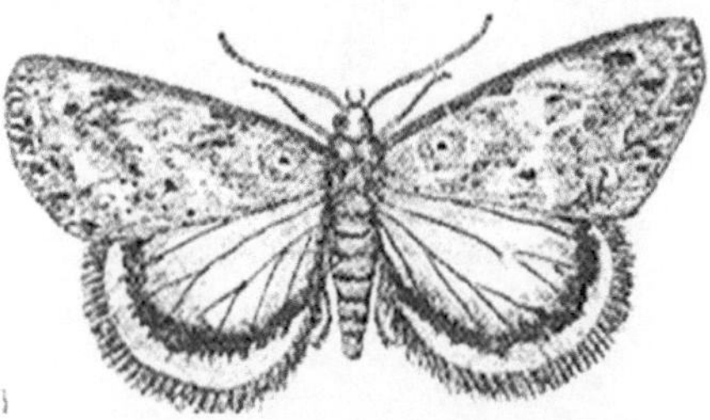

Fig. 135. American cotton boll worm
(Helicoverpa / Heliothis armigera)

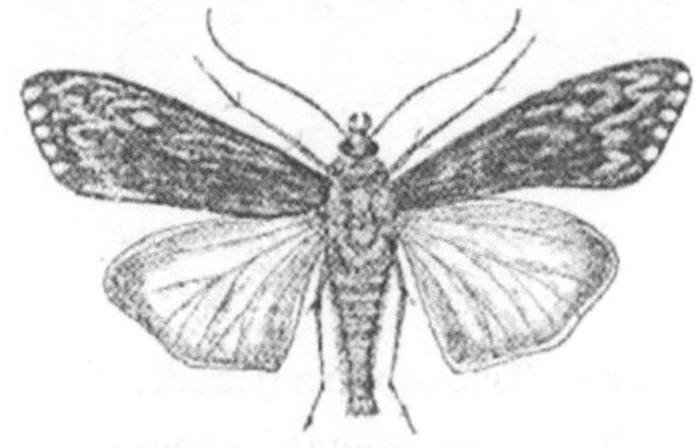

Fig. 136. The gram cutworm
(Agrotis ipsilon)

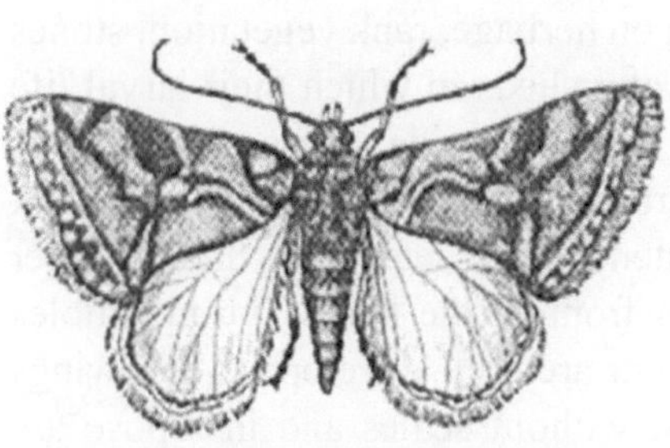

Fig. 137. The snake gourd semilooper
(Plusia signata)

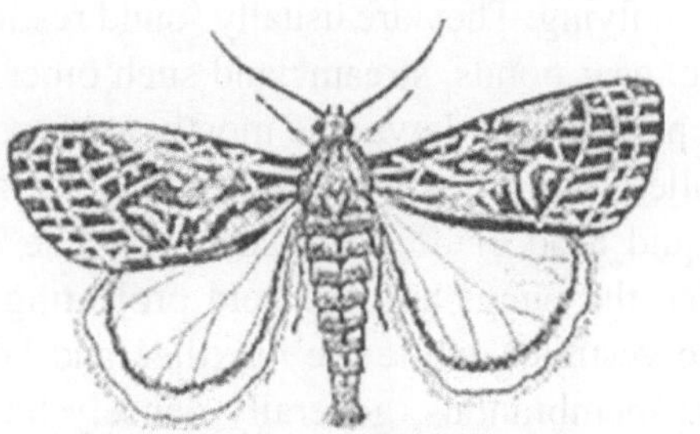

Fig. 138. The tobacco caterpillar
(*Spodoptera litura*)

Fig. 139. Castor semilooper
(Achaea janata)

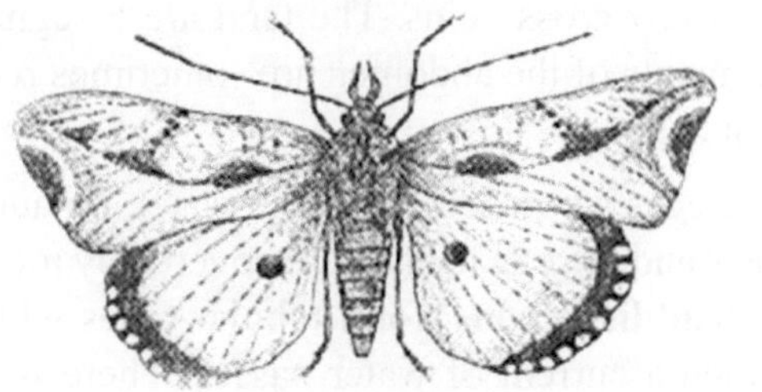

Fig. 140. The fruit sucking moth
(Othreis materna)

xxvii) **Family - Lymantriidae** (Tussock moths) - The family includes medium to large-sized, hairy moths. The males hav prominently bipectinate antennae. The females have generally short wings and a large tuft of hairs near the apex of the abdomen, which are used to cover the egg masses. The larvae are often brightly colored and thickly covered with tufts or tussocks of hairs (e-g) *Lymantria dispar* - The gypsy moth; *Psalis pennatula* - Rice yellow hairy caterpillar **(Fig. 141)**; *Euproctis fraterna* - The hairy caterpillar or tussock caterpillar of rose and castor **(Fig. 142)**.

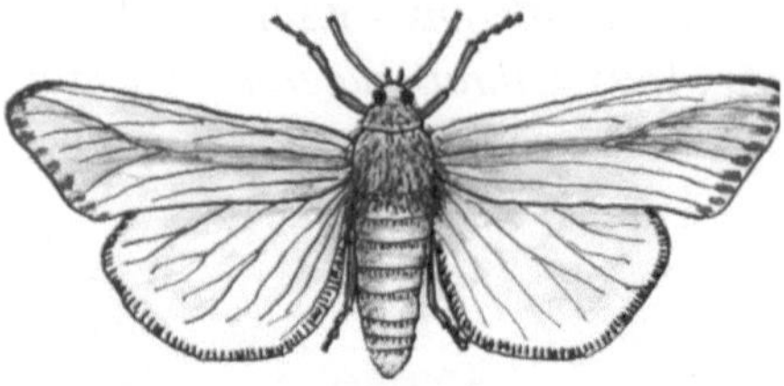

Fig. 141. Rice yellow hairy caterpillar *(Psalis pennatula)*

Fig. 142. Rose hairy caterpillar *(Euproctis fraterna)*

Order 4 - Trichoptera (Caddis flies)
('Tricho' = hair; 'ptera' = wings)

The insects belonging to this order are small to medium-sized and are moth-like in appearance and color patterns. They are poor fliers and are not generally seen flying. They are usually found resting on herbage, rank vegetation, stones etc. near ponds, streams and such other water sheds in which their larval life is passed. The larva has mostly a part of its body inside a case and hence is called **'caddle worm'**. The adults are attracted to light. They generally lick liquid food or do not feed at all. The antennae are setaceous, much longer than the wings and are held projecting in front of the head. The mandibles are vestigial, while the maxillae and labium are well developed. The wings are membranous, generally densely hairy, without scales and in repose are held roof-like over the back. The fore wings are elongated and narrow, while the hind wings are broader with a folding anal area. Wing venation simple with few cross veins. The tarsi are 5 segmented. In the female, the terminal segments of the abdomen are sometimes retractile and tubular and serve as an ovipositor.

The eggs are generally laid in water, on aquatic vegetation or on overhanging trees and the egg masses are covered by mucilage. The larvae generally known as **'caddis worms'** form tubular cases with an opening at each end, through which a current of water passes. There is a pair of hook-like appendages at the apex of the abdomen with which the larva anchors to the case and drags it during its movements. In some species, instead of portable cases, a silken cover is formed in stones etc. The larvae live submerged in water and breathe by tracheal gills or blood gills or their body wall. The larvae are mostly cruciform, but in some families they are campodeiform.

The order is divided into about two dozen families, but no Trichoptera is known to be a pest or of any economic importance (e-g) *Macronema fastuosum; Phryganea avaris; Notanotolica magna; Polymorphanisus nigricornis.*

Order 5 - Diptera (True flies and mosquitoes)

('Di' = two; 'ptera' = wings)

The Diptera are highly specialized, mostly small and soft-bodied insects. Their chief character is that they have only a single pair of membranous wings, the hind pair of wings being reduced to stumps called **halteres'**, which function as organs of equilibrium or balance. They are found in flowers feeding upon nectar, in decaying organic matter, in heaps of refuse, dung, excreta etc. Some are predacious on other insects, while a few others suck blood of man and livestock. They are also responsible for causing or spreading some of the serious diseases such as, malaria, sleeping sickness, yellow fever and **'surra'** in human beings. The larvae of some families are phytophagous and are serious crop pests. Some are parasitic, while a few others are very useful in the biological control of pests. Some are parasitic on mammals. Some aid in pollination.

The head of the adult is often large and mobile, mostly occupied by the large compound eyes. The eyes on the two sides of the head may be contiguous (**'Holoptic'**) or set apart from each other (**'Dichoptic'**). This character may differ between the sexes in some families. Three dorsal ocelli are also present. The antennae lie close to each other in the antennal grooves and exhibit many modifications. The mouthparts show a wide range of structural adaptations depending upon the different feeding habits. They may be either piercing and sucking type or sponging and lapping type, while in some others the mouthparts are poorly developed or non-functional. The mandibles are absent, except in the females of most of the blood-sucking forms and the maxillae are rarely complete. The maxillary palpi are important for classification purpose. The labium is in the form of a proboscis.

The mesothorax is more prominent than the pro- and metathoracic segments, which appear as bands. Tarsi are mostly 5 segmented. The wing venation is is generally simple. The first abdominal segment is generally reduced . In flies, only four or five segments are distinctly seen. The posterior abdominal segments are mostly retractile and form the ovipositor in the female . Metamorphosis is complete and the larvae are apodous and are called maggots. Some forms are viviparous. The larvae differ structurally depending on their food habits. Some are truly aquatic such as, mosquito larvae. They have no true legs. Aquatic larvae have accessory respiratory organs in the form of gills. Pupation takes place either by the casting of the larval skin or the last larval skin hardens forming a barrel-shaped puparium as in Muscidae.

The Diptera are divided into three suborders viz., Nematocera, Brachycera and Cyclorrhapha.

Suborder - Nematocera. In this suborder, the antennae are usually long, longer than the head and thorax together. They are mostly many segmented and arista is not developed. In the larval stage , the head is well developed and has horizontal biting mandibles. Pupae are generally free. The suborder includes about 18 families and only the important ones are dealt with below :-

i) **Family - Culicidae** (Mosquitoes) - Mosquitoes are of utmost public health importance and are found in all parts of the world. The adults of many species are carriers of some of the very serious diseases such as, malaria, filariasis, yellow fever and dengue. They are very slender, delicate, small insects with an elongated piercing proboscis consisting of 6 stylets. The entire body, legs and wings of adults of most of the forms are almost covered with scales. No ocelli are present. The males posses plumose antennae, while the females have pilose antennae. The females live longer than the males. The females are blood-suckers, while the males feed on , liquid food such as juices etc. When at rest, the culicines sit with the proboscis and the long axis of the body in a straight line, while in the anophilines the abdomen is usually inclined upwards.

   The eggs are laid singly or in masses on the surface of water. The larvae and pupae are aquatic and very active. The larvae have a big thorax and feed mostly non minute algae. But a few species are carnivorous. The last abdominal segment bears four tracheal gills. The anophiline larvae float horizontally just below the water surface, while the culicine larvae float head downward at an angle (e-g) *Anopheles culicifacies)* - Transmits malaria **(Fig. 143)**; *Culex pipens fatigans)* - Transmits filariasis and encephalitis **(Fig. 144)**; *Aedes aegypti* - transmits yellow fever, dengue and encephalitis **(Fig. 145)**.

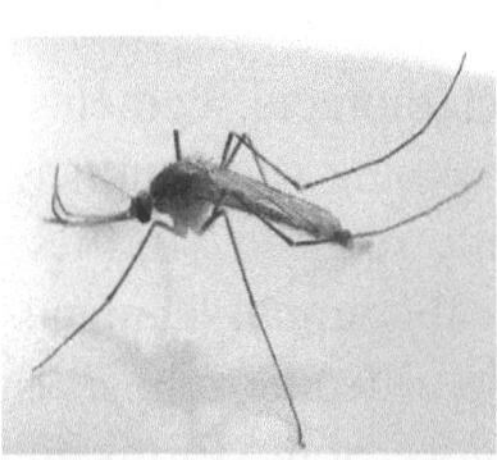

Fig. 144. Filariasis mosquito *(Culex pipens fatigans)*

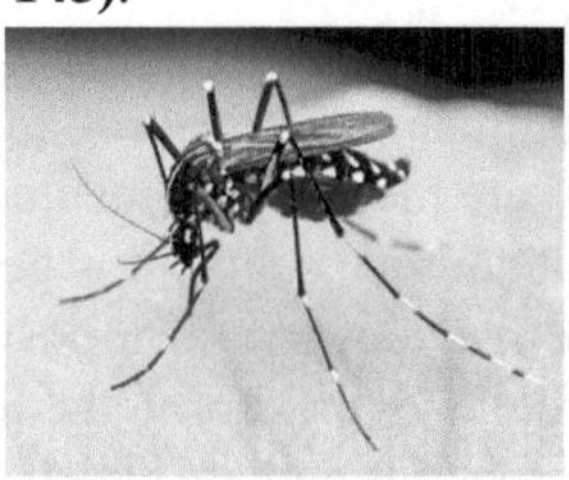

Fig. 145. Dengue mosquito *(Aedis aegypti)*

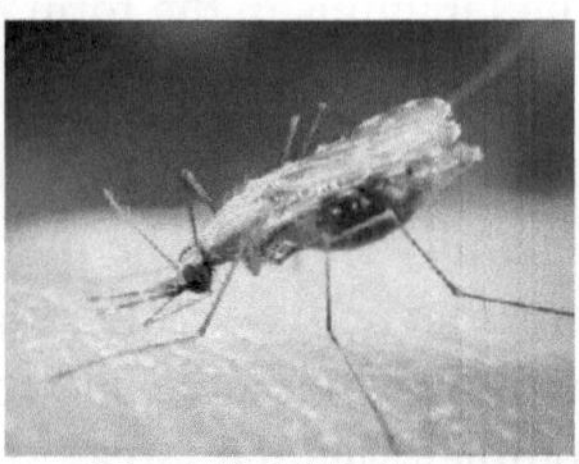

Fig. 143. Malaria mosquit *(Anopheles culicifacies)*

ii) **Family - Cecidomyiidae** (Gall midges or gall gnats) - The insects of this family are minute, two - winged, soft - bodied, slender flies, brown or yellow-colored, with dense hairs and long, thin moniliform antennae, which bear whorls of hairs. They seldom exceed 3.0 mm. in length. They have long, narrow wings and very long, thin legs The larvae are short and narrow at both the ends. The head is small and eyes are absent. The family is of much economic importance as the larval stage of many species are pests of cultivated crops and often produce galls. The larvae of some families are predacious (e-g) *Contarinia sorghicola* - The sorghum earhead gall midge; *Asphondylia sesami* - Gingelly gall fly **(Fig. 146);** *Pachydiplosis oryzae* - The rice gall fly **(Fig. 147)**; *Erosomia indica* - Mango flower gall fly; *Dasyneura lini* - Linseed gall fly.

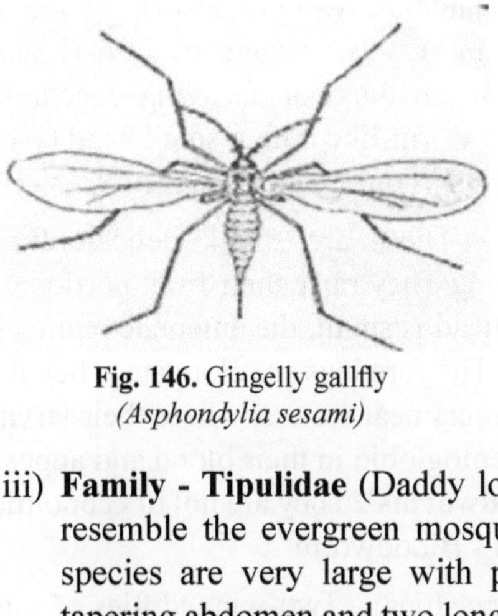

Fig. 146. Gingelly gallfly
(Asphondylia sesami)

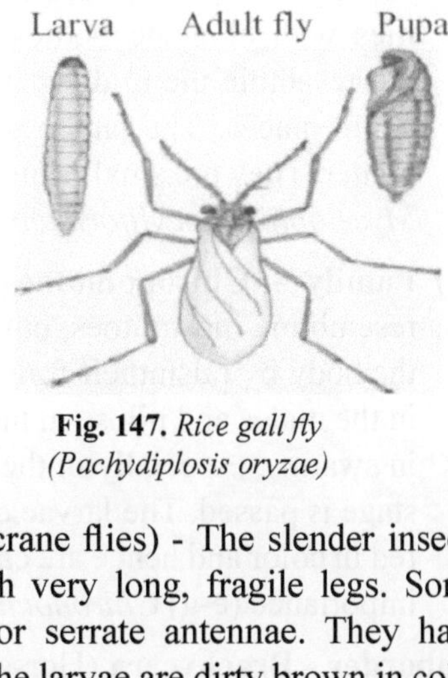

Fig. 147. *Rice gall fly*
(Pachydiplosis oryzae)

iii) **Family - Tipulidae** (Daddy long legs, crane flies) - The slender insects resemble the evergreen mosquitoes with very long, fragile legs. Some species are very large with pectinate or serrate antennae. They have tapering abdomen and two long wings. The larvae are dirty brown in color and have a tough skin, which gives them the name of leather jacket. They live among grass, roots or in water. They are elongated and hemicephalous with the head retracted in the prothorax. No species is a recorded pest of any crop plants. Few larvae are predacious (e-g) *Conosia irrorata*; *Tipula illustris*.

iv) **Family - Psychodidae** (Sand flies, moth flies) - They are minute, not more than about 4.0 mm. in length, mostly sand-colored insects found in dark, moist places such as, bathrooms. The body and wings are covered with coarse hairs or scales resembling those of moths. Some are known to suck blood from man and other vertebrates. Unlike common mosquitoes, the males of some species are also blood-suckers. The larvae are usually aquatic. They feed on decaying vegetation and animal remains (e-g) *Phlebotomus argentipes* - The insects not only suck blood from human beings, but also spreads a few diseases such as, **'kala azar'**, oriental sore etc.; *Phlebotomus minutus* - Sand fly or owl midge **(Fig. 148).**

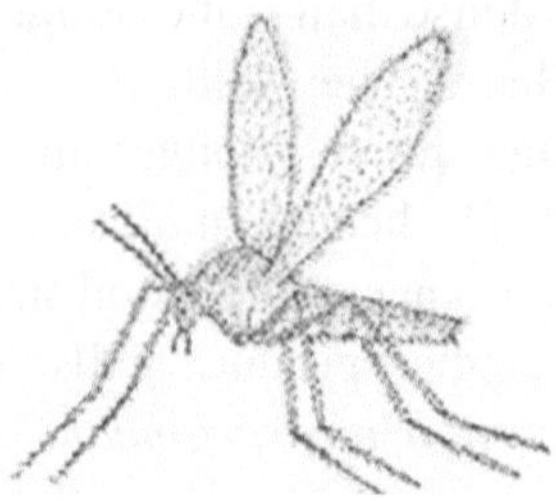

Fig. 148. Sand fly (*Phlebotomus minutus)*

v) **Family - Mycetophilidae** (Fungus gnats) - These are small, delicate flies with elongate bodies and long antennae without whorls of hairs, but resemble the midges. They occur in very large numbers in dark and damp places. The larvae generally live on fungi or decaying vegetable matter. They are small, smooth and are worm-like with a small head (e-g) *Mycetophila ancyliformans* - Pest of mushrooms.

vi) **Family - Chironomidae** (Midges) - These are small, delicate flies resembling mosquitoes, but while resting They raise their front portion of the body by raisintheir fore legs. The head is small, the antennae plumose in the males and pilose in the females. The fore legs are elongate. They fly in swarms, especially in the evening hours near waters, where their larval stage is passed. The larvae contain haemoglobin in their blood and appear red in color and hence are called **'bloodworms'.** They are not of economic importance (e-g) *Chironomus vicarius* - Bloodworm.

Suborder - Brachycera (Horse flies, robber flies) - Two-winged flies of this suborder have antennae shorter than the thorax and the 3 segmented antennae often have arista or stylet terminally. The larvae have incomplete head, usually retractile and have vertically biting mandibles. Pupae mostly obtect and free.

i) **Family - Tabanidae** (Horse flies, gad flies) - The insects of this family are medium to large - sized , stoutly built, active. two-winged, fast flying flies and have piercing proboscis in the females, with which they suck blood from the hosts, at the same time transmit a few diseases. The males feed on nectar. The eyes, which are very large and laterally extended, are contiguous in the males and separated in the females. Eggs are laid in masses on vegetation near or over water. The larvae are predators. Tabanids transmit diseases such as, **'Tularemia'** and **'Anthrax'** in animals (e-g) *Tabanus rubidus* - Pest of horse and also transmits **'surra'** disease in horses ; *Tabanus striatus* - The common gadfly **(Fig. 149).**

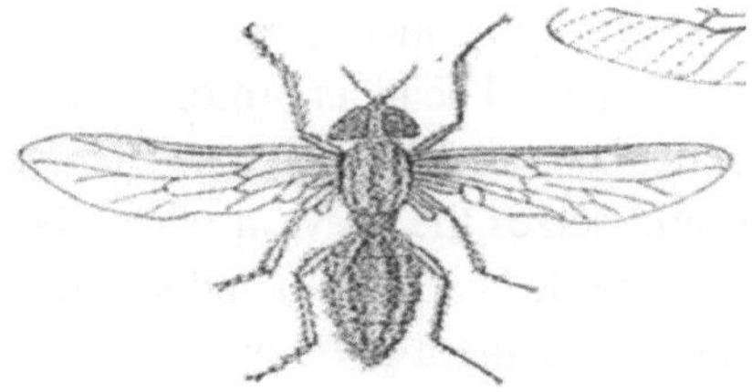

Fig. 149. The common gad fly *(Tabanus triatus)*

ii) **Family - Asilidae** (Robber flies) - This is a large family consisting of predacious insects, catch their prey with their bristly legs having powerful claws. The compound eyes are prominent and very effective for hunting prey on the wing. They feed even on the distasteful *Danais* butterflies, stinging Hymenoptera, other Diptera, grasshoppers etc. They pierce the victim with their hard and strong proboscis, making them senseless and then feed on the body fluids. They are not good fliers. They produce a somewhat loud, buzzing sound while flying. The adults of some forms are very large with a long abdomen and the body covered with bristles and have powerful legs. Some are short and bee-like. They vary in size from 3.0 - 40.0 mm. The larvae ar predators or scavengers. They are cylindrical with a small head and live in soil, wood or leaf mould (e-g) *Philodicus femoralis*; *Allocotosia surata*; *Efferia aestuans*.

Suborder - Cyclorrhapha - These flies Belong to the more advanced families of the order Diptera. Antennae are generally 3 segmented, with the two basal segments smalland the third segment large. Arista dorsal in position. Larvae have vestigial head. Pupae are exarate. The adults emerge from the puparium through a circular opening at one end of the puparium.

i) **Family - Syrphidae** (Hover flies, flower flies) - These beautifully colored-flies have smooth, polished, blue or black bodies with stripes, spots or bands of yellow. The abdomen has distinct yellow and black markings. They are found abundantly among flowering plants during bright sunshine. Often they are poised motionless in the air, their wings vibrating very rapidly and hence are known as **'hover flies'**. They feed on the pollen and nectar of flowers. Many species resemble various bees, especially bumble bees. The larvae generally live in rotting vegetable or animal matter. Some are aquatic, while some live in the nests of bees and wasps, mostly as scavengers. The larvae of several syrphids feed exclusively on aphids, coccids and other Homoptera. The head of the larva is greatly reduced and possesses a pair of short, fleshy sensory processes. Some aphid feeding larvae have long tails (e-g) *Eristalis arbustomum* - The larvae feed on soft-bodied insects such as, aphids, coccids etc. ; *Volucella signata* - The larvae spend their lives in the nests of bees and wasps; *Syrphus torvus*.

ii) **Family - Pipenculidae** (Big-headed flies) - The small, predatory, smoky-dark flies have large, spherical head, enormous eyes, and long wings. They hover about a great deal and the female poises herself suspended in the air before she darts at and seizes the prey, in whose body she lays her eggs. The big eyes set in a mobile head, small and strong legs with claws, sharp and strong ovipositor make them good hunters of insects such as, jassids and fulgorids. The larvae are endoparasitic (e-g) *Pipunculus major.*

iii) **Family- Drosophilidae** (Vinegar gnats, fruit flies) - These small, chubby, two-winged flies with large, light eyes are very common in India. They feed on fruit juices, vinegar, beer, wine etc. They breed prolifically in decaying fruits. They have an extremely short life history and in some cases, the egg to egg stage is completed in about a week's time. They breed throughout the year particularly during rains. The flies are extensively used for research work, especially in the field of genetics. The larvae of some species are leaf miners (e-g) *Drosophila melanogaster*.

iv) **Family - Tephritidae / Trypetidae** (Fruit flies) - The larvae of these flies are phytophagous, many of them living in fleshy fruits, while some are leaf miners and some others live in flower heads of Compositae. Some of them form galls. The adults are small to moderately big, more or less marked with yellow. The wings generally have dark markings. The females have a long ovipositor.

The long, white eggs are laid just under the skin of fruits by the female fly, which hatch in two or three days. The young maggots, which are dirty white or pale yellowish in color, tunnel into the fruit, feed on the pulp and becom full-grown in about two weeks. The grown- up maggots leave the fruit, fall to the ground and pupate in the soil. The maggots are blunt at the anal end, have conspicuous spiracles and taper towards the head end. The pupal stage lasts for about a week (e-g) *Bactrocera* (*Dacus*) *cucurbitae* - The Cucurbit fruit fly **(Fig.150)**; *Bactrocera(Dacus) dorsalis* - The mango fruit fly; *Bactrocera (Dacus) zonatus* - The pomegranate fruit fly.

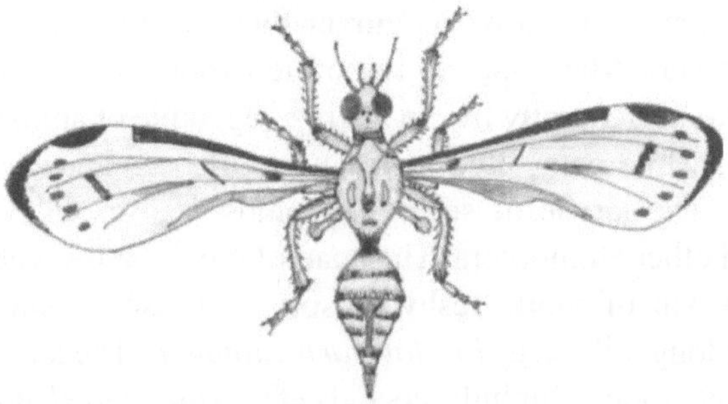

Fig. 150. The cucurbit fruit fly *(Bactrocera (Dacus) cucurbitae)*

v) **Family - Agromyzidae** - The flies of this family are very small in size and are generally grayish in color. Some species are blackish or dark grayish. The abdomen of some agromyzid flies is generally grayish- white and shining like silver. The larvae are broader at the posterior end and tapering at head end. They are mostly leaf miners or live inside stems. A few live inside fungi (e-g) *Melanagromyza obtusa* - The red gram pod fly **(Fig. 151)**; *Siphunculata funicola* - The common eye fly of man **(Fig. 152)**; *Cryptochaetum icerya* - Parasitic on cottony cushion scale *(Icerya purchasi)*.

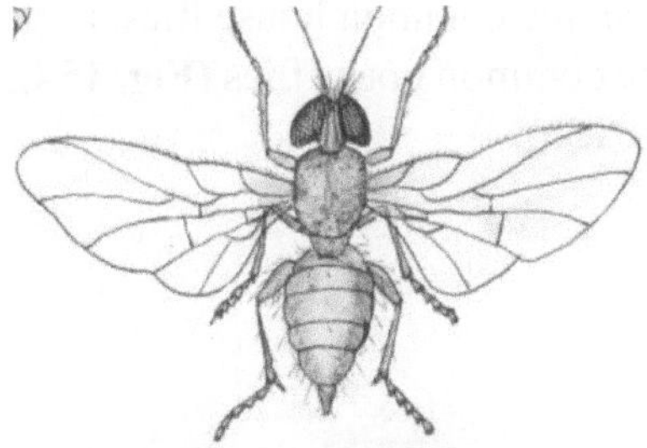

Fig. 151. The red gram pod fly *(Melanagromyza obtusa)*

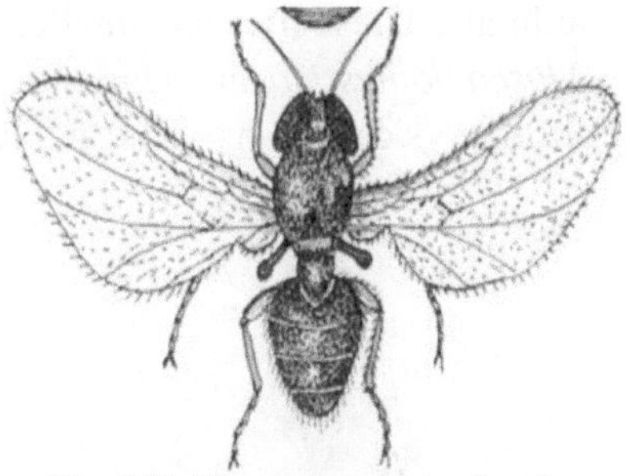

Fig. 152. The common eye fly of man *(Siphunculata funicola)*

vi) **Family - Tachinidae** (Bristle flies) - They are mostly beneficial insects, being parasitic on other insects including many crop pests, especially those belonging to the order Lepidoptera. The adult flies are stout, bristly insects usually with gray striped bodies. The females lay their eggs on the body of the host caterpillars. The young larvae, which hatch out from the eggs bore their way into the body tissues of the caterpillars, feed on the inner tissues and in due course kill them. Tachnids attack insects and non-insects off different species such as, spiders, wood lice, centipedes etc. indiscriminately (e-g) *Sturmiopsis inferens* - parasitic on sugarcane top borer caterpillars **(Fig. 153)**; *Sturmia sericariae* - Parasite on mulberry silk worms.

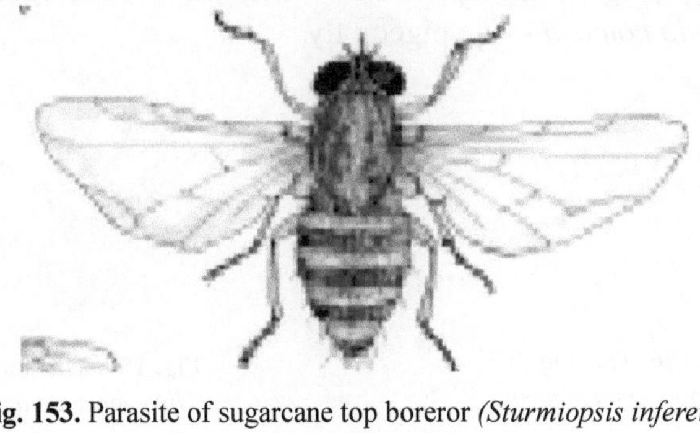

Fig. 153. Parasite of sugarcane top boreror *(Sturmiopsis inferens)*

vii) **Family - Muscidae** (House flies) - The family comprises of small to somewhat large flies. Some of them such as, the common house flies spread Diarrhoea, Dysentery, Typhoid, Cholera, Anthrax, Conjunctivitis and some other serious diseases. Many species are phytophagous and some are carnivores. The flies have short antennae and thick, bristly bodies with two wings and two halteres or balancers, which are vestigial and modified hind wings. While house flies do not bite, the stable flies, horse flies, Tse tse flies etc. are biting flies. They are found mostly in cattle sheds and both the sexes suck and feed on the blood from cattle, horses and other animals. These flies are smaller in size than the common house flies. (e-g) *Musca domestica* and *Musca nebulo* - The common house flies **(Fig. 154)**; Stomoxys calcitrans - The stable fly **(Fig. 155)**.

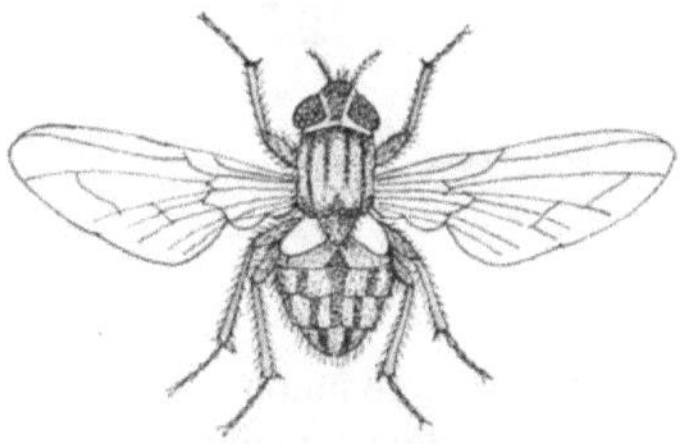

Fig. 154. The common house fly *(Musca domestica)*

Fig. 155. The stable fly *(Stomoxys calcitrans)*

viii) **Family - Hippoboscidae** - The members of this family are blood-sucking parasites common on cattle, other animals and birds and are known as cattle flies. Dog flies etc. The flies are flattened dorso-ventrally and of tough and leathery consistency, with the head sunk into the anterior part of the thorax. The legs are short and stout. Wings may be present or absent. The palpi form a sheath for the proboscis. Eggs are not laid directly, but the young larvae develop to an advanced state inside the body of the parent and pupate immediately after release (e-g) *Hippobosca capensis* - The dog fly **(Fig. 156)**; *Hippobosca maculata* - The horse fly **(Fig. 157)**; *Ornithomyia comosa* - The pigeon fly.

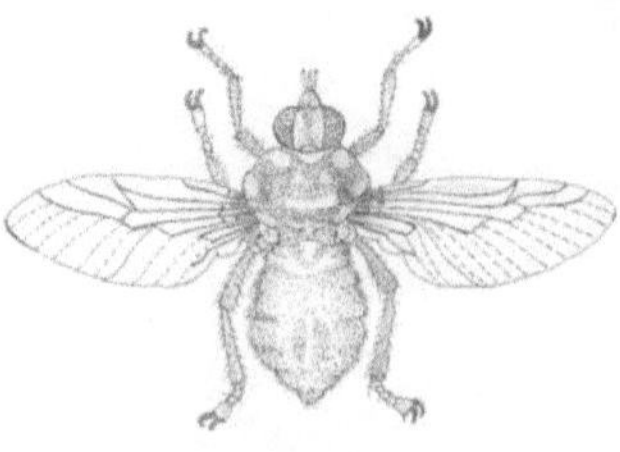

Fig. 156. The dog fly *(Hippobosca capensis)*

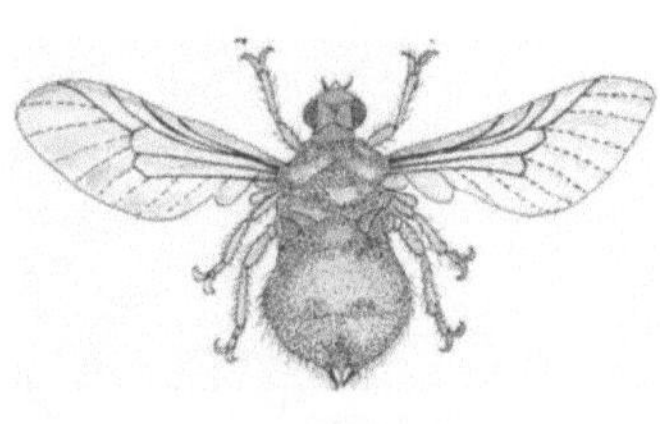

Fig. 157. The horse fly *(Hippobosca maculata)*

Order 6 - Siphonaptera or **Aphaniptera** (Fleas)

('Siphon') = tube; 'aptera' = wingless)

All members of this order are ectoparasitic in nature and possess piercing and sucking-type of mouthparts. They are wingless, very small and active insects. The bodies of fleas are compressed laterally. Though they have no eyes, they are voracious blood-suckers. They possess strong hind legs and the characteristic shape of the body is excellently fitted for gliding easily and freely among hairs, fur or feathers of mammals and birds, which form their natural hosts. The heavily chitinized body is usually covered with spines and bristles, which are sharply inclined backwards. The thoracic segments are free. Compound eyes are absent, but two ocelli are usually present. Antennae are short and stout and repose in grooves. They like warmth and therefore they leave the host soon after it dies and its body becomes cold. They usually walk, but can jump up to about 20 cm. high and 30 cm. horizontally.

Fleas are notorious transmitters of **'Bubonic plague'** caused by *Pasteurella (Bacillus) pestis* and the vector of the disease in India is *Xenopsylla cheopis* **(Fig. 158)**, the intermediate host being the domestic rat. Squirrels and some other rodents are also hosts of fleas. *Pulex irritans* is another human flea, but its favorite host is the pig. Both these belong to Family - Pulicidae.

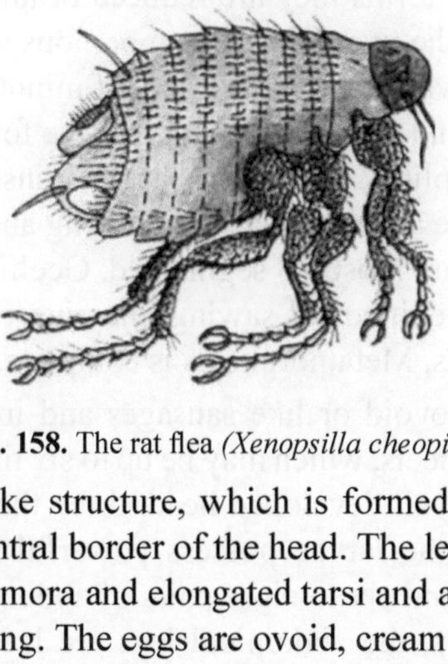

Fig. 158. The rat flea *(Xenopsilla cheopis)*

The head has a comb-like structure, which is formed by a row of powerful spines on the latero - ventral border of the head. The legs have large, flattened coxae, short and stout femora and elongated tarsi and are adapted for clinging to the hairs and for leaping. The eggs are ovoid, cream or whitish in color and are laid loose on the body of the host, but not gummed to the hair. They readily fall off and are found in the haunts of the hosts. Fleas hide in cracks of floors or in the litter found in grain godowns etc. The larvae are active , vermiformand apodous. They feed on organic debris. They have a well developed head, but lack eyes and legs. The pupa is enclosed in a cocoon. The life cycle of *enopsylla cheopis* is completed in about three weeks.

The female must feed on the blood of its normal host before it can lay fertile eggs. *Ctenocephalides canis* and *C. felis* are commonly found in dogs and cats respectively.

Order 7 - Hymenoptera (Ants, bees, wasps)

('Hymeno' = membrane; 'ptera' = wings)

This is the third largest order consisting of highly specialized insects, but the specialization is slightly less than that of Diptera. Like termites, many species of ants and wasps have various castes and are social insects, living in nests in colonies. However, a few species of wasps and bees are solitary in nature. The order is of great importance on account of the highly evolved phenomenon of parasitism. The other important members are the honeybees, which are directly beneficial to mankind, because they provide honey and bee wax and indirectly beneficial to man, because they are chiefly involved in pollination. Polyembryoni is often associated with insects of his order. Another remarkable feature of this order is reproduction by parthenogenesis, which is more common.

The adult insects of this order are minute to medium-sized, with well sclerotized body, either naked or hairy. The head is hypognathous, quite prominent and free. Compound eyes are large and well developed and widely spaced, while in some forms they are reduced or absent. The mouthparts are mandibulate-type. Of the two pairs of membranous wings, the hind wings are smaller than the fore wings and have a row of minute hooks on their anterior margin by which the hind wings get attached to the fore wings, which is known as hamulate wing-coupling . Abdomen basally constricted to form to form a narrow waist. Antennae are geniculate, fairly long and usually consist of 10 or more segments. Tarsi are mostly 5 segmented. Ocelli are absent. Ovipositor is well developed and modified for sawing, piercing or stinging. Males do not have stinging apparatus. Metamorphosis is complete.

The eggs are usually ovoid or like sausages and in many families they are carried at the tip of pedicels, which may be up to six times the length of the egg. A typical larva has a well developed head, three thoracic and ten abdominal segments. In general the larvae possess powerful biting mouthparts, three pairs of thoracic legs and six to eight pairs of abdominal legs as in saw flies. However, in the borers, the abdominal legs are lacking and in wasps, bees etc., the thoracic legs are also absent and the larvae resemble maggots. The parasitic forms are more degenerate. Ocelli are absent. The antennae are also reduced or absent. The mandibles are reduced to simple, pointed spines and the maxillary and labial palpi are also reduced or absent.

Among some parasitic forms, there is hypermetamorphosis and there may be up to ten larval forms. However, the final instar in all cases is the ovoid, maggot-like form. The pupal stage in many cases is preceded by a prepupa, which is somewhat intermediate between the larva and the pupa. The prepupa after moulting becomes the pupa, in which generally the wings and legs are free. The pupa may be enclosed in a cocoon or free as in ants and bees. A few species are pests of cultivated crops.

The order is divided into two suborders, viz., **'Symphyta'** and **'Apocrita'** or **'Petiolata'**. The members of the suborder Symphyta are all the more primitive ones. The abdomen is broadly sessile, with no marked constriction at its base. The trochanter is 2 jointed. The fore tibia mostly have 2 spurs. Ovipositor adapted for sawing and not for stinging. Larvae with thoracic and generally abdominal legs. In this suborder only the family Tenthredinidae is important. The members of the suborder Apocrita ar mostly ecto- or endoparasites of other insects. The abdomen is deeply constricted or petiolate between the first and second abdominal segments. The fore tibia have only one spur. Ovipositor is adapted for piercing in the parasitic forms and usually for stinging in some forms. Larvae are apodous. There are many important families, which are either parasitic on several important crop pests or beneficial to man in other ways.

i) **Family - Tenthredinidae** (Saw flies) - The insects of this family are small to medium-sized, brightly colored, wasp-like insects often found on foliage or on flowers. Antennae mostly 9 segmented. Parthenogenesis is common in this family. The eggs are usually laid in young shoots or in leaves. The ovipositor is saw-toothed and during oviposition, the blade of the saw is moved up and down alternately until a cavity is formed and then the eggs are laid in the cavity. The larvae are eruciform and resemble those of the Lepidoptera with only one pair of ocelli. But three pairs of thoracic legs and 6 - 8 pair of abdominal leg without crochets are present. The larvae are all plant feeders. The vast majority live exposed and are external feeders, while a few are borers or leaf miners. Pupation gradually takes place in an elongate - oval silken cocoon and in a few cases inside an earthen cell(e-g) *Athalia lugens proxima* - The mustard saw fly **(Fig. 159)**.

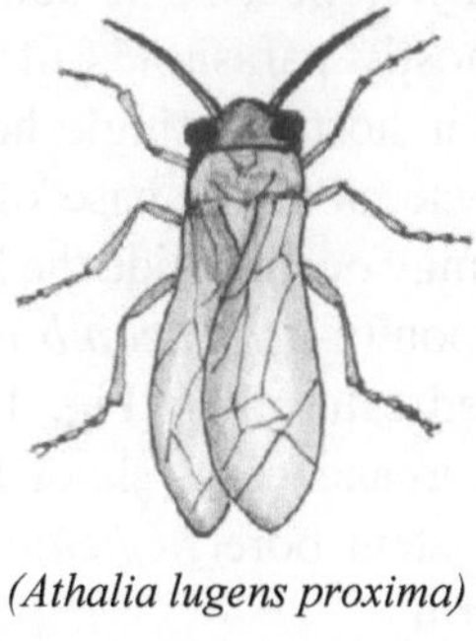

Fig. 159. The mustard saw fly *(Athalia lugens proxima)*

ii) **Family - Ichneumonidae** (Ichneumon flies) - The members of this family generally called as Ichneumonids are rather large, slender and wasp – like insects. They have long, filiform antennae, with 16 or more segments and moderately long legs. Some species are apterous or brachypterous and resemble the braconids. They are black, yellow or reddish-yellow in color and are more active on warm, sunny days. The two sexes are almost identical. The abdomen of the female is long and slender and the ovipositor is very long, sometimes extending up to 1.5 cm. The young larvae have a long tail at he anal end of the body. The full-grown larva pupates inside a silken cocoon.

Most of the Ichneumonids are parasitoids or less frequently hyper parasitoids and attack a great variety of hosts, especially of Lepidopera and most of them are internal Parasites of the immature stages of the host (e-g) *Isotima javensis* - Larval parasitoid of rice stem borer, sugarcane top borer and pupal parasitoid of sugarcane early shoot borer, internode borer and top borer **(Fig. 160)**; *Isotima dommermani* - Pupal parasitoid of sugarcane top borer; *Melcha ornalipennis* - Larval parasitoid of sugarcane top borer *(Eriborus trochanteratus)*.

Fig. 160. Ichneumonid parasitoid *(Isotima javensis)*

iii) **Family - Braconidae** - These are generally called as Braconids. They are minute insects, which are attracted to flowers and light. They resemble the Ichneumonids in many respects. The abdomen is sessile, sub-sessile or petiolate. They are mostly parasitoids of Lepidoptera. Eggs are laid on or in the body of their hosts. A single host larva may harbor about 100 - 150 parasitoid larvae as in the case of *Apanteles* parasitizing the larva of *Pieris.* Pupation may occur inside the host caterpillar or outside its body within a silken cocoon (e-g) *Bracon brevicornis* - Larval parasitoid of coconut black - headed caterpillar **(Fig. 161)**; *Microbracon chilonis*- Larval parasitoid of sugarcane early shoot borer; *Apanteles ruficrus* - Larval parasitoid of rice stem borer; *Chelonus narayani* - Parasitoid of American cotton boll worm.

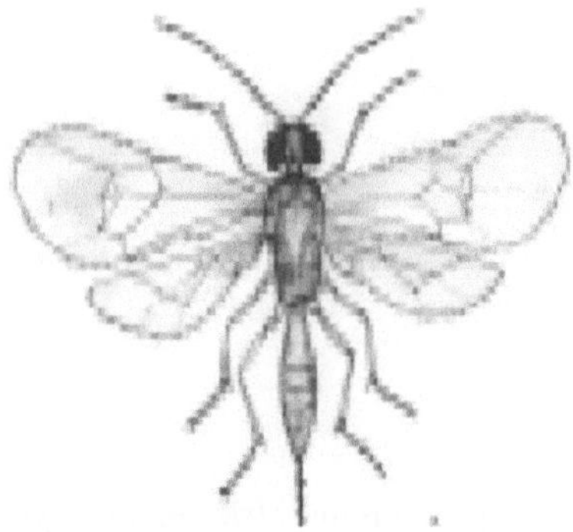

Fig. 161. Bracon parasitoid *(Bracon brevicornis)*

iv) **Family - Chalcididae** - Most of the species of this family are small to medium-sized wasps, either parasitoids or hyper parasitoids and are of great economic importance as a natural means of control of many pests belonging to Lepidoptera, Diptera and Coleoptera. The characteristic feature of these insects is that the hind coxae are five to six times larger than the fore coxae, with the hind tibial spurs larger than the mid tibia l spurs and the hind femora larger with a row of short teeth beneath. The wing venation is reduced to a single anterior vein. Ovipositor short and strong. The hyper- parasitoids attack Ichneumonids and Tachinids (e-g) *Brachymeria nephantid* - Larval parasitoid of coconut black - headed caterpillar **(Fig. 162)**.

Fig. 162. Chalcid parasitoid *(Brachymeria nephantidis)*

v) **Family - Eulophidae** - They are commonly known as Eulophids. They are minute, pupal parasitic wasps and are mostly parasitic on coccids, aphids and some of the leaf mining Lepidoptera. The fore wings are narrower with pubescence on the wing lamina, but the hairs are not arranged in rows. Ovipositor present at the tip of the abdomen (e-g) *Tetrastichus pyrillae* - Parasitic on sugarcane *Pyrilla* ; *Trichospilus pupivora* - Pupal parasitoid of coconut black - headed caterpillar **(Fig. 163)** *Aphelinus mali* - Parasitoid of apple wooly aphid.

Fig. 163. Pupal parasitoid *(Trichospilus pupivora)*

vi) **Family - Trichogrammatidae** - The family includes minute, fringe - winged species, some be ing the smallest among the insects. All the species are egg parasitoids particularly of Lepidoptera. Th adults have geniculate antennae. The head, which is short, is concave behind. The fore wings are broad and the tarsi are usually 3 segmented. The wings have microscopic hairs, which are usually arranged in rows. A single egg may give rise to several individuals. Some species are reared in large numbers under laboratory conditions and released in sugarcane fields for the control of borer pests. (e-g) *Tirchogramma chilonis* - Egg parasitoid of sugarcane internode borer and American cotton boll worm **(Fig. 164)**; *Telenomus manolus* - Egg parasipoid of groundnut red hairy caterpillar; *Telenomus beneficiens* - Egg parasitoids of rice stem borer, sugarcane early shoot borer and sugarcane top borer.

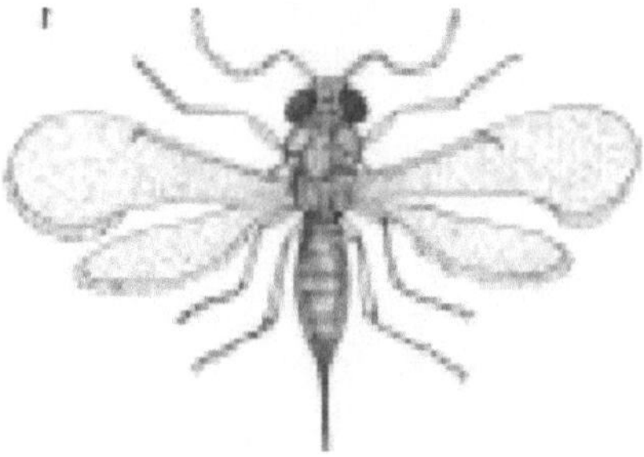

Fig. 164. Egg parasitoid *(Trichogramma chilonis)*

vii) **Family - Bethylidae** - Generally called as Bethylids, these are mostly small, ant-like' Black insects with often reduced wings in the females. Some species are dimorphic in each sex. The head is elongate and depressed with often punctiform eyes and antennae inserted close to the clypeus. They are mostly parasitic on the larvae of Lepidoptera and Coleoptera (e-g) *Perisierola nephantidis* - Larval parasitoid of coconut black-headed caterpillar **(Fig. 165)**.

Fig. 165. Bethylid larval parasite *(Perisierola nephantidis)*

viii)**Family - Formicidae** (Ants) - They are chiefly characterized by the basal one or two segments of the abdomen being in the form of separate nodes. The antennae are usually elbowed and have a long scape and a number of small joints forming the flagella. In the males of some species, the scape is very short. The legs are long and most species can run actively.

Ants occur practically everywhere in terrestrial habitats. They are generally small insects, black or brown in color. They are typically social insects living in colonies in which the various forms are adapted for special functions and there is well organized division of labor. The commonly found forms consist of the queen, males and females with two or three forms of workers and the soldiers with large head and mandibles. The nests are in the soil or on trees. The ants generally work in light. The queen is larger than the members of the other castes, usually winged, which are shed after the nuptial flight. The queen usually starts a colony and does most of the egg-laying in the colony. Winged males, which maybe found are smaller than the queen, short-lived and die soon after mating.

In general ants are scavengers, the workers bringing to the nest the food for the entire community. The food consists of dead insects and any other animals or vegetable matter, sap of plants etc. Some species known as harvesters store seeds of grasses, small millets, rice, wheat etc. in their nests. Some suck the honeydew secreted by aphids, coccids and other Homoptera. Ants have different means of defense. Many species can sing, while many others bite and a few give off a foul-smelling secretion (e-g) *Oecophylla smaragdina* - The red tree ant is often a nuisance on fruit trees **(Fig. 166);** *Monomorium destructor* - It is troublesome in the Household; *Monomorium salomonis* - Carnivorous ants that feed on living insects of which termites are the common hosts *Camponotus compressus* - The black ants that often shelter colonies of phids, coccids etc. and use them as cows for their honeydew; **(Fig. 167).**

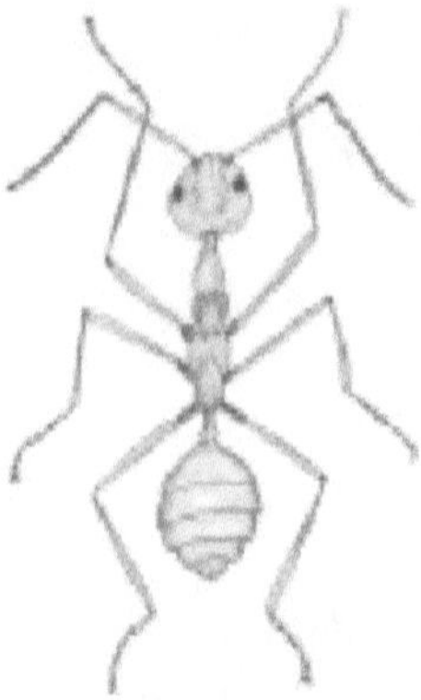

Fig. 166. Red tree ant
(Oecophylla smaragdina)

Fig. 167 - Black ant
(Camponotus compressus)

ix) **Familt - Vespidae** (Wasps) - Many species of this family are social in habit, while a few are solitary forms. The members of this family have usually long and slender petiole, moderately long legs, wings folded longitudinally when at rest and the fore tibia having a comb for cleaning the antennae and the middle tibia having two terminal spurs and simple claws.

They are often yellow or red with black markings and well developed clypeus and dentate mandibles. The eyes are reniform. The glossa and paraglossa terminate in sclerotized pads. The wing venation do not reach the apical margin. The abdomen is sessile and conical with well developed sting. These insects are more active in warm and hot countries. Many of them posses warning coloration. Most of the species have social habits and the colony includes queen, males and workers. The nests generally do not last for more than one season.

The wasps are generally predacious on Lepidopterous larvae, which are stored in the cells of the nests after stinging and paralyzing them. Each cell may contain up to a dozen caterpillars. A single egg is laid in each cell from a suspensory filament from the roof of the cell, which is subsequently sealed. Nests of wasps are generally made of papery material made out o chewed vegetable fiber. The cells are generally hexagonal in form and the nest hangs with the opening downward.

The food of wasps generally consists of caterpillars, bugs, mantids, grasshoppers, beetles etc., which are brought into the nests by the parents. Some wasps feed on fruit juice and nectar. Some wasps sting powerfully (e-g) *Vespa cincta* - The yellow banded, stinging wasp; *Vespa orientalis* - The social wasp **(Fig. 168)**.

Fig. 168. Social wasp *(Vespa orientalis)*

x) **Family - Sphecidae** (Solitary wasps, digger wasps) - They are mostly fossorial, solitary wasps, usually large, shining black or black and orange or red in color. Some forms build single mud cells on rocks, trees, buildings, furniture etc. or nest underground in sandy places. They are also predacious, mostly on Lepidopterous larvae, bugs, Orthopterans, spiders etc. They sting the hosts to induce rapid paralysis and keep them in the nests as food for their developing larvae. Some hosts die as a result of stinging, but remain undecomposed for several days. These wasps are generally known as thread -waisted wasps, because of the long and very slender petiole between the thorax and abdomen. The adults are commonly found on flowers (e-g) *Eumenes conicus* – Solitary mud wasp; *Ampulex compressa* – Solitary wasp; *Ammophila procera* - The digger wasp **(Fig. 169)**.

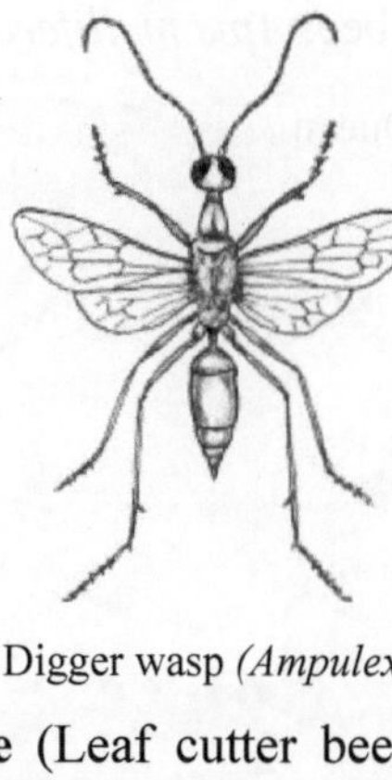

Fig. 169. Digger wasp *(Ampulex compressa)*

xi) **Family - Megachilidae** (Leaf cutter bees) - These are insects, which build their nests in rotten wood or in the soil with neatly cut oval or round pieces of leaves or flower petals. The female bee holds the leaf with her legs, while she cuts with her mandibles. A few forms build mud cells. Each bee makes a solitary nest preparing one cell at a time fills it with honey and pollen and lays on it. (e-g) *Megachile anthracina* - The black leaf cutter bee of rose, pigeon pea etc. *M. lanata* - Leaf cutter bee **(Fig. 170)**.

Fig. 170. Leaf cutter bee *(Megachile lanata)*

xii) **Family - Apidae** (Honeybees) - This family includes a number of small to medium-sized, solitary and parasitic bees and the social honeybees. Honeybees are man's most valuable domesticated insects and are associated with the human race for more than 4000 years, They exhibit polymorphism and most of them live in colonies with queen, drones and workers. The workers, which are the working force do all kinds of work such as, collection of nectar, pollen, concentration of honey, nursing the young ones and other members, defense of colony, construction of wax combs, secreting royal jelly or bee's milk etc. They have large compound eyes, geniculate antennae and the workers have pollen baskets in their metathoracic legs (e-g) *Apis dorsata* - The giant rock bee; *Apis Indica* - The Indian honeybe **(Fig. 171)**; *Apis florea* - The little honeybee; *Melipona irridipennis* - The dammer bee; *Apis mellifera* - The European honeybee.

Worker Queen

Drone

Fig. 171. The Indian honey bee *(Apis indica)*

xiii)**Family - Evanidae** (Ensign wasps) - These are small wasps with short and compressed gaster (distal portion of the abdomen) attached o the propodeum by long, slender petiole, carried almost like a flag. They are parasitic on ootheca of cockroaches (e-g) *(Evania appendigaster).*

Order 8 - Coleoptera (Beetles, weevils)

('Coleo' = Sheath. 'ptera' = wings)

The order is the largest one in the animal kingdom, numbering about 2,50,000 recorded species, which is about 40 per cent of the class insecta.

Being of concealed habit, they are not so conspicuous as Lepidoptera. They vary in length from less than 1.0 mm. to 15.0 cm. They are cosmopolitan in nature and are found in land, water or air. They are found in soil, humus, decaying plant or animal matter etc. Large numbers are found living on plants, feeding on leaves, roots, boring into stems, branches etc. Some live on wool, hides, furniture, timber, tobacco, drugs, stored grains, oilseeds, stored foods and several other products. Some species are predacious and rarely a few are parasitic. Some live in the nests of social insects.

The head is prognathous or hypognathous and heavily sclerotized. The vertex and frons are prolonged anteriorly to form a rostrum in weevils and the mouthparts are borne at the apex of the rostrum and the antennae are also carried forward in the rostrum. The head is freely movable. The mouthparts are biting and chewing type and the mandibles are well developed. Compound eyes are mostly present, but are wanting in a few species. Ocelli are also present. The antennae are of varied forms, either serrate, moniliform, filiform or lamellate, but never setaceous. Antennae are usually 11 segmented. The prothorax is the largest of the thoracic segments and mobile. Metathorax is much reduced. The pronotum consists of a single sclerite and entirely visible from above. One of the most distinctive feature of this order is the structure of the wings. The fore wings are modified into horny or leathery elytra (sing. - elytron), which generally meet to form a straight line like a mid-dorsal suture when in repose. The elytra serve as a protecting sheath. In many species, the elytra are abbreviated, not covering the abdomen completely, thus exposing the terminal abdominal tergite. (pygidium) beyond the tips of the elytra. They are not useful in flight and during flight they are merely kept stretched outside on each side. The hind-wings are membranous and useful in flight. They are folded longitudinally and transversely beneath the elytra during repose or often reduced in some forms. The abdomen has 8 or 9 visible segments. Tarsal segments vary from three to five.

Eggs are of two types; the soft, oval eggs laid in concealment or the harder, variously shaped eggs without ornamentation are laid openly. The eggs are laid singly or in batches. The larvae are campodeiform or scarabeiform or eruciformand are generally called grubs. The larvae of weevils are apodous. Larvae have well developed head and biting and chewing type of mouthparts. In some species, hypermetamorphosis is noticed. Normally metamorphosis is complete. The larvae are devoid of sucker feet. Pupation takes place openly or in a cocoon of mud, anal excretion or fibers, but never in silken cocoons. Pupae are adecticous, exarate and rarely obtect. A large number pupate in underground earthen cells.

The order comprises of many suborders, but the two suborders viz., Adephaga and Polyphaga are important :

Suborder - Adephaga - Hind coxal cavities divide the first visible abdominal sternite. The hind wings have two medial-cubital cross veins. Prothorax has pleural sutures. The thoracic legs of larvae have tarsi and two claws.

i) **Family - Cicindelidae** (Tiger beetles) - The members of this family are small to medium-sized, generally 1.25 - 2.50 cm. long and brightly colored, green, brown or black with spots or bands of white or pale yellow. They have prominent eyes and large, serrated mandibles. They are among the most vicious and fierce of insects, particularly in the larval stages. The beetles are also adapted for predacious life. They can run rapidly with their long, slender legs and with their mouthparts they quickly seize and firmly clasp the prey. They can fly fast, but only for short distances. Most of the species are diurnal in habit. They are commonly seen in damp places, in rice fields, thick vegetation, in forests or on river banks. The larvae are found in vertical burrows in the earth, which may extend 30 cm. or more in depth. They remain at the entrance of the burrow with their broad head ;and thorax extending out and catch any moving insects outside. The captured prey is dragged down to the bottom of the burrow and then eaten (e-g) *Cicindwela vigintiguttata Cicindela sexpunctata* - Tiger beetle **(Fig. 172)**.

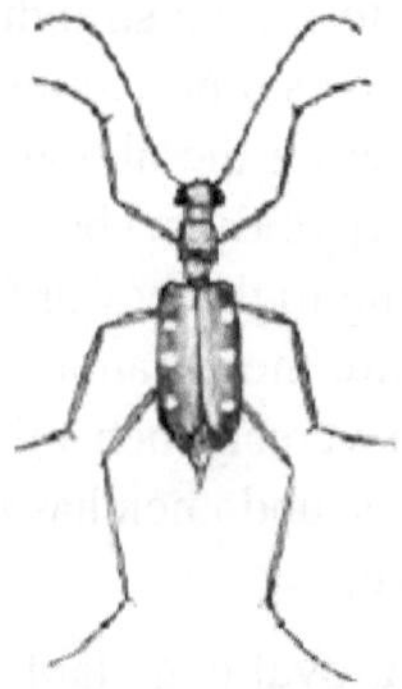

Fig. 172. Tiger beetle *(Cicindela sexpunctata)*

ii) **Family - Carabidae** (Predacious ground beetles) - In general body form and coloration, these insects are more akin to Cicindelidae, but the body is usually oval, broader and more flattened and are found under stones, logs, leaves, barks, debris or running on the ground. These carnivorous beetles are usually black, but often have a bronze or violet tinge, narrow head, long, thin antennae and long legs. The legs are either long adapted for running or short and thick, adapted for digging. The antennae project prominently in front of the head. The elytra in some species are soldered together and hence they can not fly. Most of them exude foul-smelling liquid from glands opening above the anus. The larvae, which are campodeiform or long - legged type, are slender, elongate and active with large head and long, caliper-like mandibles. They are predacious on caterpillars of other insects and so are beneficial. Pupation takes place in the soil (e-g) *Scarites indicus* - Predacious ground beetle **(Fig. 173)**; *Carabus nemoralis*.

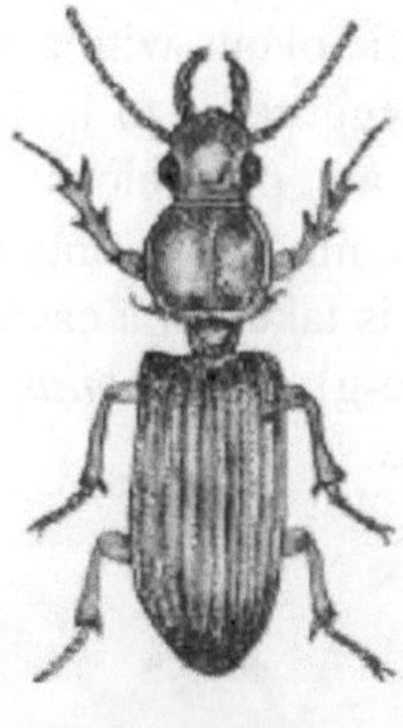

Fig. 173. Predatory beetle *(Scarites indicus)*

iii) Family - Ditiscidae (True water beetles, predacious diving beetles) - A family of aquatic beetles, which are found in both running and standing waters. The adults are capable of prolonged flight and can exist in land also. The larvae, as well as the adults are carnivorous and feed on a variety of small aquatic insects including small fishes. Some species are very large, while some are rather small. The body form and legs are modified to suit aquatic mode of life The body form is oval without any projections or angles, the broad head fitting tightly into the thorax. The hind legs are flattened, oar-like and fringed with bristles for swimming. They give out an unpleasant smelling fluid from the anus. The antennae are filiform. The elytra cover the entire abdomen and the wings are large and functional. The beetles obtain air at the surface of the water, but can remain submerged for long periods, because they carry air in a chamber provided beneath the elytra. The air is used for respiration during submergence. The male is distinguished by the broadened fore tarsi, which form cup-like suckers for gripping the female when mating. The grubs are elongated with a large, flat head and long tapering body. The larva comes to the surface of water with tail upwards and takes in air supply through two spiracles. The larval mandibles are sickle-shaped. Pupation takes place in the mud near the water shed (e-g) *Dystiscus marginalis*; *Hydroporus aper*; *Copelatus indicus.*

Suborder - Polyphaga - In the insects belonging to this suborder, the hind coxal cavities do not divide the first abdominal sternite. Hind wings have no cross veins. No clear sutures are found on the prothorax. Larval thoracic legs have no tarsi and only one claw is present.

i) **Family - Hydrophilidae** (Water scavenger beetles) - The hydrophilids are both aquatic and terrestrial . They are more globose, rounded, smooth, black or dull black colored, rather small and not exceeding 1.25 cm. in length. They are primarily vegetarians The eggs are laid in a hollow case, which is attached to aquatic plants with a mast - like projecting process. Some species fasten the egg cases to their own bodies. The larvae are similar to dytiscids and are predacious, while some are vegetarians. Pupation takes place in the mud . The obtain air by coming to the surface of water, head up and air is taken in a cavity on the lower surface of the body through spiracles (e-g) *Hydrophilus olivaceus* - Water scavenger **(Fig. 174)**; *Hydrous piceus.*

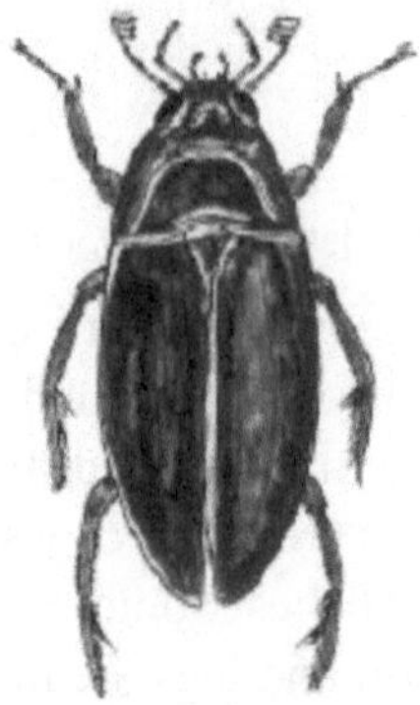

Fig. 174. Water scavenger beetle *(Hydrophilus olivaceus)*

ii) **Family - Scarabeidae** (Cockchafers, white grubs, root grubs, unicorn beetles, rhinoceros beetles etc.). This is a large family comprising of beetles of various sizes, mostly somber, dark brown-colored. However, some are very beautifully colored in metallic blue or green. They are convex insects, heavy-bodied, oval or elongate. The head projects forward as a flat plate, beneath which the mouthparts are located. Some such as, the unicorn beetles and rhinoceros beetles are very large and some of which are armed with large horns or protuberances extending forwards from the thorax. Sexual dimorphism is common in some species. The elytra are hard and rough and cover the abdomen completely. The legs are large and powerful. Tarsi 5 segmented. The antennae are 8 - 10 segmented and are lamellate. Because of certain differences in the morphological characters, the members belonging to this family have been placed under different sub-families viz.- **Melolonthinae** (White grubs, root grubs, cock chafers), **Dynastinae** (Unicorn beetles, rhinoceros beetle), **Scarabinae** (Dung beetle), **Cetoninae** (Rose flower cutting beetle), **Rutelinae** (Leaf chafer) etc.

Sub-family - Melolonthinae (White grub (June beetle), cockchafer) - White grubs remain under the soil, cut, bite and feed on the roots, rootlets and underground stalk of crops such as, sugarcane, finger millet etc. Rarely the grubs bore into the stem and feed on the inner tissues. The beetles come out of the soil during dusk and mate. The female beetle lays eggs on the moist soil below the plant canopy. The young grubs hatching out of the eggs, burrow into the soil, reach the roots, bite and feed on the roots. Full grown grub is whitish in color, long, smooth, shining, fleshy with a dark head, wrinkled body segments and curved ventrally like a **'C'** in its natural position. It pupates under the soil andmerges as an adult beetle. The beetle is cylindrical in shape and black in color. (e-g) Sugarcane white grub *(Holotrichia consanguinea)* - **Fig. 175.**

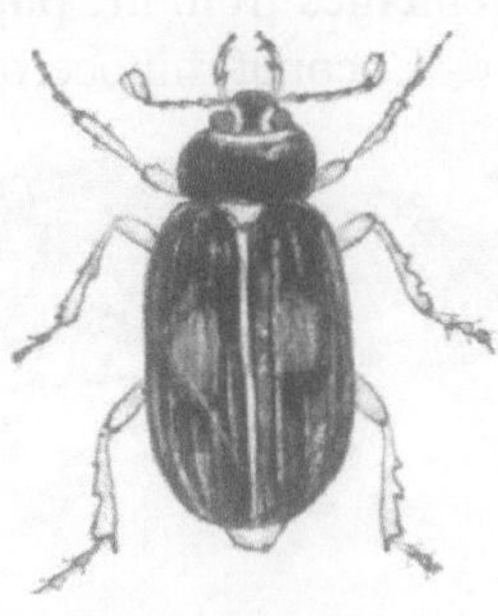

Fig. 175. Sugarcane white grub *(Holotrichia consanguinea)*

Cockchafers are usually smaller, oval in shape, stout and glossy, bright-colored and the elytra do not cover the abdomen completely. The small head has well sclerotized labrum. The beetles feed on the foliage of host plants such as, vegetable crops by night and hide in the soil during the daytime. They are attracted to light. The larvae feed on the roots of host plants and are often very destructive. They are scarabeiform, fleshy, dull white in color, with a brown head, the body curved in an arch and the apical segments large and smooth. Pupation takes place in the soil in a mud cell. A few species have very large scutellum (e-g) *Melolontha vulgaris* - A serious pest of vegetable crops.

Sub-family - Dynastinae (Unicorn beetles, Rhinoceros beetle) - The beetle is black or Reddish - brown in color, stoutly built with elongated cylindrical body and measures 3.5 - 5.0 cm. in length. The dorsal side is overed by a smooth, shining, black, thick and hard elytra. The ventral side is saw dust - colored and hairy. It has well developed wings and can fly long distances. On the dorsal side, near the tip of the head, the beetle has a pointed horn bent slightly backwards, which is longer in the male and much shorter or rudimentary in the female and hence the insect is referred to as the **'rhinoceros beetle'.** There is a is a depression on the dorsal side of the first thoracic segment. The beetles fly by night and are attracted by light.

The adult beetle burrows through the unopened, tender fronds, biting and chewing the fibrous portions and feeding on the juice and throws out the remnants as a dry, fibrous mass through the burrow. This dry fibrous mass and the faecal matter of the beetle are seen at the mouth of the burrow.

The female beetle lays yellowish-white, oval eggs in manure pits or heaps of organic debris at a depth of 5.0 - 15.0 cm. The grubs emerging from the eggs feed on the rotting organic matter and grow. The full-grown grub is dirty white in color, smooth, shiny, stout and fleshy with wrinkled body segments. It has a light brown, small head and a darker tail end. In its natural position, it is always curved up ventrally in the form of a **'C'**. The grown-up grub burrows to a depth of 30.0 - 100.0 cm., construct a pupal chamber in the manure pit and pupates inside it. The beetle emerges from the pupa after a period of 10 - 60 days (e-g) *Oryctes rhinoceros* - Coconut rhinoceros beetle **(Fig. 176)**.

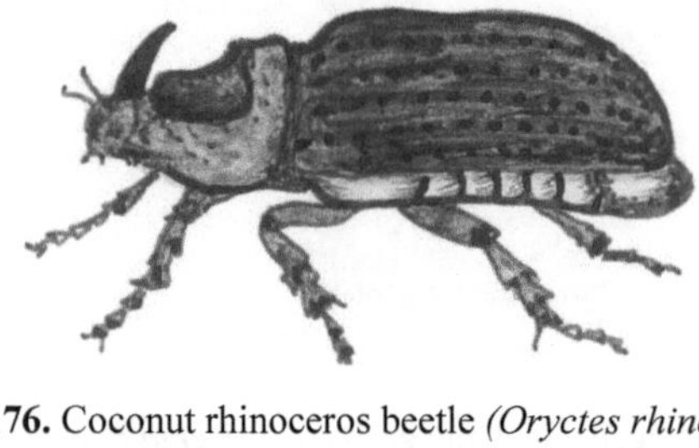

Fig. 176. Coconut rhinoceros beetle *(Oryctes rhinoceros)*

Sub-family - Rutelinae (Shining leaf chafer / chafers of the plains) - The shining, bright-colored beetles bite and feed on the foliage of vegetable crops. The grubs, which live under the soil bite and feed on the roots of rice, pearl millet, and other cereals, thus cause severe damage (e-g) *Anamola varians; Anamola biharensis* **(Fig. 177)**.

Sub-family - Cetoninae (Rose flower cutting beetle) - The rose flower cutting beetles cut and damage rose flower buds and flowers (e-g) *Oxycetonia versicolor* - Rose flower cutting beetle **(Fig. 178)**.

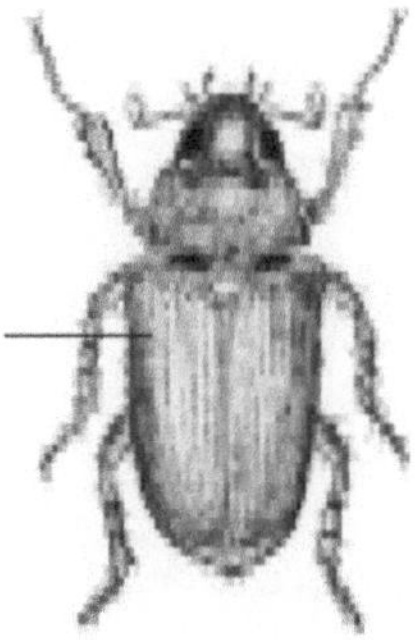

Fig. 177. Shining leaf chafer *(Anamola biharensis)*

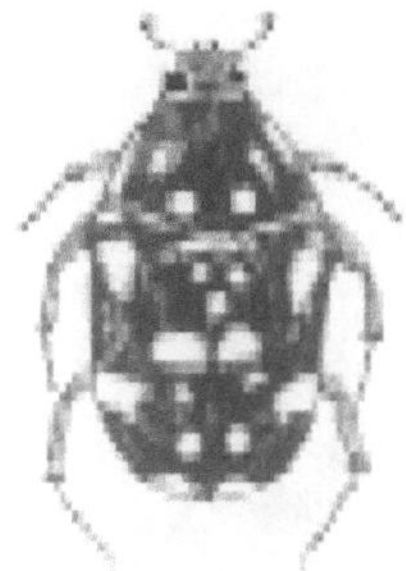

Fig. 178. Rose flower cutting beetle *(Oxycetonia versicolor)*

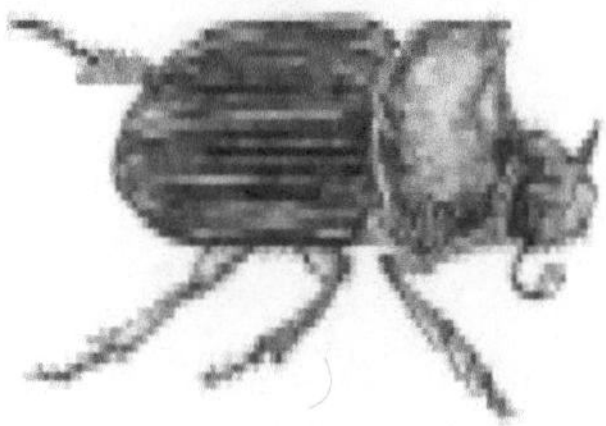

Fig. 179. Dung roller beetle *(Heliocopris bucephalus)*

Sub-family - Scarabinae (Dung roller beetles) - The beetles are round or oval and often very convex beetles with membranous mandibles and live almost entirely on dung or dead insects. The larvae are often found in dung balls buried to a depth of 7.5 - 17.5 cm. below the soil surface by the mother beetle. He dung ball is hollow and a single, white egg is fixed inside. The larva feed from inside the ball and pupates there itself. In some forms, there are groups of egg balls, which are guarded by the parents (e-g) *Heliocopris bucephalus* - Common dung roller beetle **(Fig. 179)**.

iii) **Family - Buprestridae** (Jewel beetles, metallic wood-boring beetles) - Some species of this family are delicate and less than 6.0 mm. long, while others are robust, long and nearly 2.5 cm. in length. They are one of the most brilliantly colored groups, many forms being bright green with red reflections. Some species are covered with efflorescence produced from a secretion in the skin. The integument is hard and strong and the head partly sunk in the thorax, which is strongly flexed to the abdomen. The antennae are serrate, short and can be readily concealed under the head. The legs are short and folded under the body when at rest. The tarsi are 5 segmented. The wings are large and functional in flight. The metallic green elytra are accurately applied to the body. They are good fliers and are diurnal in habit. They feed on the leaf tissues. The larvae are flat-headed borers, some boring into branches or woody stems, while others mine the leaves or feed on the tissues beneath the bark of trees. The larvae are legless, with the thoracic segments swollen, the abdomen very long and slender, giving a clubbed appearance. The head is small and entirely drawn into the thorax. Abdomen 9 segmented and there are no anal processes. Pupa is naked and is found in the larval burrow (e-g) *Sphenoptera gossipii- The cotton stem borer*; *Buprestes parsina* - Borer pest of forest trees **(Fig. 180)**; *Sphenoptera perotetti* - Ground nut stwem borer.

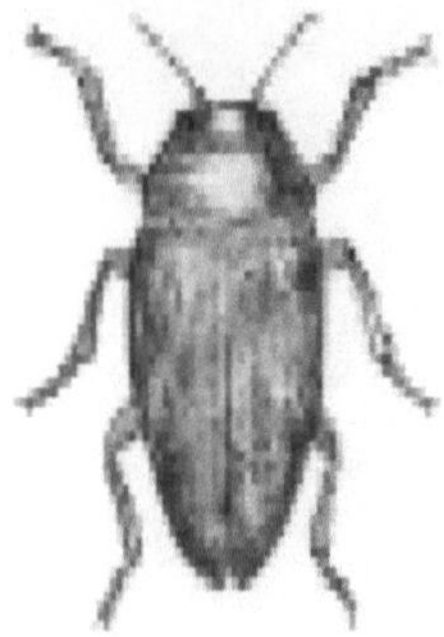

Fig. 180. Jewel beetle *(Buprestes parsina)*

iv) **Family - Elateridae** (Click beetles, wire worms) - The family consists of both dull brown or yellow - colored, small forms and the bright-colored larger forms, which are about 1.25 cm or more in length. The body is elongated, usually with parallel sides. The pronotum is rounded anteriorly and sharply pointed posteriorly Antennae are usually serrate. The head is small and embedded in the solid prothorax, which is large and powerful, fitting into the elytra and is mobile. The beetles leap with a clicking noise with each jump. Clicking is effected by the flexible union of the prothorax and mesothorax and a prosternal spine that fits into a groove on the mesosternum. Tibia posses two spurs. The larvae are cylindrical and elongated, with three pairs of legs. They appear like rusted wire and hence are called **'wire worms'**. A few are pests of wild plans (e-g) *Agrypnus fuscipes* - Click beetle **(Fig. 181).**

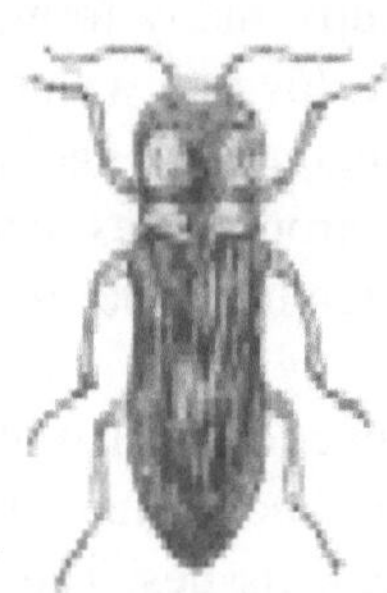

Fig. 181. Click beetle *(Agrypnus fuscipes)*

v) **Family - Lampyridae** (Glow worms, fire flies) - The members of this family are luminous insects, nocturnal in habit and provided with photogenic organs near the end of the abdomen emitting light, which is stronger in the females. Light is produced by the oxidation of a substance called **'Luciferin'** in the photogenic organs in the presence of an enzyme called **'Luciferinase'**. The males are usually winged and have well developed eyes, while the females are apterous with small eyes, resembling larvae, thus exhibiting sexual dimorphism. The elytra are soft and flexible. The eggs, larvae and pupae are also luminous to some extent. The luminous larvae are called **'glow worms'**. The larvae are predacious, while the adults, which are found on vegetation are also predacious (e-g) *Diaphanes marginellus* – Glow worms.

vi) **Family - Dermestidae** (Hide and skin beetles, grain beetles) - They are mostly household pests. The adults are small to medium-sized beetles, 0.8 - 1.8 cm. in length, oval or elongate-oval and convex, with short, clubbed antennae, usually hairy or covered with scales. They are black or dull-colored, but many of them have characteristic color patterns Household species feed upon furs, hides, horns, wool, bacon, cheese etc., making holes and damaging them badly and complete their life cycle there itself. So- me are found on museum specimens, wax articles etc., while a few others are storage pests. The larvae have tufts of long hairs. The last larval skin often wholly envelops the pupa (e-g) *Trogoderma granarium* - The wheat beetle or **'Khapra beetle'**, the larvae of which cause extensive damage to stored grains such as, wheat and also sorghum, barley and pulses **(Fig. 182);** *Anthrenus flavipes* - Household pest that damages wool, fur etc.; *Anthrenus voras* - The carpet beetle that damages carpets **(Fig. 183).**

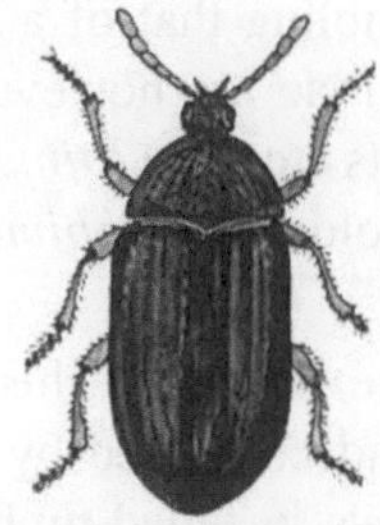

Fig. 182. 'Khapra' beetle
(Trogoderma granarium)

Fig. 183. Carpet beetle
(Anthrenus voras)

vii) **Family - Anobiidae** (Wood worms, wood borers) - The species belonging to this family are often very destructive. The beetles and grubs are both borers and they make neat, cylindrical tunnels for their abode. The adult beetles are cylindrical to oval, pubescent beetles, 1.0 - 9.0 mm. in length, with the head deflexed and usually connected from above by a hook-like pronotum. Most of them have the last three antennal segments enlarged. The beetles in the tunnels produce knocking noise with their heads, generall called the 'death watch', which is probably a sexual call. The grubs are fleshy, with larger abdominal segments (e-g) *Lasioderma serricorne* - The grubs and adults bore into cheroots, cigarettes, cured tobacco, ginger, museum specimens and insect collections; *Stegobium paniceum* - Damages books, papers, dried wood, furniture, dried vegetable matter, various drugs and sometimes cereals **(Fig. 184);** *Anobium punctatum* - Bores into timbers, wooden structures, furniture etc.

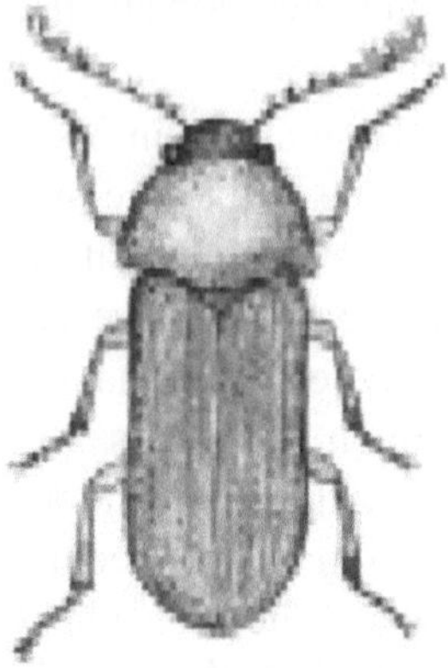

Fig. 184. Wood boring beetle *(Stegobium paniceum)*

viii)**Family - Ptinidae** (spider beetles) - These beetles are not wood borers, but are associated with stored food, dead insects and dry vegetable matter. They feed on decaying animal and vegetable matter. The beetles are small, often cylindrical and somber, dark brown, blackish or reddish in color, with long legs and rounded abdomen, resembling that of a spider. The antennae are often feebly developed. The larvae are however borers in wood, furniture, books, drugs, tobacco, biscuits etc (e-g) *Ptinus fur* - It is a museum pest; *Ptinus claviceps* - It is a household pest; *Gibbium psylloides* - It is a common household pest.

ix) **Family - Bostrychidae** (Grain beetles) - The beetles of this family are small, elongate, cylindrical beetles, with head concealed by the hood - like pronotum. They make cylindrical burrows in felled timber or dried wood and sometimes attack unhealthy standing trees. They are essentially polyphagous and starch is an essential constituent of their food. The body is often truncated posteriorly and armed with small projections. The anntennae usually have less than 11 segments and end in a terminal club. The elytra are smooth or sculptured. The larvae have a curved body, enlarged at the thorax, with well developed legs, annular spiracles and mandibles. The family is of great economic importance in forestry and wood-using industries, as they cause damage to the manufactured soft wood and sap wood articles, plywood, matches, packing cases, boards, shacks, tent poles, tool handles and sports goods. (e-g) *Dinoderus distinctus* - The bamboo borer; *Rhizopertha dominica* - The lesser grain beetle **(Fig. 185)**; *Heterobostrychus aequalis* - The powder post beetle.

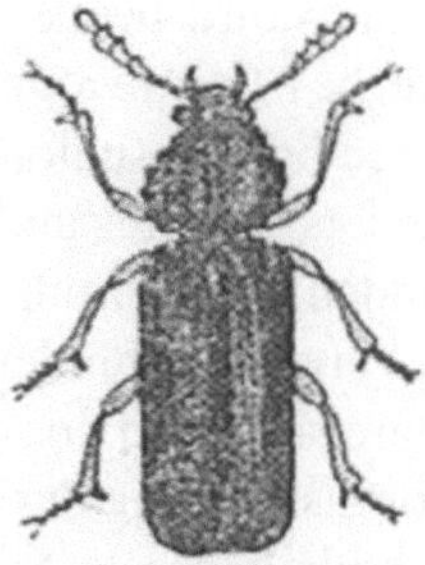

Fig. 185. Lesser grain beetle *(Rhizopertha dominica)*

x) **Family - Lyctidae** (Wood boring insects) - This is closely allied to Bostrychidae, but differs slightly in their morphological characters. The antennae are club of 2 segments, instead of three. The pronotum is flattened and not hood-like. They are rather small, less than 5.0 mm. in length, elongate, light brown to almost blue and are found both in freshly cut, as ell as in old timber, furniture etc (e-g) *Lyctus africanus, L. brunneus*- They are of much importance as pests in saw mills, wood-using factories, furniture, wooden fittings in buildings etc.

xi) **Family - Cucujidae** (Flat bark beetles, cucujid beetles) - The beetles are minute to small, elongated beetles, flat, reddish, brownish or yellowish in color and are usually found on cut logs under the bark. Many are predacious on mites and small insects. Some are pests of stored products and stored grains, The antennae are long, filiform and sometimes with a small club at the terminal end (e-g) *Oryzaephilus surinamensis* - The saw-toothed grain beetle, which has a row of sharp, teeth-like projections on each side of the prothorax. It burrows holes in grains, dried fruits etc. and feeds on them **(Fig. 186)**; *Cryptolestes pusillus, C. minutus* - Feeds on coarse biscuits and other dried food products.

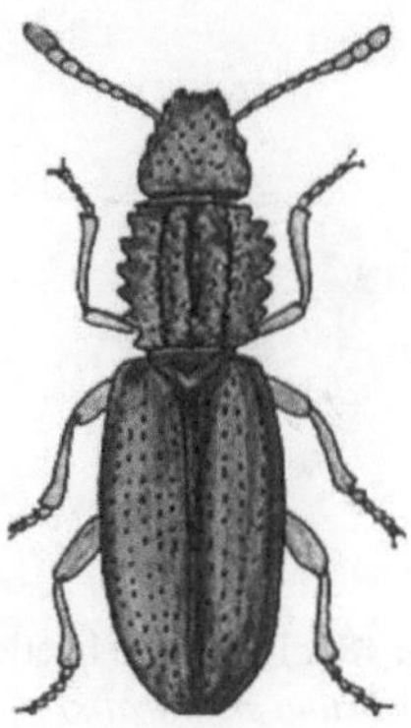

Fig. 186. Saw-toothed grain beetle *(Oryzaephilus surinamensis)*

xii) **Family - Coccinellidae** (Lady bird beetles) - This is a large family of important beetles, which can be easily recognized by their oval or rounded, hemispherical form. They are brightly colored, black, yellow, red, brown or their mixtures and with spots or bands and are about 6.5 mm. long. The head is small and nearly hidden by the prothorax, which fits smoothly into the rounded elytra. The legs are short, hidden under the body and adapted for running. The antennae are short and not clubbed. The vast majority of the coccinellids are carnivorous, feeding during the larval and adult stages upon aphids and other soft-bodied insects. Species of *Coccinella, Rodolia, Menochilus* and *Brumus* are some such predacious insects. A single larva may consume about 500 aphids in its life time. The adults are even more voracious feeders. The Phytophagous ones such as, *Epilachna* species are also important, as they cause serious damage to many cultivated crops and vegetables. When disturbed, many of the coccinellids discharge a bitter, amber-colored fluid. Many species are gregarious in habit.

Eggs are laid in clusters openly on the plants. A single female may lay up to 1,000 eggs. They are cigar-shaped and are laid on one end. The larvae are active, campodeiform, broader in the middle and tapering on both the ends. The body segments have spines or tubercles bearing hairs. There is an anal foot, which assists in locomotion. Most larval forms are black, spotted with yellow, some red and and a number of them have waxy processes similar to those of the mealy bugs, which they eat. Pupation takes place openly on plants, the larva fixing itself with its anal foot and the pupa remaining often partly enveloped by the arval skin. The beetles hibernate in the adult stage (e-g) *Epilachna vigintioctopunctata* - The egg plant spotted beetle **(Fig. 187)**; *Coccinella septum punctata* - Predatory on sorghum stem borer and sugarcane leaf hopper; *Brumus suturalis* - Predatory on sorghum stem borer **(Fig. 188)**; *Rodolia cardinalis* - Predatory on the scale insect *(Icerya purchasi)* and the citrus scale insects **(Fig. 189)**; *Menochilus sexmaculatus* - Predatory on aphids.

Fig. 187. Lady bird beetle *(Epilachna vigintioctopunctata)*

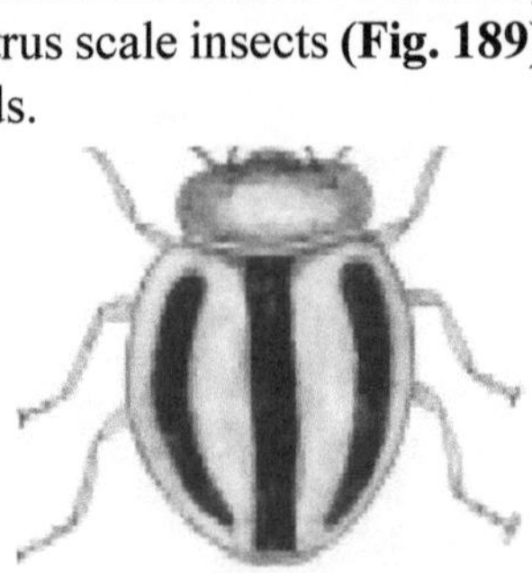

Fig. 188. Lady bird beetle *(Brumus suturalis)*

Fig. 189. Lady bird beetle *(Rodolia cardinalis)*

xiii)**Family - Tenebrionidae** (Darkling beetles, false ground beetles, mealworm beetles) - They are generally dark brown, ground beetles found most abundantly in deserts and dry places, feeding on dead and dried leaves, while a few are pests of stored products. The body is hard, often flattened and elongated, the elytra fitting closely, while some forms are aperous. The antennae are short. The larvae are elongate and cylindrical and the thoracic legs are adapted for running. The larvae usually feed on vegetable matter, but eat locust hoppers also (e-g) *Tribolium castaneum* - The red flour beetle that attacks different grains, flour, oil cakes, pulses, dried fruits, museum specimens etc. **(Fig. 190).**

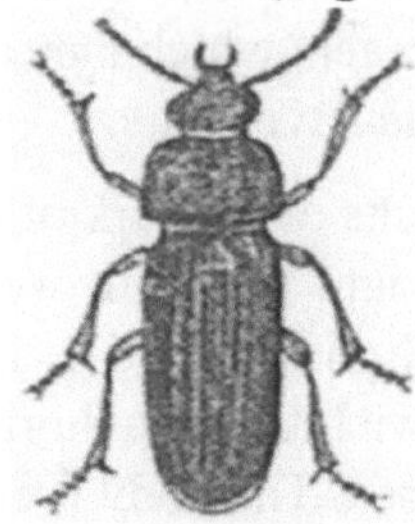

Fig. 190. Red flour beetle *(Tribolium castaneum)*

xiv)**Family - Meloidae** (Oil beetles, blister beetles) – The members of this family are generally quite conspicuous in appearance, sometimes with warning colors, 1.25 - 2.50 cm. long and moderately stout. Some members are green, blue, brown or dull-colored. The head is joined to the prothorax by a distinct, narrow neck and is strongly deflexed. The legs are long. The elytra only loosely cover the upper surface of the body. The wings are used in flight. The beetles are diurnal and feed on flowers and leaves and are injurious to many crops. When disturbed, the beetles of some genera secrete an oily acid containing cantharidin, which causes blisters on the skin. Species of *Mylabris, Epicauta* and *Lytta* yield large quantities of this drug. The larvae are whitish, **'C'**- shaped, cylindrical and usually apodous. They prey upon the eggs of Orthoptera and Hymenoptera. Hypermetamorphosis is common. The first stage larva is campodeiform, which later becomes eruciform or scarabeiform (e-g) *Mylabris phalerata*; *Lytta tenuicollis* **(Fig. 191)**; *Epicauta atkinsoni.*

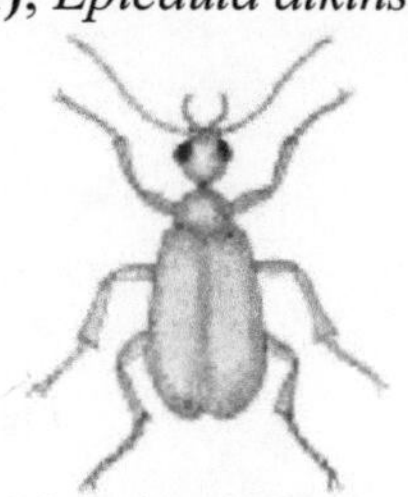

Fig. 191. Blister beetle *(Lytta tenuicollis)*

xv) **Family - Cerambycidae** (Longicorn or long-horned beetles) - This family differs from of this order in having very long antennae, usually at least two-third as long as the whole body, capable of being flexed backwards and are usually inserted on prominent tubercles. It is a large family of large insects ranging from less than 1.25 cm. to more than 2.5 cm. in length, with robust body and hard integument. They are usually somber and cryptically colored and some have bright warning coloration. Some species are serious pests of fruit trees and coffee plants. They are mostly nocturnal in habit. The head is distinct, with large eyes and powerful mouthparts, especially the mandibles. The elytra cover a pair of large membranous wings and closely applied to the abdomen, which they cover completely. The legs are long nd pubescent. All tibia have spurs.

Eggs are laid singly in cracks of the bark of trees, which hatch into legless larvae. The larvae tunnel into the hard, woody tissues, eating out large, winding galleries in which they live. The larva is elongate, yellowish in color, with a hard and brown head. The thorax is swollen and the abdomen narrower, but well developed. The body is covered with a fine pubescence and the thoracic and abdominal segments have dorsal sclerotized plates. The abdominal legs are lacking and the thoracic legs generally reduced. Locomotion takes place with the aid of Dorsal and ventral, segmentally arranged abdominal swellings.. The pupa is found in the larval tunnel and the head and tail regions are covered with frass to form a chamber or in a white cocoon. The larval life may extend up to 3 years or more (e-g) *Batocera rufomaculata* - The mango stem borer **(Fig. 192)**; *Plocaederus ferrugineus* - The cashew stem borer; *Sthenias grisator* - The grapevine stem girdler **(Fig. 193)**.

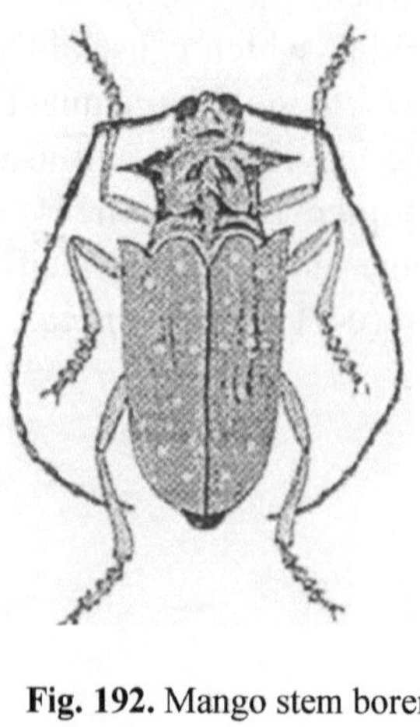

Fig. 192. Mango stem borer *(Batocera rufomaculata)*

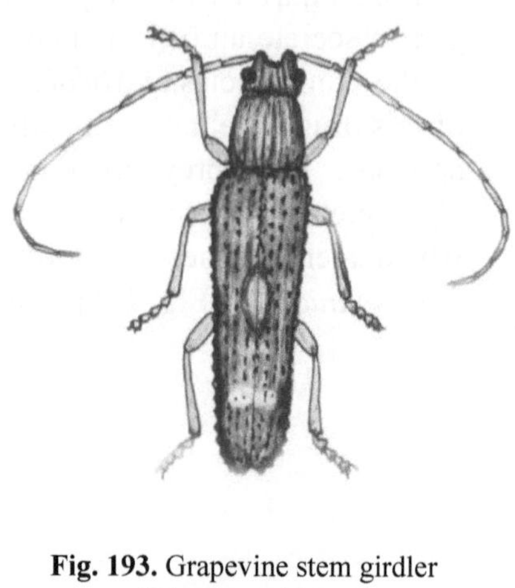

Fig. 193. Grapevine stem girdler *(Sthenias grisator)*

xvi)**Family - Bruchidae** (Pulse beetles) - This is also a large family and the larvae mostly live in leguminous seeds such as, grams, beans, lentils etc., while some attack coconuts. And palm nuts. These small, short beetles, about 6.5 mm. in length are somber-colored and inconspicuous. The body is covered with fine hairs. The head is small, with a blunt rostrum and with short, pectinate or serrate antennae. The elytra are striated, truncate and not covering the pygidium. The legs are short and the hind femora are thickset and often toothed. The abdomen is also thickset.

The eggs are small, oval in shape, yellow in color and are laid in a semi-liquid condition, which stick to the grains on hardening. On hatching, the larva eats through the inner wall of the egg shell into the seed coat and then into the seed itself, where it feeds on the cotyledons..The larva is eruciform white and curved and in the first instar may have three pairs of functional legs and a series of thoracic spines, which help the larva to bore into the seed and establish itself. After the first moult, all these appendages disappear. The full-grown larva is eruciform, grub-like and with a thick body. Pupation takes place within the seed or outside in a cocoon made of excretory matter. Hypermetamorphosis is present. These beetles breed rapidly and up to six generations may appear in a year (e-g) *Bruchus chinensis* - The pulse beetle **(Fig. 194)**.

Fig. 194. The pulse beetle *(Bruchus chinensis)*

xvii)**Family - Chrysomelidae** (Leaf cutting beetles) - The family comprises of a very large and varied forms than any other family of Coleoptera. All are brightly colored, herbivorous and diurnal in habit. The beetles are mostly smooth, without hairs and frequently with metallic coloration. The hind legs are not thickened and toothed as in as is found in Bruchidae. The larvae are soft, fleshy, with short legs and usually brightly colored. Pupation takes place in the soil. (e-g) *Raphidopalpa foveicollis* - The red pumpkin beetle **(Fig. 195)**; *Dicladispa armigera* - The rice hispa; *Scelodonta strigicollis* - The grapevine leaf beetle; *Leptinotarsa decimlineata* - The potato beetle.

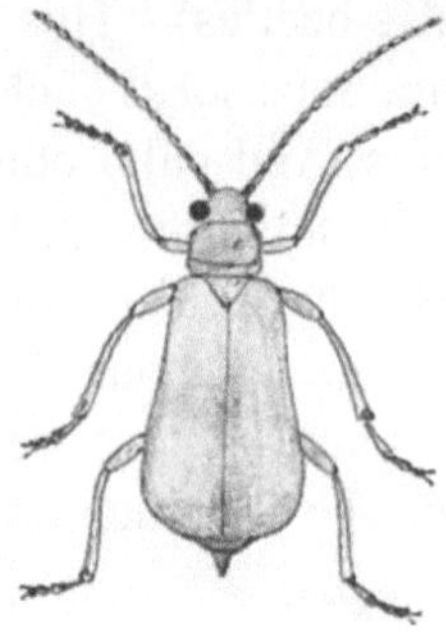

Fig. 195. Red pumpkin beetle *(Raphidopalpa foveicollis)*

xviii) **Family - Curculionidae** (Weevils or snout beetles) – The most important and characteristic feature of this large family is that the head is drawn into a distinct rostrum or snout in the vast majority of the forms. The antennae are clubbed and elbowed and the palpi are reduced and rigid. The beetles vary in length from 0.3 - 5.0 cm. They may be of different colors, generally dull brown, gray, black, rich brown or green. The body is clothed in scales in many species. The body is often short and thickset. The rostrum or snout is of varied forms, which in the females of some species helps in boring holes for laying eggs or putting the eggs deep in the holes made with the ovipositor. The minute mouthparts are situated at the apex of the snout. The abdomen is completely covered by the elytra.

The eggs, which are small, oval and smooth or soft, white and elongate, are laid singly and in large numbers spread over a number of plants. The larvae are white, soft, legless, with a distinct brown head and stout, much wrinkled body, which is fleshy and slightly curved. They are internal feeders. A few species are aquatic in the larval and adult stages. The majority of species pupate in the soil or inside the host plant.

A few genera of this family lack rostrum and posses short antennae. They are cylindrical in shape and many species bore into the bark and stem or the roots.

Weevils have the habit of feigning dead at the approach of danger. They fold the legs and antennae close to the body and drop to the ground and become extremely hard to find (e-g) *Sitophilus oryzae* - The rice weevil; *Sitophilus granarium* - The grain weevil; *Rhynchophorus ferrugineus* - The red palm weevil **(Fig. 196)**; *Pempherulus affinis* - The cotton stem weevil; *Sternochetus mangiferae* - The mango nut weevil; *Xyleborus fornicatus* - The tea shot-hole borer; *Myllocerus subfasciatus* - The brinjal ash weevil; *Myllocerus pusthulata* - The ash weevil **(Fig. 197)**; *Apoderus tranquebaricus* - The mango leaf-twisting weevil **(Fig. 198)**.

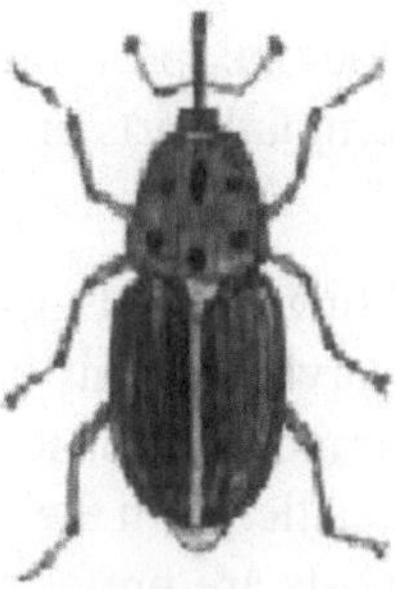

Fig. 196. Coconut red palm weevil *(Rhynchophorus ferrugineus)*

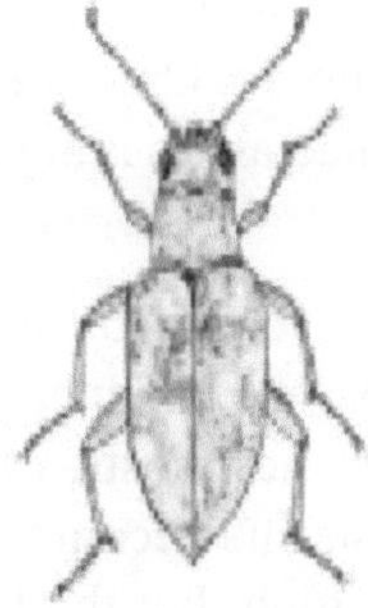

Fig. 197. Ash weevil *(Myllocerus pusthulata)*

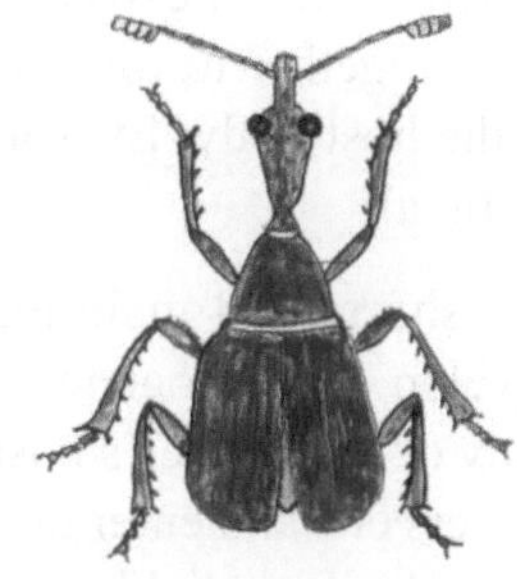

Fig. 198. Mango leaf twisting weevil *(Apoderus tranquebaricus)*

Order - 9. Strepsiptera (Stylopids)

('Strepsi' = twisted ; 'ptera' = wings)

The members of this order are very small and minute and brown or black in color. All are endoparasitic.in the larval stage. The larvae are called 'triungulins' The females always remain in the puparium, which protrude from the body of the host and are known as 'stylops'. The males are free-living and are winged. The hosts include some Homoptera and Hymenoptera.

Fig. 199. Stylopid *(Eoxenos laboulbeni)*

In the males, the head is transverse, with protruding compound eyes and ocelli are absent. The antennae are peculiar in that the third segment is flabellate, giving it a branched appearance and all the antennal segments are covered with sensory organs. The mouthparts are of biting type, but are very much reduced. The metathorax is very much enlarged and is nearly half of the body length. The anterior wings in males are reduced and club-like, resembling the halteres of Diptera, while the hind wings are large, fan-shaped and have radiating veins without cross veins. The abdomen is 10 segmented and no cerci are present. The female, which is wingless, is larviform, apodous and enclosed in the persistent larval exoskeleton, except in a few forms. The head and thorax are fused together to form a cephalothorax. Only the cephalothorax protrudesout of the host body. Eyes and antennae are lacking and the mouthparts are also vestigial.

The short-lived male fertilizes the female by just alighting on it. Eggs can develop parthenogenetically also. The minute, young larvae remain in the body of the mother's host and reach new hosts, when their mother's host visits flowers etc. and enter into new hosts and become endoparasitic. Both the sexes of the host are liable for parasitization, but the female hosts are preferred. A single individual of the host may have many stylop parasites. Jassids, wasps and bees are parasitized by many Strepsiptera (e-g) *Eoxenos laboulbeni* - Parasite of predatory wasp -*Polistes hebraeus* **(Fig. 199)**.

Appendices

Appendix 1. Main characteristics of important classes of Phylum- Arthropoda

Crustacea	**Insecta**	**Arachnida**	**Chilopoda**	**Diplopoda**	Onychopora
1. Body division-2, Cephalothorax and abdomen.	Body division-3, Head, thorax and abdomen.	Body division-2, Cephalo-thorax and abdomen.	Body division-2, Head and body.	Body division-2, Head and body.	Body division 2, Head and body.
2. Antennae-2 pairs	Antennae-1 pair.	Antennae - none.	Antennae-1 pair.	Antennae-1 pair	Antennae-1 pair.
3. Legs-usually 5 pairs, but may be many in some cases.	Legs-3 pairs, one pair in each thoracic segment.	Legs-4 pairs.	Legs-many, one pair per segment	Legs-many, one or two pairs per segment	Legs-many, one pair per segment.
4. Wings-absent.	Wings-2 pairs, one pair in each of the meso- and metatho-rax	Wings-absent	Wings-absent.	Wings-absent.	Wings-absent.
5. Develop-ment-ametabolous.	Develop-ment-metabolous.	Develop-ment-ametabolous.	Develop-ment-ametabo-lous.	Develop-ment-ametabo-lous.	Develop-ment-ametabolous.
6. Habitat-chiefly marine and fresh water, rarely terrestrial.	Habitat-terrestrial and fresh water, rarely marine.	Habitat-terrestrial	Habitat-terrestrial	Habitat-terrestrial.	Habitat-terrestrial.
(e-g) Crabs, prawns	Grasshop-pers, flies, butterflies, beetles, bees, cockroaches	Spiders, mites, ticks, scorpions	Centipedes	Millipedes	Peripatus

Appendix 2. Scientific classification of Class - Insecta

Class - Insecta

Sub class

- Apterygota or Ametabola
  - **Orders**
    - * Protura
    - *Thysanura
    - * Collembola
- Pterygota or Metabola
  - **Division**
    - Exopterygota or Heterometabola
      - **Sub division**
        - **Paurometabola**
          - **Orders**
            - * Dictyoptera
            - * Orthoptera
            - * Phasmida
            - * Dermaptera
            - * Embioptera
            - * Isoptera
            - * Psocoptera
            - * Mallophaga
            - * Siphunculata (Anoplura)
            - * Hemiptera (Rhynchota)
            - * Thysanoptera
        - **Hemimetabola**
          - **Orders**
            - * Odonata
            - * Ephimeroptera
            - * Plecoptera
    - Endopterygota or Holometabola
      - **Orders**
        - * Neuroptera
        - * Mecoptera
        - * Lepidoptera
        - * Trichoptera
        - * Diptera
        - * Siphonaptera
        - * Hymenoptera
        - * Coleoptera
        - * Strepsiptera

Appendix 3. Economic classification of insects

INSECTS

- Insects of no economic importance
- Insects of economic importance
  - Injurious insects
    1. Pests of cultivated plants (Crop pests)
    2. Pests of stored grains, stored products, household articles, furniture etc.
    3. Pests of man, cattle, other domestic animals, poultry, other birds etc.
    4. Household and disease carrying Insects
  - Beneficial insects
    - **1. Productive insects**
      - (i) Silk
      - (ii) Honey
      - (iii) Lac
      - (iv) Dyes and chemicals
      - (v) Food
      - (vi) Drugs
      - (vii) Ornaments
      - (viii) Other products
    - **2. Helpful insects**
      - (i) Parasitoids
      - (ii) Predators
      - (iii) Pollinators
      - (iv) Weed killers
      - (v) Scavengers
      - (vi) Soil ameliorators
      - (vii) For research purposes
      - (viii) For recreation

Appendix 4: Different types of metamorphosis in insects

Type of metamorphosis	Example	Egg stage	Feeding, moulting and growing stages	Non feeding transformation stage	Full grown, reproducing stage
Ametabola	Silver fish			Nil	
Paurometabola	Grasshoppers			Nil	
Paurometabola	Bugs			Nil	
Paurometabola	Thrips			Nil	
Hemimetabola	Dragon flies			Nil	
Holometabola	Moths				
Holometabola	Flies				
Holometabola	Beetles				
Holometabola	Bees				

Colour Plates

Order- Thysanura
(*Silver fish*)

order-Orthoptera
(Grass Hopper)

Order:Dictyoptera
(*Prayng mantis*)

Order- Gryllidae
(*Cricket*)

order-Phasmida
(*Leaf insect*)

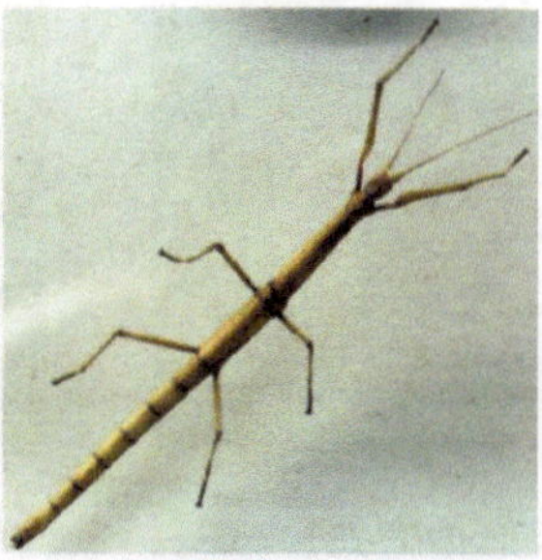

Order: Phasmida
(*Stick insect*)

Order:Dermaptera
(Earwig)

order-Siphunculata
(Human head louse)

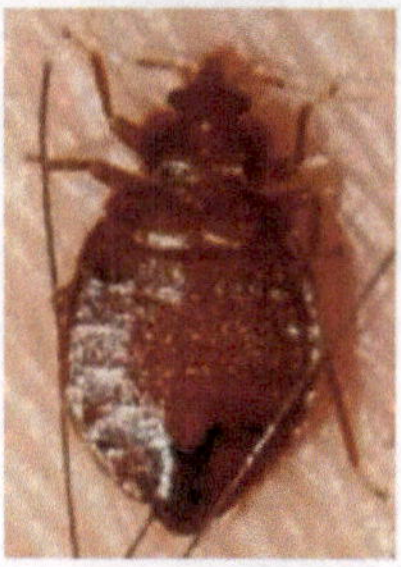

Order - Siphunculata
(Bed bug)

Order - Hemiptera
(*Cicadas*)

Order – Thysanoptera
(*Rice leaf thrips*)

Order - Lepidoptera
(*Cabbage butterfly*)

ler – Lepidoptera
3lue butterfly)

Order – Lepidoptera
(*Citrus leaf miner*)

Order – Lepidoptera
(*Diamond black moth*)

er – Lepidoptera
; caterpillar moth)

Order – Lepidoptera
(Hawk moth)

Order – Lepidoptera
(Gram cutworm moth)

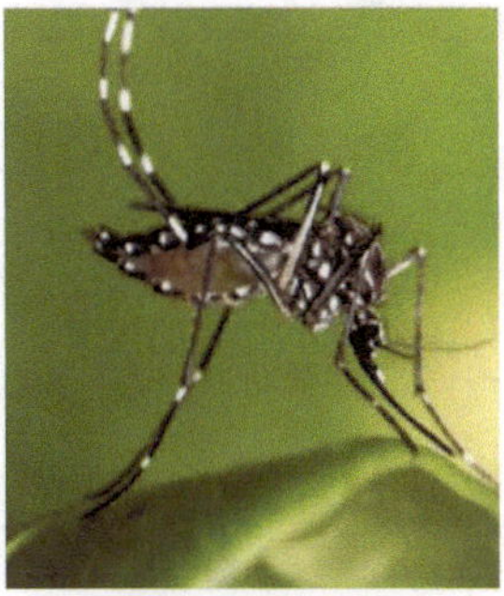

Order – Diptera
(*Aedes mosquito*)

Order – Diptera
(*Sesamum gall fly*)

Order – Diptera
(*Common house fly*)

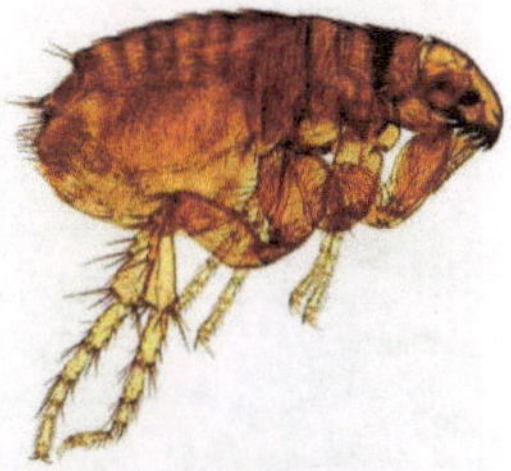

Order – Siphonaptera
(Rat flea)

Order – Hymenoptera
(Red ant)

Order - Hymenoptera
(Rose leaf cutter bee)

Order – Coleoptera
(Tiger beetle)

Order – Coleoptera
(Dung roller beetle)

Order – Coleoptera
(Rhinoceros beetle)

Order – Coleoptera
(Khapra beetle)

Order – Coleoptera
(Wood borer beetle)

Order – Coleoptera
(Saw toothed beetle)

Order – Coleoptera
(Lady bird beetle)

Order – Coleoptera
(Blister beetle)

Order – Coleoptera
(Ash weevil)

Order – Coleoptera
(Lady bird beetle)

Order – Coleoptera
(Mango stem borer)

Order – Coleoptera
(Mango leaf twisting weevil)

References

Anon. 1997. Hand book of Agriculture. Director, Director of Publicaion and Information on Agriculture, Krishi Anusandhan Bhavan, Pusa, New Delhi - 110 012. pp. 1303.

Hem Singh Pruthi. 1969. Textbook on Agricultural Entomology. Indian Council of Agricultural Reseach, New Delhi. pp. 977.

Lewin Devasahayam, H. 2014. Elements of Entomology. New India Publishing Agency, Pitam Pura, New Delhi - 110 012. pp. 448.

Ramakrishna Ayyar, T.V. Handbook of Economic Entomology for Suth India, Controller of Stationary and Printing, Government of Madras. pp. 516.

Ross H. Arnett and Richard I. Jacques. 1981. Simon and Schulter's Guide to Insects. Simon and Schulter Inc., New York. pp. 512.

Saxena, R.C. and Srivasta, R.C. 2009. Entomology at a glance. Agrotech Publishing Academy, Udaipur - 313 002. pp. 464.

Vasantharaj David, B and Kumaraswamy, T. 1975. Elements of Economic Entomology. Popular Book Depot, 29-potters Street, Saidapet, Chennai -600 015. pp. 507

Index

A
Abductor muscles - 60
Absorption - 76, 80, 81, 82, 83, 84
Accessory glands - 80, 93, 105, 106, 108
Acephalus larvae - 127
Acrididae - 34, 110, 120, 146, 148
Acrosome - 108
Acrotrophic type - 105
Adenotrophic viviparity - 33, 109
Adephaga - 216
Aedeagus - 73, 74, 106, 108
Aestivation - 13
Afferent neurones - 88
Agromyzidae - 203
Air sacs - 98, 100, 113, 114
Alary - 86, 87, 146, 148
Aleyrodidae - 160, 172
Alder flies - 178
Ameiotic - 110
Ametabola - 123, 136
Amphipneustic - 103
Anatomy - 75, 133
Ant lions - 28, 180
Ants - 11, 12, 14, 15, 17, 23, 27, 29, 30, 31, 64, 85, 103. 116, 128, 137, 153, 169, 193, 206, 211
Anemotropism - 8
Anisoptera - 139
Anisozygoptera - 140
Anobiidae - 223
Anoplura - 158
Antennifer - 46
Anthrax - 200, 204
Antidiuretic hormones - 92
Ants - 2, 11, 12, 14,.15, 17, 23, 27, 29, 30, 31, 64, 95, 96, 103, 116, 128, 169, 178, 205
Aorta - 85, 86, 87, 91, 92
Aphaniptera - 205
Aphidiae - 172
Apidae - 214
Apneusic - 102
Apocrita - 207
Apodemes - 39, 43
Apodous larvae - 127, 216
Apolysis - 41
Apomietic - 110
Apomistic - 110
Apposition image - 118, 119
Apterigota - 64, 123, 136, 139
Apterous - 64, 120, 152, 157, 159, 160, 173, 177, 188, 208, 223
Arctidae - 194
Asilidae - 48, 201
Assassin bugs - 161, 168
Arachnida - 3, 133, 134, 233
Araeopidae - 171
Archaea - 134
Arolium - 61
Arthropoda - 2, 3, 4, 134, 233
Atrium - 73, 74, 97, 99, 108
Auditory organs - 6, 111, 113, 146, 147, 149
Auditry hairs - 115
Auricular valve - 86, 87

B
Bacteria - 27, 35, 79, 85, 134, 155
Bag worms - 188
Bark and stem borers - 188
Basal membrane - 40, 66, 79
Batesian mimicry - 12
Beetles - 11, 12, 13, 15, 16, 22, 23, 24, 28, 45, 65, 67, 120, 130, 135, 212, 215
Belostomatidae - 166
Bethylidae - 210
Bethylids - 210
Big-headed flies - 202
Binocular eyes - 119
Biotic agencies - 4
Biotic potential - 10
Bipolar neurone - 111, 113, 115
Bittacidae - 181
Black ants - 9, 22
Blastoderm - 122
Blattaria - 143

Blister beetles - 11, 22, 227
Blues - 186
Bombicidae
Bombycidae - 192
Bostrychidae - 224
Brachycera - 127, 198, 200
Braconidae - 208
Braconids - 208
Brain - 3, 44, 87, 89, 91, 92, 93, 118, 128
Brain hormones - 92
Bristle flies - 203
Bruchidae - 229
Bubonic plague - 205
Buccal cavity - 77
Buprestridae - 127
Bursa copulatrix - 105
Bursicon - 42

C

Cabbage butterfly - 10, 185, 186
Caddis flies - 196
Caddle worm - 196
Caelifera - 146, 148
Campaniform sensilla - 66, 111
Carnivorous - 5, 10, 45, 143, 145, 181, 198, 211, 217, 226
Campodeiform larvae - 127
Carabidae - 217
Castes of termites - 153
Caterpillars - 8, 9, 11, 21, 25, 64, 76, 127, 168, 182, 186, 188, 189, 191, 192, 193, 194, 195, 203, 212, 217
Cecidomyiidae - 199
Centipedes - 2, 135, 203, 233
Central nerve cord - 44, 85, 89, 90, 91, 92, 97
Central sinus - 85
Cerambicidae - 127
Cerci - 39, 72, 73, 74, 82, 111, 115, 127, 136, 137, 139, 141, 142, 143, 144, 145, 147, 148, 149, 151, 152, 153, 157, 158, 160, 166, 177, 179, 181, 231
Cercopidae - 84, 169
Cervical region - 59
Cetoninae - 219, 220
Chalcididae - 209
Chermidae - 172
Chemo receptors - 46, 111, 115, 116
Chemotropism - 7
Chilopoda - 3, 134, 135, 233
Chironomidae - 200
Chitin - 2, 3, 37, 38, 39, 40, 41, 50, 71, 80, 84, 98, 102, 134, 205
Chitinase - 41
Chordotonal organs - 110, 113, 115
Chorion - 104, 105, 106, 121
Chrysalid - 21, 22, 130, 184
Chrysomelidae - 229
Chrisopidae - 179
Cibarium - 44, 53, 76
Cicadellidae - 170, 172
Cicadidae - 160
Cicindelidae - 216, 217
Cimicidae - 163
Cirixidae - 167
Claspers - 102, 127
Classification - 61, 63, 133, 134, 136, 139, 183, 197
Cleavage energid - 122
Click beetles - 222
Closed type - 88
Clypeus - 43, 44, 49, 51, 52, 53, 55, 210, 212
Coccidae - 129, 160, 175
Coccinellidae - 226
Cochlididae - 189
Cock chafer - 218, 210
Cockroaches - 8, 10, 14, 45, 72, 135, 143, 215, 233
Coelom - 3
Coleoptera - 5, 33, 34, 45, 48, 49, 83, 84, 109, 110, 120, 125, 126, 127, 209, 210, 215, 229
Collateral glands - 103, 105, 106, 108
Collumbola - 64, 71, 83, 84, 102, 138
Colon - 79
Common chemical receptors - 116
Common oviduct - 103, 105
Common thrips - 177
Complete metamorphosis - 66, 124, 125
Compound eyes - 37, 44, 45, 46, 89, 116, 117, 118, 119, 125, 127, 138, 139, 140, 142, 143, 144, 150, 152, 153, 154, 155, 157, 160, 172, 174, 177, 181, 183, 197, 201, 205, 214, 215, 231
Conductivity - 89
Copulatory pouch - 72
Corbicula - 63
Coreidae - 164
Cornea - 117, 118
Corneogen layer - 117
Corpora allata - 92, 93, 128
Corpora cardiaca - 92, 93

Cortical area - 121
Cossidae - 188
Crabs - 134
Crambidae - 190
Crane flies - 100
Cranium - 43
Cremaster - 130
Crochets - 64, 207
Crop - 18, 77, 78, 81, 82, 90
Crustacea - 3, 134, 233
Cryptophagidae - 187
Culicidae - 198
Curcujidae - 225
Curcujid beetles - 225
Curculionidae - 1230
Crystalline cone - 117, 118, 119
Cyclorrhapha - 93, 126, 198, 201

D

Dactilopiidae - 175
Daddy long legs - 199
Darkling beetlwa - 227
Delphacidae - 171
Dermal receptors - 116
Dermaptera - 34, 150
Dermestidae - 223
Deuteroceribrum - 89
Deutoplasm - 121
Diapause - 13, 93
Diaspididae - 176
Diastole - 86, 87
Dichoptic - 197
Dictyoptera - 104, 143
Digestion - 24, 41, 42, 76, 77, 78, 81, 83, 155
Digger wasps - 213
Diplopoda - 134, 135, 233
Diptera - 33, 34, 48, 49, 59, 60, 66, 67, 69, 84, 93, 98, 104, 109, 110, 115, 125, 127, 130, 197, 198, 201, 206, 209, 231
Ditiscidae - 217
Diving beetles - 217
Dog fly - 204
Dorsal diaphragm - 75, 85, 86
Dorsal commissure - 99
Dorsal vessel - 84, 85, 86
Drosophilidae - 202
Dung beetle - 13, 23, 219\
Dynastinae - 219

E

Earth measurers - 193
Ecdysal glands - 92, 93
Ecdysis - 41, 42, 98, 122, 128
Ecdysone - 41, 42, 93, 128
Ecdysteroids - 92
Eclosion - 42, 92, 129
Eclosion hormones - 92
Efferent neurones - 88
Egg cells - 104
Egg depositor - 71, 102
Egg guide - 73
Egg pod - 11, 105, 106
Ejaculatory duct - 73, 106, 108
Ejestion - 76, 84
Elateridae - 222
Embiidae - 15, 152
Embioptera - 151
Embryo - 24, 33, 34, 105, 109, 122, 126
Embryonic development - 37, 108, 121
Embryonic primordium - 122
Emperor moths - 192
Empodium - 61, 64
Endochorion - 106
Endocrine system - 91, 92, 93
Endoparasites - 16, 27, 207
Endopterygota - 178
Endoskeleton - 39
Energid - 122
Ensifera - 145
Ensign wasps - 215
Ephemeroptera - 74
Epicranial suture - 44, 45
Epidermis - 2, 37, 38, 39, 40, 41, 42, 49, 51, 66, 68, 93, 98, 115, 130, 131
Epiphallus - 73
Epipharynx - 51, 52, 53, 54, 55, 57, 58, 59
Eruciform larvae - 127
Eukarya - 134
Eucephalus larvae - 127
Eulophidae - 209
Eulophids - 209
Eupteroptidae - 192
Evanidae - 215
Evertible glands - 11
Exochorion - 106
Exopterygota - 123, 138, 142
Exoskeleton / Integument - 2, 3, 5, 7, 10, 37, 38, 39, 41, 59, 102, 120, 122, 134, 205

Extrinsic muscles - 62
Exuvia - 23, 42, 122
Eye fly - 203

F

Faceted eyes - 45
Fecundity - 5, 8, 10, 24,
Felt pads - 97
Femoro alary - 146, 148
Fertilization - 3, 33, 105, 106, 108, 109, 121, 122
Filter chamber - 82
Flagellum - 46, 47, 48, 49, 115
Fleas - 45, 53, 54, 205
Flower cutting beetle - 219, 220, 221
Flower flies - 201
Flower thrips - 177
Follicle - 104, 105, 106, 107
Food channel - 53, 54, 55, 56, 58, 160
Formicidae - 211
Frenulum - 68, 69, 184, 188, 190, 194
Frons - 43, 44, 46, 215
Fruit flies - 8, 15, 31, 202
Fulgoridae - 171
Fungus gnats -

G

Gad flies - 200
Galleriidae - 189
Gall gnats - 199
Gall midges - 21, 199
Gaster - 72, 215
Gastric caecae - 79, 81
Gelechiidae - 187
Gena - 43, 44, 115
Genital chamber - 103, 105
Genital aperture - 38, 71, 72, 73, 74, 106, 139
Genital atrium - 73, 74, 108
Genital ducts - 105, 108
Genitalia - 38, 71, 72, 74, 102, 106, 123, 137, 142, 143, 145, 157, 177, 181
Geometers - 193
Geometridae - 193
Geotropism - 8
Germarium - 104, 105, 107
Germ band - 122
Gerridae - 162
Giant silk worm moths - 192
Gizzard - 77, 78
Glands - 33, 38, 41, 51, 80
Glia - 92
Glial cells - 88, 92
Glow worms - 120, 222, 223
Gonocoxite - 73
Gonopore - 103, 106
Gracillariidae - 187
Gradual metamorphosis - 123, 124
Grain beetles - 223, 224
Grubs - 3, 9, 21, 23, 127, 216, 218, 219, 220, 223
Gryllidae - 120, 146, 147, 148
Gryllotalpidae - 148
Gula - 43, 44, 50
Gustatory receptors - 116

H

Haematomyzidae - 158
Haematophagous - 5
Haematopinidae - 159
Haemocoel - 2, 34, 64, 83, 84, 85, 86, 87, 97, 109, 136
Haemocoelus viviparity - 109
Haemocytes - 85
Haemolymph - 41, 83, 85, 102
Haemocoel - 2, 34, 64, 83,84, 85, 86, 87, 109, 136
Hairy caterpillars - 9, 25, 194
Haplodiploidy - 110
Hard scale - 176
Hatching - 13, 22, 33, 109, 121, 122, 123, 144, 166, 169, 219, 229
Hawk moths - 193
Haustellate type - 49
Heart - 75, 84, 85, 86, 87, 92, 99
Heliotropism - 9
Hemelytra - 67, 160, 161, 162, 163, 164, 165
Hemerobiidae - 178
Hemicephalus larvae - 127
Hemimetabola - 138
Hemimetabolous development - 124
Hemipneustic - 101
Hemiptera - 45, 49, 52, 67, 82, 84, 105, 109, 115, 120, 160, 161, 167
Herbivorous - 10, 135, 147, 150, 229
Hermaphroditism - 34, 110
Hesperiidae - 186
Heterobolous development - 123
Heterometabola - 125, 138, 142
Heteroptera - 67, 97, 160, 161
Hexapoda - 134, 135

Hibernation - 13, 15, 24, 31
Hippoboscidae - 204
Histoblasts - 130
Histolysis - 130, 131
Hodotermitidae - 156
Holometabola - 125, 178
Holometabolous development - 125
Holopneustic - 101
Holoptic - 197
Homoptera - 82, 96, 160, 161, 168, 201, 211, 231
Honeybees - 5, 17, 22, 25, 27, 29, 30, 95, 206, 214
Honey comb moths - 94, 189
Hooked spines - 69, 177
Hormonal control mechanism - 81
Horn worms - 193
Horse flies - 200, 204
House flies - 204
Hover flies - 23, 201
Humeral bristles - 68
Humuli - 69
Hydrofuge - 46, 162
Hydrophilidae - 218
Hydrophilus - 46, 218
Hydrotropism - 9
Hygro receptors - 46, 111, 116
Hymenoptera - 5, 13, 15, 31, 40, 60, 69, 72, 83, 96, 102, 104, 110, 115, 125, 127, 153, 201, 206, 227, 231
Hyper metamorphosis - 126, 178, 206, 216, 227, 229
Hyperparasites - 16
Hypognathous - 45, 143, 206, 215
Hypopharynx - 38, 49, 51, 52, 53, 54, 58, 76 77, 80, 81, 89, 96, 116, 143, 150

I

Ichneumonidae - 34, 62, 119, 208
Ichneumonids - 208
Ilium - 79, 82
Imaginal buds - 130
Imago - 122, 129, 130, 139
Incomplete metamorphosis - 66, 124, 125
Indotermitidae - 156
Ingestion - 49, 76
Insecta - 3, 64, 69, 133, 134, 135, 136, 182, 215, 233
Instars - 11, 15, 31, 41, 42, 65, 66, 122, 123, 125, 126, 128, 129, 137, 159, 179, 206, 229
Instincts - 110
Integument - 38, 40, 42, 65, 76, 83, 97, 102, 111, 112, 115, 221, 222
Inter neurones - 88
Interspecific - 16, 27, 95
Intima - 17, 77, 79, 98, 99, 114
Intraspecific - 13, 14, 15, 27, 29, 30, 31
Intrinsic muscles - 62
Iris - 117
Isoptera - 48, 104, 139, 153

J

Jassidae - 170
Jewel beetle - 23, 221
Johnston's organ - 113, 115
Jugal bristles - 68
Jugal lobes - 69
Juvenile hormone - 42, 92, 03, 128

K

Kalotermitidae - 156
Khapra beetle - 0, 223

L

Labidurudae - 151
Lace wings - 178
Lacciferidae - 175
Lady bird beetles - 16, 226
Laeral ocelli - 46, 117
Lacaniidae - 175
Lampyridae - 120, 222
Leaf chafer - 219, 220, 221
Leaf cutter bees - 213
Leaf cutting beetles - 229
Lepidoptera - 5, 14, 29, 34, 42, 48, 59, 60, 68, 94, 98, 101, 102, 110, 115, 120, 126, 127, 130, 181, 182, 184, 194, 203, 207, 210, 215
Lepismatidae - 137
Life cycle - 22, 24, 29, 32, 64, 91, 93, 121, 123, 145, 148, 172, 173, 205, 223
Light producing organs - 120
Ligula - 50, 172
Limacodidae - 189
Lobsters - 134
Longicorn beetles - 12
Loopers - 127, 193
Lophopidae - 171
Lycaenidae - 186
Lygaeidae - 167
Luciferin - 222

Luciferinase - 222
Lyctidae - 225
Lymantriidae - 196

M

Machlidae - 137
Mallophaga - 158
Macropterous - 6, 21
Maggots - 7, 22, 23, 102, 127, 148, 197, 202, 206
Malpighian tubules - 3, 38, 79, 82, 83, 84, 85, 131, 136, 143, 178
Mantidae - 143, 144
Mantid flies - 178
Mantispidae - 126, 178
Mantodea - 143, 144
Margarodidae - 174
Mealy bugs - 11, 42, 174, 175, 226
Mechano receptors - 111, 113
Mecoptera - 68, 101, 181
Median ocellus - 46
Megachilidae - 213
Melolonthinae - 219
Meiotic - 110
Meloidae - 126, 227
Melolonthinae - 219
Membracidae - 169
Menoponidae - 158
Meroistic type - 105
Mesenteron - 76, 78, 79, 81, 82
Measuring worms - 193
Metabola - 138, 142, 178
Metamorphosis - 91, 93, 121, 122, 123, 125, 126, 129, 130, 133, 136, 139, 141, 142, 145, 150, 157, 158, 160, 181, 182, 197, 206, 216, 227, 236
Metapneustic - 102
Metarbelidae - 188
Micropyle - 106, 121
Midges - 21, 23, 199, 200
Millipedes - 2, 135, 233
Mimicry - 6, 12, 22
Miridae - 162
Mies - 15, 21, 30, 135, 176, 179, 225, 233
Mitochondria - 88
Monophagous - 5, 10
Morphology - 37, 43, 133
Moth flies - 199
Mosaic image - 118
Mosquitoes - 14, 49, 51, 53, 64, 101, 102, 113, 116, 121, 135, 197, 198, 199, 200
Motor neurones - 88
Moulting - 3, 9, 37, 40, 41, 42, 84, 91, 93, 122, 128, 207
Moulting fluid - 40, 41, 42, 93, 122
Moulting hormone - 93
Mullarian mimicry - 12
Muller's organ - 114
Muscidae - 48, 198, 204
Mutualism - 17, 29, 155
Mycetophilidae - 200
Myrmeleontidae - 180

N

Nabididae - 168
Naiads - 102, 124, 139, 141, 142
Nematocera - 127, 198
Nemopteridae - 179
Nepidae - 166
Nerve cells - 91, 02, 111, 113
Nervous system - 3, 7, 75, 81, 88, 89, 90, 91, 92, 110, 117, 134, 136
Neural control mechanism - 81
Neuron - 88, 91, 111, 113, 115, 118
Neuroptera - 22, 84, 101, 104, 126, 127, 178
Neurosecretory cells - 91, 92
Noctuidae - 194
Nocturnal moths - 191, 194
Notonectidae - 167
Nucleus - 41, 88, 105, 108, 112, 122
Nymphalidae - 60, 185
Nymphs - 8, 33, 42, 101, 102, 108, 123, 124, 139, 140, 141, 144, 145, 148, 150, 154, 161, 165, 168, 169, 171, 172, 174, 177

O

Occiput - 44
Ocelli - 45, 46, 89, 116, 117, 127, 139, 143, 144, 150, 152, 157, 158, 159, 160, 162, 163, 165, 168, 177, 183, 184, 197, 198, 205, 206, 207, 215, 231
Odonata - 34, 60, 70, 71, 98, 124, 138
Oesophagus - 52, 53, 77, 81, 82, 89, 91, 92
Olfactory receptors - 116
Oligopneustic - 101
Oligophagous - 5, 10
Oligopod larvae - 127
Oligotomidae - 152
Ommatida - 45
Ommatidium - 116, 117, 118, 119

Omnivorous - 10, 143, 147, 148
Onychophora - 134
Oocyte - 91, 92, 103, 104, 105, 110
Oogenesis - 105
Oogonia - 104
Opisthonagthous - 45, 160
Orthoptera - 48, 49, 59, 67, 83, 104, 115, 120, 145, 146, 213, 227
Ostia - 86
Ostial valve - 87
Ovarioles - 33, 105, 104, 109
Ovaries - 34, 102, 103, 104, 106, 107, 109, 110
Oviducts - 34, 102, 103, 105, 109
Oviparity - 33, 108, 109, 160, 173
Ovipositor - 71, 73, 102, 106, 116, 139, 140, 142, 143, 145, 146, 147, 148, 150, 151, 160, 162, 196, 197, 202, 203, 206, 207, 208, 209, 230
Ovoviviparity - 109, 160
Oysters - 134

P

Paedogenesis - 34, 110
Panchaetothripidae - 177
Panoistic type - 104
Panorpidae - 181
Papilionidae - 69, 184
Parameres - 74
Parasitoids - 16, 22, 27, 28, 110, 208, 209, 210
Parental care - 13, 14, 151
Parthenogenesis - 34, 109, 110, 172, 173, 177, 206, 207
Paurometabola - 142
Paurometabolous development - 123
Pedicel - 47, 48, 113, 115, 172, 179, 206
Pediculidae - 159
Penis - 73, 74, 102, 106, 108
Pentatomidae - 165
Perikaryon - 88
Peripatus species - 134, 233
Periplasm - 121, 122
Peripneustic - 101
Peritrophic membrane - 79, 81
Petiolata - 207
Phallobase - 73, 74, 108
Phallus - 73, 74, 106, 108
Pharate pupa - 128
Phasic muscles - 62
Phasmida - 149
Phasmidae - 150
Pheromones - 14, 30, 43, 92, 94, 95, 175
Pherophoridae - 192
Philopteridae - 168
Phioeothripidae - 177
Photo receptors - 111, 116, 117
Phototropism - 7
Phthiridae - 159
Phycitidae - 190
Phyllidae - 150
Physical receptors - 111, 115
Phytophagous - 5, 45, 135, 145, 146, 148, 161, 162, 197, 202, 204, 226
Phytotoxemia - 162
Pieridae - 185
Pigment cells - 117, 118
Pilifers - 55
Pipenculidae - 202
Planta - 64
Plasma - 62, 85, 120
Plecoptera - 124, 142
Plume moths - 192
Plutellidae - 187
Polyembryony - 34, 110
Polyphaga - 216, 218
Polyphagous - 5, 10, 16, 20, 31, 32, 173, 195, 224
Polypneustic - 101
Polypod larvae - 127
Polytrophic type - 104
Post embryonic development - 121, 122
Post gena - 43, 44
Praying mantis - 28, 31, 63, 143, 145
Predators - 9, 13, 15, 16, 22, 24, 28, 31, 32, 40, 135, 149, 169, 200, 201
Predatory beetles - 217
Prepupa - 125, 177, 207
Proboscis - 55, 56, 57, 58, 163, 168, 171, 182, 183, 192, 193, 194, 197, 198, 200, 201, 204
Proctodaeum - 40, 76, 77, 82, 83
Prognathous - 45, 127, 150, 152, 215
Propneustic - 101
Propodium - 72
Prothoracic glands - 41, 92, 93
Protocerebrum - 89, 91, 92
Protopod larvae - 126
Protura - 64, 71, 102, 136
Proventricular plate - 78
Proventriculus - 78, 81, 82
Pseudococcidae - 174

Pseudo halteres - 67
Pseudoplacental viviparity - 33, 109
Pseudotracheae - 50, 58
Psillidae - 172
Psocidae - 157
Psocoptera - 157
Psychidae - 188
Psychodidae - 126, 199
Pterygota - 133, 138, 142, 178
Ptinidae - 224
Pulsating organs - 87
Pulse beetles - 229
Pulvilli - 61, 64
Pupa adecticous - 130
Pupa coarctata - 130
Pupal stage - 11, 66, 124, 126, 129, 139, 131, 160, 172, 177, 202, 207
Pupa libera - 130
Pupa obtecta - 139
Puparium - 11, 93, 128, 130, 198, 201, 231
Pyloric valve - 76, 79
Pyralidae - 191
Pyraustidae - 191
Pyrrhocoridae - 163

Q

Queen bee - 96, 116, 120
Queen substance - 96

R

Receptacle - 80, 105
Rectal papillae - 80, 83
Red gram pod fly - 203
Reduviidae - 161
Repellent - 8
Reproductive system - 34, 40, 75, 91, 102, 103, 106, 110
Retina - 117, 118, 119
Retinaculum - 68, 69
Retinula - 117, 118
Retiinular cells - 117
Rhabdome - 117, 118, 119
Rhabdomere - 117, 118
Rheotropism - 8
Rhinoceros beetle - 218, 219, 220
Rhinotermitidae - 156
Robber flies - 26, 28, 200, 201
Rostrun - 52, 57, 171, 172, 215, 229, 230
Rutelinae - 219, 220

S

Salivarium - 76
Sand flies - 23, 199
Sarcolemma - 62
Saturniidae - 192
Saw flies - 127, 206, 207
Scape - 47, 113, 211
Scarabeidae - 48, 126, 218
Scarabeiform larvae - 127
Scarabinae - 219, 221
Schoenobiidae - 191
Sclerites - 37, 38, 43, 44, 57, 59, 65, 71, 72
Scolopale - 111, 112, 113, 115
Scoloparia - 112
Scolopidia - 113, 145
Scolopophorous organs - 112
Scorpion flies - 127, 181
Scorpions - 135, 233
Secretagogue mechanism - 81
Semiloopers - 127, 195
Seminal vesicle - 74, 106, 108
Sensilla basiconia - 115
Sensilla coeloconica - 115
Sensilla placoidea - 115
Sensilla styloconica - 115
Sensilla trichoidea olfactoria - 115
Sense organs - 22, 28, 46, 88, 89, 110, 111
Sensilla - 46, 66, 94, 111, 113, 114, 115, 116
Sensitivity - 88, 89, 111
Sensory neurones - 88, 111, 115
Setae - 40, 64, 96, 111, 128, 177, 179, 185, 186
Sex pheromones - 04
Shrimps - 134
Sieve plates - 97
Simple eyes - 37, 43, 45, 116, 117, 138, 174
Siphonaptera - 126, 205
Siphunculata - 158
Skin beetle - 23, 223
Skippers - 67, 181, 182, 183, 184, 186
Slug caterpillar - 189
Snout beetles - 230
Snout moths - 190
Social insects - 13, 14, 15, 27, 29, 30, 31, 94, 95, 119, 128, 206, 211, 215
Social symbiosis - 17, 29
Soothsayer - 145
Sound producing organs - 113, 119, 146, 160, 168
Spermatheca - 74, 103, 105, 106

Spermathecal glands - 103, 105
Spermatids - 107
Spermatocytes - 107
Spermatogenesis - 107
Spermatogonia - 107
Spermatozoa - 33, 34, 106, 107, 108, 110
Sperms - 105, 108, 109
Sphecidae - 213
Sphingidae - 193
Spiden - 135, 203, 213, 233
Spider beetle -
Spiracles - 6, 38, 59,72,91, 97, 99, 100, 101, 102, 119, 120, 127, 130, 136, 202, 218, 219, 224
Spiracular sound organs - 120
Spittle - 84, 169
Stadia - 120
Stadium - 42, 122
Stemmata - 117
Sterile females - 5
Stigma - 139, 178
Stigmata - 38, 72, 97, 136
Stimulus - 14, 29, 56, 88, 93, 110, 111
Stinging apparatus - 71, 206
Stomodaeal feeding - 155
Stomodaeal valve - 78, 81
Stomodaeum - 40, 76, 77, 79, 91
Strepsiptera - 49, 60, 66, 67, 126, 127, 129, 231, 232
Stridens - 120
Stridulatory organs - 119, 146, 148, 150
Stylopidae - 231
Sub oesophageal ganglion - 89, 90, 92
Superposition image - 118, 119
Supra oesophageal ganglion - 89
Surra - 197, 200
Swallow tails - 184
Symbiosis - 17, 28, 29
Symphyta - 204
Synapse - 88
Syrphidae - 201
Systole - 86, 87

T
Tabanidae - 200
Tachinidae - 203
Taenidia - 98
Taxes - 7
Tegmina - 67, 143, 145, 146, 147, 148, 149, 150
Tenebrionidae - 227
Tenthredinidae - 127, 207
Tentorial pits - 45
Tentorium - 43, 44, 45
Tephritidae - 202
Teragriidae - 188
Terminal filament - 104
Termitarium - 14, 30, 95
Termitidae - 155, 156
Testes - 102, 106, 107, 108
Tettigonidae - 120, 146
Thermo receptors - 46, 111, 116
Thermotropism - 8
Thigmotropism - 8
Thripidae - 177
Thysanoptera - 33, 69, 109, 176
Thysanura - 64, 83, 136
Tiger beetle - 216
Tiger moths - 194
Tingidae - 165
Tipulidae - 127, 199
Tormogen cells - 111
Tracheae - 7, 38, 40, 58, 66, 69, 90, 91, 97, 98, 99, 100, 102, 107, 136
Tracheoles - 7, 65, 97, 98, 100, 101
Trichogen cells - 111
Trichogrammatidae - 210
Trichoid sensilla - 111
Trichoptera - 68, 127, 129, 196, 197
Trypetidae - 202
Tritoceribrum - 89, 91
True viviparity - 33, 109
Trypanasomiasis - 161
Tuleremia - 200
Tunica propria - 103
Tussock moths - 196
Tympana - 120, 146
Tympanal organs - 72, 113, 114, 146, 148
Tympanum - 113, 114, 120, 146

U
Urticating hairs - 11

V
Vagina - 33, 74, 76, 103, 105, 106
Valiidae - 168
Vasa deferentia - 106, 107, 108
Ventral sinus - 85

Ventricle - 86, 87
Vertex - 43, 44, 215
Vespidae - 212
Vibrating membranes - 120
Vinegar gnats - 202
Visceral muscles - 61, 62
Vitellarium - 104
Vitellin membrane - 105, 121
Vitellogenesis - 105
Viviparity - 33, 34, 108, 160, 173

W

Wasps - 5, 12, 13, 14, 16, 21, 22, 27, 28, 29, 30, 31, 64, 95, 103, 116, 128, 137, 153, 169, 170, 206, 214
Water scavenger beetles - 23, 1218
Weevils - 17, 29, 45, 127, 215, 216, 230
White ants - 14, 17, 29, 30, 95, 103, 153
White grubs - 217, 218
Wing cells - 69
Wire worms - 222
Wood borers - 16, 31, 223, 224
Wood boring beetle - 221

Z

Zygoptera - 140